ANNALS OF THE NEW YORK ACADEMY OF SCIENCES
Volume 1066

CELL INJURY
MECHANISMS, RESPONSES, AND REPAIR

Edited by Raphael C. Lee, Florin Despa, and Kimm J. Hamann

The New York Academy of Sciences
New York, New York
2005

Copyright © 2005 by the New York Academy of Sciences. All rights reserved. Under the provisions of the United States Copyright Act of 1976, individual readers of the Annals are permitted to make fair use of the material in them for teaching or research. Permission is granted to quote from the Annals provided that the customary acknowledgment is made of the source. Material in the Annals may be republished only by permission of the Academy. Address inquiries to the Permissions Department (editorial@nyas.org) at the New York Academy of Sciences.

Copying fees: For each copy of an article made beyond the free copying permitted under Section 107 or 108 of the 1976 Copyright Act, a fee should be paid through the Copyright Clearance Center, Inc., 222 Rosewood Drive, Danvers, MA 01923 (www.copyright.com).

♾ The paper used in this publication meets the minimum requirements of the American National Standard for Information Sciences—Permanence of Paper for Printed Library Materials, ANSI Z39.48-1984.

Library of Congress Cataloging-in-Publication Data

Cell injury : mechanisms, responses, and repair / edited by Raphael C. Lee, Florin Despa, and Kimm J. Hamann.
    p. ; cm. — (Annals of the New York Academy of Sciences ; v. 1066)
Based a conference held in May 2004 at the University of Chicago, Chicago, Ill.
Includes bibliographical references and index.
ISBN 1-57331-616-4 (cloth : alk. paper) — ISBN 1-57331-617-2 (pbk. : alk. paper)
    1. Pathology, Cellular. 2. Cell physiology. 3. Cell death. I. Lee, R. C. (Raphael Carl), 1949– . II. Despa, Florin. III. Hamann, Kimm Jon. IV. New York Academy of Sciences. V. Series.
    [DNLM: 1. Cell Survival—physiology—Congresses. 2. Cells—pathology—Congresses. 3. Cell Cycle—physiology—Congresses. 4. Cell Cycle Proteins—physiology—Congresses. 5. Membrane Proteins—physiology—Congresses. W1 AN626YL v.1066 2005 / QU 375 C3927 2005]
Q11.N5 vol. 1066
[RB113]
500 s—dc22
[611'.01815]

2005035413

GYAT / PCP

ISBN 1-57331-616-4 (cloth)
ISBN 1-57331-617-2 (paper)
ISSN 0077-8923

ANNALS OF THE NEW YORK ACADEMY OF SCIENCES
Volume 1066
December 2005

CELL INJURY
MECHANISMS, RESPONSES, AND REPAIR

Editors
RAPHAEL C. LEE, FLORIN DESPA, AND
KIMM J. HAMANN

This volume is the result of a seminar series entitled **Cell Injury: Responses and Repair** held between March 31st and June 2nd, 2004 at the University of Chicago, Chicago, Illinois.

CONTENTS

Foreword. *By* THOMAS K. HUNT	vii
Introduction. *By* RAPHAEL C. LEE, FLORIN DESPA, AND KIMM J. HAMANN	ix

Part I. Cell Structure and Integrity

Biological Water: Its Vital Role in Macromolecular Structure and Function. *By* FLORIN DESPA	1
Thermal Stability of Proteins. *By* JOHN C. BISCHOF AND XIAOMING HE	12
The Physics of the Interactions Governing Folding and Association of Proteins. *By* WEIHUA GUO, JOAN-EMMA SHEA, AND R. STEPHEN BERRY	34
Molecular Crowding Effects on Protein Stability. *By* FLORIN DESPA, DENNIS P. ORGILL, AND RAPHAEL C. LEE	54

Part II. Modes of Cell Injury

Mechanical Cell Injury. *By* KENNETH A. BARBEE	67
Cell Injury by Electric Forces. *By* RAPHAEL C. LEE	85
Electroconformational Denaturation of Membrane Proteins. *By* WEI CHEN	92

Heat Injury to Cells in Perfused Systems. *By* DENNIS P. ORGILL, STACY A. PORTER, AND HELENA O. TAYLOR 106

Cryo-Injury and Biopreservation. *By* ALEX FOWLER AND MEHMET TONER .. 119

Oxidative Reactive Species in Cell Injury: Mechanisms in Diabetes Mellitus and Therapeutic Approaches. *By* LEONID E. FRIDLYAND AND LOUIS H. PHILIPSON 136

Part III. Cellular Responses to Injury

The Mechanisms of Cell Membrane Repair: A Tutorial Guide to Key Experiments. *By* RICHARD A. STEINHARDT 152

The Role of Ca^{2+} in Muscle Cell Damage. *By* HANNE GISSEL 166

Protein Denaturation and Aggregation: Cellular Responses to Denatured and Aggregated Proteins. *By* STEPHEN C. MEREDITH 181

Thermally Induced Injury and Heat-Shock Protein Expression in Cells and Tissues. *By* MARISSA NICHOLE RYLANDER, YUSHENG FENG, JON BASS, AND KENNETH R. DILLER 222

Cellular Response to DNA Damage. *By* JOHNNY KAO, BARRY S. ROSENSTEIN, SHEILA PETERS, MICHAEL T. MILANO, AND STEPHEN J. KRON 243

Autophagy. *By* AMEETA KELEKAR 259

Magnetic Resonance Imaging of Changes in Muscle Tissues after Membrane Trauma. *By* HANNE GISSEL, FLORIN DESPA, JOHN COLLINS, DEVKUMAR MUSTAFI, KATHERINE ROJAHN, GREG KARCZMAR, AND RAPHAEL LEE 272

Part IV. Therapeutics for Cell Injury

Na^+-K^+ Pump Stimulation Improves Contractility in Damaged Muscle Fibers. *By* TORBEN CLAUSEN 286

Multimodal Strategies for Resuscitating Injured Cells. *By* JAYANT AGARWAL, ALEXANDRA WALSH, AND RAPHAEL C. LEE 295

Membrane Sealing by Polymers. *By* STACEY A. MASKARINEC, GUOHUI WU, AND KA YEE C. LEE 310

A Surfactant Copolymer Facilitates Functional Recovery of Heat-Denatured Lysozyme. *By* ALEXANDRA M. WALSH, DEVKUMAR MUSTAFI, MARVIN W. MAKINEN, AND RAPHAEL C. LEE 321

Index of Contributors 329

Financial assistance was received from:

- THE UNIVERSITY OF CHICAGO

The New York Academy of Sciences believes it has a responsibility to provide an open forum for discussion of scientific questions. The positions taken by the participants in the reported conferences are their own and not necessarily those of the Academy. The Academy has no intent to influence legislation by providing such forums.

CELL INJURY
MECHANISMS, RESPONSES, AND REPAIR

THIS BOOK BELONGS TO:
Medical Library
Christie Hospital NHS Trust
Manchester
M20 4BX
Phone: 0161 446 3452

Foreword

THOMAS K. HUNT

Emeritus Professor of Surgery, University of California, San Francisco, San Francisco, California 94143 USA

At the time my obsession with wound healing began 40 years ago, collagen, epithelization, and a little angiogenesis were the whole field. I tried to visualize how individual cells might react to injuries. Do they recover or do they die? If they recover, do they regenerate or do they bear scars as tissues do? On searching the literature, I found little and, lacking the courage to answer such questions, I stayed on the beaten path. Surely, though, I thought, when I smashed my thumb with a hammer, cells must suffer as much as connective tissue. Do hammered cells just collapse like oversqueezed balloons? Can they recover at all or do they die at the slightest trauma? Is there a patch for punctured cell membranes? (This book says that there may be one.) Aside from inflammation, do injured cells influence healing? Can injured cells incite unwanted scar directly, or is inflammation a necessary intermediary? For a number of good and bad reasons, wound healers have skipped past those questions in the rush to clarify the issue of growth factors arising from coagulation and inflammation. The diversity and subtleties of injuries were overlooked. This book attends to a number of overlooked opportunities.

Unfortunately, the course we took, though a productive one, tended to isolate the injury-induced deposition of vascularized connective tissue ("wound healing") from the rest of its genre (arteriosclerosis, diabetic retinopathy, ischemic injury, and so on). The more we see of wound healing, the more we must concede that vascularized scar is the final common pathway of many human diseases and has many origins, many of which are not preceded by injuries in the usual sense.

Diabetic retinopathy is an instructive case. It is scar tissue in the retina in which the vascular element is more than usually apparent, probably because, as opposed to other scars, we can see it through an ophthalmoscope. There is no mechanical damage. Clearly, there is loss (death?) of normal cells and replacement of the normal stroma due to scar. Where or what is the injury? Is it the result of normal cells taking another phenotype, that is to say, being misdirected to producing scar by their environment? Inflammation is minimal, so what is the origin of the signals that induce angiogenesis and connective tissue deposition? It seems to me that injury must have occurred.

We have almost forgotten the diversity of injury and we do not know how much mechanical, electrical, or "metabolic" injury is necessary to make an individual cell complain enough to incite its surrounding tissues to do something about it. Does the

Address for correspondence: Dr. Thomas K. Hunt, Wound Healing Laboratory, Department of Surgery, Room HSW-1652, University of California, San Francisco, CA 94143-0522. Voice: 415-476-0410.
wound@itsa.ucsf.edu

"complaint" arise from hypoxia, as some will say, or from lactate accumulation, as I believe? There is no evidence that hypoxia precedes the scar. We have ignored the fact that non-inflammatory cells release angiogenic factors and cytokines that stimulate collagen and proteoglycan deposition! Are injured but still viable cells the source of unwanted connective tissue deposition?

On the one hand, we need to know how to save cells that, though injured, have reparative capacities or will resume their original functions. On the other hand, we need to know how sick a cell has to be in order to incite the deposition of vascularized scar in the course of trying to save itself. Is the scar just the result of normal attrition and replacement in the diabetic environment?

At the time my interest began, only a few brave souls puzzled over the fate and functions of pre-existing, presumably injured, cells in wound sites. During many, but not all, of those years, were it not for Raphael Lee, I would scarcely have thought of how injured cells repair themselves, much less of how cells are injured absent an obvious trauma. Finally, he has got the concept on paper and in one place! To my knowledge, this is the first compendium on repair of injured cells, and he has put it together in a context in which "injury" and "repair" can be seen in their broader contexts.

I like the first sentence from Agarwal, Walsh, and Lee: "Biologists commonly consider a wound to be an acquired defect in the structural integrity of tissues." It is true. We are careless about that, and have tended to see wounds as an anatomic "fracture" of connective tissue that needs to be stuck together again as rapidly as possible. We see the glue as deposition of coagulation proteins and later the deposition of new, "connective" tissue. Though it may be a fine point, this view makes the tacit assumption that the hallmark of an injury is what happens after mechanical trauma, rather than as a protean process that pervades multicellular life and follows the inevitable injuries that also afflict individual cells. I suspect that in time, we will strip off layer after layer of inflammatory stimuli, metabolic events, and mechanical or electrochemical influences in search of the lowest common denominator that we hope will be the quintessence of "injury." I suspect, however, that there is no such point.

After all, there is no point in evolution at which "healing" became possible. Rather, repair of life's weak and often broken spots has always borrowed from already existing normal life processes. Cells were re-adhering to each other on the way to multi-cellular life before collagen even evolved. As long as life creates substance, there will be collisions and exchanges of mechanical forces. As long as life depends upon oxygen, carbohydrate, and, minerals there will be electrons that will go astray and injure the inner workings of cells. Fridlyand and Philipson have described that process in a remarkably brief and informative chapter. The consequences of such subtle injuries as a localized rupture of the cell membrane are discussed, and evidence for the possibility that a lipid patch may limit the extent of injury is summarized. This is truly a new idea! Can cells be given a head start on repair?

While the authors have sought to deal mainly with repair, they have by necessity also examined "injury." They have expanded the scope of injury from simple mechanical or electrical wounds all the way to incineration in the fire of carbohydrate metabolism. If you want to know how much rough handling a cell can stand, you would do well to read on. If you are brave enough to attempt an understanding of the full spectrum of injury and repair, you really *must* read on!

Introduction

RAPHAEL C. LEE, FLORIN DESPA, AND KIMM J. HAMANN

Pritzker School of Medicine, The University of Chicago, Chicago, Illinois 60637 USA

When the subject of responses to injury or wound healing arises, the discussion usually pertains to reparative processes at the tissue or organ system level. Until recently, relatively little attention has been paid to the healing of wounded cells. Although much is known about the responses of individual cells to injury, and about their repair processes, there has not been a collective synthesis published that integrates the interdisciplinary aspects of the cellular healing responses. This *Annals* volume represents the first endeavor to bring this subject into focus.

Each of the many and various molecular processes involved in cell repair are the subject of active research efforts scattered over numerous biomedical science research fields. When viewed collectively, it becomes clear that cellular wound-healing activities are highly organized and complex. By comparison, the reparative processes involved in tissue wound-healing reflects the outcome of complex coordinated events involving many cells and cell types. Reparative processes at the cell level are also complex and coordinated, involving highly orchestrated series of molecular events designed to detect and repair injured components of the cell. As opposed to healing of tissue injury, which often occurs by replacement of damaged tissue with scar, cellular wound-healing processes are more regenerative and, when successful, the repair is more precise.

Cell Injury: Mechanisms and Repair is concerned chiefly with describing the processes of injury and healing at the molecular level. In the spring of 2004, a conference was organized at The University of Chicago to bring together experts on the various aspects of cell injury and repair, to share information and consider each aspect of the healing response in light of all the other processes that are simultaneously occurring in cells while they are healing and responding to injury. The symposium has since evolved into a graduate-level core course in molecular medicine and pathology at The University of Chicago. Like the original symposium, this book is organized in four sections, which progress from basic structure and physical integrity of the mammalian cell to modes of cell injury and cellular responses to ways in which we may be able to utilize our understanding of these types of injury and subsequent responses for therapeutic strategies that limit injury or enhance repair.

Part I of this *Annals* volume focuses on the structural factors which are deterministic of cell integrity and the physicochemical modes of cell injury. It is essentially a materials-science approach to cell injury. The chapters review basic aspects of mammalian cell structure, including not only the biophysical nature and responses

---

Address for correspondence: Dr. Raphael C. Lee, Plastic and Reconstructive Surgery, University of Chicago, 5841 S. Maryland Avenue, MC 6035 J641, Chicago, IL 60637. Voice: 773-702-6302; fax: 773-702-1634

r-lee@surgery.bsd.uchicago.edu

of specific cell components such as the plasma membrane but also interactions with the extracellular and intracellular "matrix of life," water. This pertains to basic determinants of protein stability, protein assemblies and organelles. Thus, information about the energetics of passage through intermediate steps leading to aggregation of unfolded proteins and about the role of the biological solvent (water) as an active player in all these molecular events is discussed in the context of their role in the pathogenesis of cell injury. Physical and chemical aspects of interactions within and between proteins are reviewed and the effects of temperature and molecular crowding on these interactions are discussed.

In Part II, several different biophysical modes of cell injury are reviewed in a series of chapters that examine electrical injury to cells such as electroporation of the lipid bilayer and electrical denaturation of membrane proteins, as well as the effects of temperature extremes on cells. In these latter chapters, effects of excessive heat on individual cells and their components, as well as the effects of freezing and thawing on cells in both cryo-injury and biopreservation attempts are considered. Thus, the chapters in this section give the reader an overview of the types of direct cell injury which promote cellular responses and for which we are currently seeking and testing therapeutic strategies.

Part III of this volume is devoted to the healing responses of cells. In the opening chapter of this section, a tutorial overview of endogenous and therapeutic mechanisms of cell membrane repair is presented, giving the reader an introduction to key experiments in the elucidation of these concepts. Subsequent chapters in this section review the roles of endogenous substances, including calcium and heat shock proteins, in responses to cell injury. Molecular mechanisms involved in the induction of and the cellular response to DNA damage are also detailed in this section. Considerations of genetic syndromes and the clinical phenotypes resulting from aberrations in DNA repair are included. This part of the text concludes with a treatise on autophagy, a relative newcomer to the spectrum of endogenous protective responses to injury and stress, and discusses the pathological implications of deregulation of the autophagic response in mammalian cells.

The final components of the text deal with therapeutic strategies to rescue injured cells by augmenting the cell's natural healing responses. Many of the strategies discussed are those we considered when resuscitating damaged tissues and organs. These include inhibition of injurious factors such as reactive oxygen species, as well as direct repair of membranes through the use of specific polymers or through stimulated enhancement of endogenous repair mechanisms. By distinguishing cellular wound-healing process from tissue and organ wound-healing processes, it is hoped that the therapeutic goals will be better defined, and that this will result in more effective clinical resuscitation efforts.

The editors would like to thank Sandra Marijan for her enormous help coordinating the development of the seminar series and the text. We would also like to thank Dr. Julian Solway, the Chairman of the Committee of Molecular Medicine at the University of Chicago, for his support and for making this effort possible.

Biological Water

Its Vital Role in Macromolecular Structure and Function

FLORIN DESPA

Department of Surgery, MC 6035, The University of Chicago, Chicago, Illinois 60637, USA

ABSTRACT: Water in tissues and cells is confined by intervening cellular components and is subject to structural effects that are not present in its bulk counterpart. The structuring effects lower the dielectric susceptibility of water molecules and induce a "red shift" of their relaxation frequency. This is also a source of polarization fields that contribute to the effective interactions between macromolecules. The behavior of water molecules at hydrophilic sites is different from that at hydrophobic sites, and this dissimilar behavior promotes the anisotropy of the hydration shell of proteins. The anisotropy of the hydration shell is essential for the enzyme function, but it is also important in detecting denaturation of the protein (i.e., proteins expose their hydrophobic parts to water during unfolding). The most significant differences between biological and ordinary water will be presented along with how this information can be used to decipher patterns in dynamical behavior of biological water and to detect possible structural changes of the cellular components.

KEYWORDS: biological water; protein dynamics; injuries

INTRODUCTION

Water is the critical substance for production of biochemical energy (photosynthesis) and the most common product of the metabolic processes as well. Water represents the matrix of life on Earth. Because life on Earth is so tightly connected with water, many human achievements based on water and aqueous solutions became a matter of fact. "As the fish forgets the water in the ocean." we often neglect the essential role of water in our life. The water content of the living cell (TABLE 1) is about 70%, making the molarity of the human body less than 1 mole (for an average molecular weight < 10 kDa).

However, the water control in a human body is rigorous. On one hand, a deficiency in hydration of less than 5% is usually fatal. On the other hand, an increase of the water content in cells and tissues over the physiological limit changes the protein activity and may trigger also cell malfunctioning and death.

Address for correspondence: Florin Despa, Pritzker School of Medicine, MC 6035, The University of Chicago, 5841 S. Maryland Avenue, Chicago, IL 60637. Voice: 773-702-5767; fax: 773-702-1634.

  fdespa@uchicago.edu

TABLE 1. The composites of a mammalian cell

Cell Component		% Weight
Water		70
Inorganic ions (Na^+, K^+, Cl^-, Ca^{2+}, Mg^{2+}, etc.)		1
Metabolites		3
Macromolecules	Proteins	18
	RNA	1.1
	DNA	0.25
	Polysaccharides	2
Lipid bilayer	Phospholipids	3
	Glycolipids plus cholesterol	2

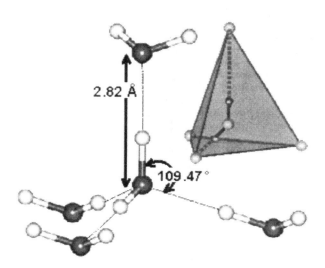

FIGURE 1. Three-dimensional structure of bulk water.

From a chemical point of view, water has a very simple structure (FIG. 1) in comparison to the complicated architectures of other biological molecules, such as the amino acids.

Despite its simplicity, water has unusual thermodynamic parameters (melting and boiling points, vaporization, and fusion heat), higher than expected for liquids composed of hydrogen and oxygen. In addition, water shows abnormal structural properties: maximal density at 4°C decreases its viscosity with a pressure up to about 1,000 atm.

Water in tissues and cells (biological water) rarely is thicker than a few molecular layers and mostly confined by intervening cellular components. This water is markedly different from the bulk counterpart. Although our knowledge about biological water is incomplete, all theories of cell biochemistry have explicit or implicit as-

TABLE 2. The energy (E) of common bonds in vacuum and water

Bond	E (vacuum) [kcal/mole]	E (water) [kcal/mole]
Covalent	90	90
Ionic	80	1
Dipolar	4	1
van der Waals	1	1

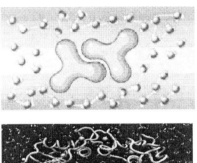

hydrophobic interactions

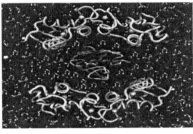

crowding

FIGURE 2. Water mediates nonspecific interactions in biological systems, such as interactions between hydrophobic molecules (*top*) and crowding effects (*bottom*). Crowding effects are manifest on the dynamics of the protein in the center, which is obstructed by the surrounding proteins.

sumptions about the physical properties of this water.[1] Most of them consider biological water as a solvent which rescales the strength of Coulomb interactions (ionic and dipolar) between macromolecules with respect to vacuum (TABLE 2). Also, it is admitted that this solvent mediates the hydrophobic interactions and plays a role in setting the level of cellular crowding (FIG. 2).

STRUCTURAL EFFECTS IN BIOLOGICAL WATER: PAIR CORRELATION APPROXIMATION

Water at a macromolecular interface is subject to structural effects which are not present in its bulk counterpart.[2] Jacobson,[3] more than fifty years ago, suggested in a general manner that these structuring effects actually expand beyond the first hydration layer and may give rise to long-range hydration structures; the details of his explanation can now be formulated in a more quantitative manner.[2] At the interface

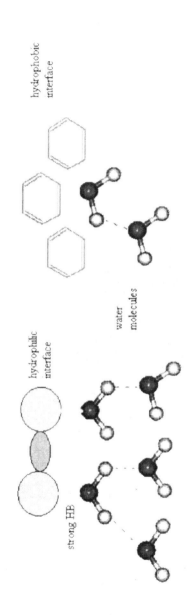

FIGURE 3. The dynamics of water at interfaces. The perturbation of the HB exchange between water molecules at these interfaces lead to a correlation in pairs of this water.

with a macromolecule, the free rotation of a water molecular dipole ($\vec{d}$) is likely to be obstructed by local geometric constraints, strong interactions with surface electric charges, or by a hydrophobic effect (FIG. 3).

This reduces the number of possibilities of hydrogen bond (HB) exchange of this water molecule with other water molecules from its vicinity. The depletion (f) of the HB exchange (say, from m possibilities for a molecule in bulk water, to $m - f$ possibilities at the interface) lowers the entropy of the water molecule and leads to extended lag times for the reorientation of $\vec{d}$. This enhances the probability that one water dipole (i) joins the slowly-fluctuating dipole of a neighbor (j) and creates a relatively long-lived dipole pair (ij). The interspace r_{ij} between dipoles in a pair and, therefore, the spatial ordering of water molecules, ranges between the typical interdistance of bulk water molecules a_0 [$a_0 = (\frac{3}{4}\pi n)^{1/3}$; n is the density of bulk water] and a critical distance (r_c ($r_c > a_0$). The formation of water dipole pairs with the largest interspace (r_c) is favored by the large decrease in entropy, while pairs separated by short distances ($\sim a_0$) correspond to small changes of the entropy. r_{ij} is random within its range ($a_0 \leq r_{ij} \leq r_c$). Consequently, the vector dipole field $\vec{E}$ at each site in the correlated region is also a random variable, and so is the thermodynamic average $\langle \vec{d} \rangle$ of the water molecular dipole moment. The magnitude and distribution of $\vec{E}$ determine the departure of the properties of structured water from those corresponding to bulk water. The probability distribution of $\vec{E}$, $P(\vec{E})$, as well as the maximum most probable value of $\vec{E}$, E_s, were derived based on basic molecular principles.[2] The main assumption of the model, which is physically intuitive, consists in the fact that the librational dynamics favors the formation of structures of water molecules correlated in pairs (FIG. 3). Thus, the approach yields a quantitative description of the librational dynamics of water under the constrains of the vicinal macromolecules.

POLARIZATION EFFECTS AND DIELECTRIC SUSCEPTIBILITY OF CONFINED WATER MOLECULES

The structuring effects lower the dielectric susceptibility of water molecules and induce a "red shift" of their relaxation frequency.[2] The librational dynamics of water is also a source of polarization fields, which contribute to the effective interactions between macromolecules.[4] For the particular case of hydrated hydrophobic molecules, the polarization field in the region of correlated water molecules can induce attractions between hydrophobes (FIG. 4). The hydrophobic interaction—the apparent attraction between hydrophobic species in water—is considered a key factor in maintaining the correct folded conformation of a protein molecule and also the main cause of protein aggregation. This attraction is thought to result, in a way that is still imperfectly understood, from changes in the arrangement of hydrogen bonds between water molecules surrounding a hydrophobe.[2] This gives rise to a local polarization of the interfacial water which is shown to be strong enough to induce long-range attraction between hydrophobic molecules. The polarization fields give rise also to induction effects which make water molecules and hydrophobes actually attract each other,[4,5] but not nearly as strongly as water attracts itself! These recent results[2,4] increased our understanding about the way proteins enhance their intramolecular interactions as they fold or associate. Furthermore, the approach presented

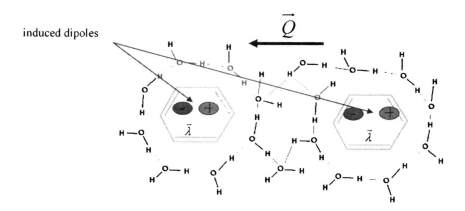

FIGURE 4. Polarization field ($\vec{Q}$) of water structured around hydrophobes. $\vec{\lambda}$ is the induced dipole by the polarization field.

above gives additional support to the idea that water confined in nanoscale hydrophobic environments has quite different solvent properties from those of the bulk liquid.[6]

Water's high dielectric constant is the reason why it is a good solvent for ions: it screens their electrical charges and so prevents them from aggregating. But in the vicinity of hydrophobic residues in a protein chain, the reduction in dielectric constant means that charged residues will interact much more strongly,[2] potentially helping to fix the protein's folds in place. Some details are given below.

Structural changes of water in the vicinity of macromolecules lead to modifications in the dielectric properties of their hydration shells. It was shown[2] that molecules experiencing high constraints ($f/m \to 1$) are characterized by a low susceptibility to follow an external electric field. From these results one can infer that, in the particular case of hydrophobic interfaces, where water molecules are constrained by the lack of HB exchange, there is a drop in the dielectric permittivity of the surrounding water. The trend of the electric susceptibility is to decrease from that of bulk water ($f < 1$) towards its value at the hydrophobic interface ($f \to m$). In return, Coulombic interactions between charged groups will systematically be enhanced in the direction of a neighboring hydrophobe. Therefore, hydrophobic residues play an active role in mediating intramolecular interactions between the polar side-chain residues of a protein.[2]

In FIGURE 5 we can see the thermodynamic effects of confinement upon the hindered rotation motion of molecular dipoles in biological water. We display the average dielectric susceptibility of biological water[2] against $\beta E_L d$, $\beta = 1/k_B T$, where $k_B \cong 1.38 \times 10^{-23}$ JK^{-1} stands for the Boltzmann constant, T is the temperature and $E_L d$ is the Lorentz energy. An increase of the temperature ($\beta E_L d \to 0$) lowers the susceptibility of the biological water. Actually, the model of biological water described above correctly predicts that the very-low-temperature susceptibility, which is not relevant to biology, is low (not shown in FIGURE 5), and then increases because of

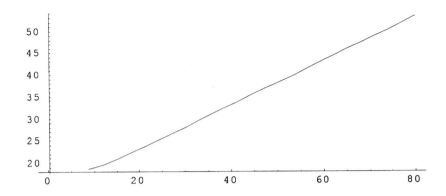

FIGURE 5. Average dielectric susceptibility χ of the hydration layer as a function of temperature ($\beta E_L d = E_L d/k_B T$) for $f/m = 0.5$.

the formal decrease of the polarization field with T, and only decreases again at relatively high temperatures.

In the above context, it is relevant to recall that the solid form of water (ice) has a higher dielectric constant than liquid water, at temperature well above $0K$. For example, the values of the static dielectric constant of ice range from 91.5 at $-0.1°C$ to 133 at $-65.8°C$.[8] These high values of the dielectric constant are a direct consequence of the ordering of ice, which reduces random fluctuations of internal fields. Nevertheless, the current view is that the degree of ordering in ice is higher than the degree of water ordering around proteins.[9,10] Therefore, in order to recover the ice-like dielectric characteristics within the present theory we need to take into account higher-order correlations between the water dipoles. It is worth mentioning here the recent progress in simulating freezing of water to a known ice structure.[11] The key result of the simulation performed by Matsumoto et al.[11] is that ice nucleation occurs once a sufficient number of relatively long-lived hydrogen bonds develop spontaneously at the same location, forming a highly correlated, compact nucleus.

THE ANISOTROPY OF THE HYDRATION WATER OF PROTEINS

It is interesting to compare the water structure at a hydrophobic site, which is, basically, a distribution of dipole pairs, with that corresponding to a hydrophilic site. A hydrophilic group, characterized by a permanent dipole of moment $\vec{\lambda}_0$, aligns neighboring water dipoles along $\vec{\lambda}_0$ in a region of space determined by the interplay between the pair-wise solvent–solute interaction and the entropy change.

The dissimilar behavior of water molecules at these two sites promotes the anisotropy of the hydration shell of a protein.[7] The anisotropy of the hydration shell is essential for the enzyme function and is part of the recognition process by other molecules or proteins. In this context we can say that a polar group is fully expressed on a protein surface when $\lambda_0 / \langle \vec{\mu} \rangle \gg 1$.

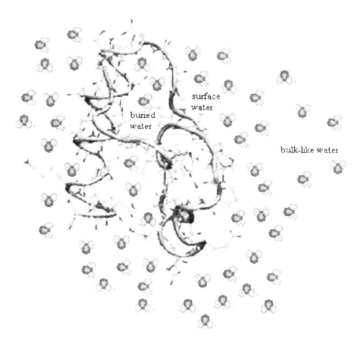

FIGURE 6. Hydration water can be divided in three main compartments: buried water, surface water and bulk-like water.

It is also interesting to compare the dynamics of water at various locations with respect to the protein structure.

Roughly, we can distinguish three water compartments (FIG. 6): bulk-like water, interface water, and water buried deep into the structure of the protein (crystallographic water).

The bulk water is characterized by a random, fluctuating, three-dimensional network of H-bonds.[12] An interruption of this structural arrangement by solute proteins changes the local properties of water in a region of space determined by the interplay between the pair-wise solvent–solute interaction and the entropy change. The water response time at this interface is fully dictated by the mobility of water molecules in the environment, but does not necessarily follow the strength of individual interactions. The equilibrium of the opposing forces determines this mobility and the highest mobility is determined by entirely balanced forces in all directions. This is the case of bulk-like water, at large distance from the protein structure. At this location, the exchange of H-bonds may occur over the entire 4π solid angle and is fast, with an exchange frequency $v_b \cong 1.710^{13}\ s^{-1}$.[13] This trade is associated with a fast and random reorientation of the individual dipoles ($\vec{d}$) of the water molecules under normal, body-temperature conditions. Water molecules at a protein–water interface can be localized by strong H-bonding with $\Delta E \cong 4k_B T$ energy per bond[4] at physiological temperature. Reorientation of the molecular dipole becomes more difficult since it requires surmounting this additional potential barrier. The reduction of the bulk relaxation frequency (v_b) of a water molecule localized by only one H-bond yields a period of $\tau_s = 3.9 \times 10^{-12}$ s. We can compute the frequency of reorientation of a molecular dipole localized by two H-bonds (corresponding to an energy barrier equal to $2\Delta E$), which gives a lifetime of about $\tau_m = 2.8 \times 10^{-10}$ s. We notice that a water molecule localized by three H-bonds is already rotationally immobilized as the one localized by four H-bonds (or more). The corresponding lifetimes computed as above are $\tau_{l'} = 2.1 \times 10^{-8}$ s and $\tau_{l''} = 1.9 \times 10^{-6}$ s, respectively. They match the time scale of the exchange between buried water molecules and bulk water via conformational fluctuations.[14,15] For protein in solution, they also match the rotational correlation time of the protein.[14,15] In the case of immobilized proteins and tissue $\tau_l \sim 10^{-6}$ s or longer, corresponds to actual residence time of buried water molecules (and, possibly, labile protons) as suggested by Foster et al.[16] and amply investigated in several other works (see, for review, Refs. 14 and 15).

It should be cautioned that although we considered that each water component has a distinct correlation time associated with it, we expect that each represents a distribution of values.

BIOLOGICAL WATER AS A PROBE IN BIOMEDICAL ENGINEERING

In above material, we have presented the most significant differences between biological and ordinary water. We now can use this information to decipher patterns in dynamic behavior of biological water and detect possible structural changes of the cellular components. This is important for various biomedical engineering applications as, for example, interpreting the dielectric and magnetic response of water in living tissues. In the following, we present the basic aspects that need to be taken into consideration for relating the dielectric and magnetic response of biological water to the state of the macromolecules.

It should be observed that the pair correlation of the water molecular dipoles changes the dispersion properties of the biological water. To probe this phenomenon, the complex dielectric constant $\varepsilon^* = \varepsilon' - i\varepsilon''$ has to be examined. Both ε' and ε'' are frequency-dependent and, for biological water, they also depend on the parameter f describing the degree of obstruction of the HB exchange of the individual water molecular dipole (see above). $f = 0$ corresponds to bulk water in which all the H-bonds per water molecule are fully satisfied. The sudden drop of the relaxation frequency of water in the vicinity of a molecular surface is a consequence of precluding H-bonds, which hinders the rotational motion of the molecular dipoles, as described in above. Therefore, ε' and ε'' characterizing the hydration shell of a hydrophobe are given by the superposition of f Debye-type contributions.[2] This is an important verifiable result which provides a way to probe the water molecules next to biomolecules. It is not hard to see that the dielectric loss, which is proportional to ε'', will now be characterized by a convolution of f separate Debye peaks. Their characteristic frequencies will systematically be "red shifted" in comparison with that of the bulk water.[2] The maximum drop of the relaxation frequency occurs next to the interface ($f/m \to 1$). This approach provides the frame for a direct comparison between the dispersion properties of bulk water and its structured counterpart, the biological water.

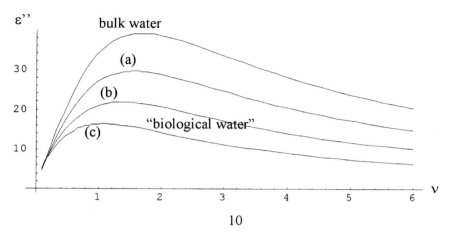

FIGURE 7. Theoretical predictions for the dielectric loss in bulk water versus biological water (a, b, and c). By increasing the volume of the structured water, the dielectric loss is "red shifted".

The things to be compared in FIGURE 7 are ε'' theoretical curves for bulk-like water and biological water (a, b and c). The curves a, b and c correspond to three different volumic partitions between bulk-like and structured water. We observe a significant departure of the dispersion properties of biological water from those characterizing bulk water. The relaxation frequency corresponding to the maximum dielectric loss in biological water is low, as we estimated above, while the peak decreases and achieves a more gradual slope. As expected, by enriching the correlated water species with low relaxation frequencies, the peaks of dielectric loss (a), (b), and (c) are systematically "red shifted" in comparison with the respective maximum of the dielectric loss in bulk water. The broad trend of the peaks is due to the superposition of the individual Debye behavior of water each type of correlated water.

Similar physical behavior can be seen in the magnetic relaxation (MR) of biological water, which is usually described by two relaxation times; spin-spin and spin-lattice relaxation times (see Koenig[14] for review). The spin-spin and spin-lattice relaxation times describe the time the water protons aligned by the magnetic field needs to come to equilibrium with the other degrees of freedom in the system. Each spin may exchange quanta of energy with the environment (the other spins in the system plus the lattice). Because spontaneous emission is very improbable at the relatively low frequencies of MR experiments, the energy exchanges are stimulated and require the spin system to couple to fluctuating fields in the sample. The relaxation occurs when a component of the fluctuation spectrum matches the nuclear Larmor frequency and stimulates a spin flip. The relaxation measurement, therefore, provides information about the dynamics at the Larmor frequencies. This information can be achieved by using a quantum statistical theory that relates the molecular motions of the biological water molecules to the intensity of the magnetic fluctuations at the required transition frequencies. In the present context, a very informative approach would be to develop a model for the fluctuations in the time domain by constructing a correlation function that depends on the details of the restricted rotations and translations at each location of the water molecules with respect to the protein structure. In such a model, the effective relaxation rate will be related to the behavior of water in each water compartment relevant to the magnetic relaxation (i.e., buried water, surface water, and bulk-like water). In this way, one can measure the extent at which the recompartmentalization of water due to structural changes of the protein alters the magnetic relaxation of surrounding water.

On the base of the above illustrative arguments, we think that further experimental investigations of far-infrared dynamics of water in protein solutions, as well as cell suspensions and tissues, might be relevant in understanding the frequency dependence of the dielectric and magnetic response of biological water. This has an immediate biomedical relevance for *in vivo* characterization of the state of proteins in cells and tissue.

REFERENCES

1. BALL, P. 1999. H_2O: A Biography of Water. Phoenix. London.
2. DESPA, F., A. FERNANDEZ & R.S. BERRY. 2004. Dielectric modulation of biological water. Phys. Rev. Lett. **93**: 228104.
3. JACOBSON, B. 1953. Hydration of deoxyribonucleic acid and its physico-chemical properties. Nature **172**: 666.
4. DESPA, F.A. FERNANDEZ & R.S. BERRY. The origin of long range attraction between hydrophobes in water. Proc. Natl. Acad. Sci. USA. In press.
5. HILDEBRAND, J.H. 1979. Is there a "hydrophobic effect"? Proc. Natl. Acad. Sci. USA **76**: 194.
6. BALL, P. 2004. Grease makes ions stickier. Nature **432**: 688.
7. PAL, S.K., J. PEON & A. ZEWAIL. 2002. Biological water at the protein surface: dynamical solvation probed directly with femtosecond resolution. Proc. Natl. Acad. Sci. USA **99**: 1763.
8. ONSAGER, L. & M. DUPUIS. 1962. The electrical properties of ice. *In* Electrolytes.: 27–46. Pergamon. Oxford.
9. LIPSCOMB, L.A., F.X. ZHOU & L.D. WILLIAMS. 1996. Clathrate hydrates are poor models of biomolecule hydration. Biopolymers **38**: 177.
10. STILLINGER, F.H. 1980. Water revisited. Science **209**: 451.
11. MATSUMOTO, M., S. SAITO & I. OHMINE. 2002. Molecular dynamics simulation of the ice nucleation and growth process leading to water freezing. Nature **416**: 409.
12. SCEATS, M.G. & S.A. RICE. 1982. Water: A Comprehensive Treatise, Vol. 7. F. Franks, Ed. Plenum. New York.
13. VENABLES, D.S. & C.A. SCHMUTTENMAER. 2001. Structure and dynamics of nonaqueous mixtures of dipolar liquids. II. Molecular dynamics simulations. J. Chem. Phys. **113**: 3249.
14. KOENIG, S.H. 1982. Classes of hydration sites at protein-water interfaces: the source of contrast in magnetic resonance imaging. Biophys. J. **69**: 593.
15. BRYANT, R.G. 1996. The dynamics of water-protein interactions. Annu. Rev. Biophys. Biomol. Struct. **25**: 29.
16. FOSTER, K.R., H.A. RESING & A.N. GARROWAY. 1976. Bounds on "bound water": transverse nuclear magnetic resonance relaxation in barnacle muscle. Science **194**: 324.

Thermal Stability of Proteins

JOHN C. BISCHOF[a,b,c] AND XIAOMING HE[d]

Departments of Mechanical Engineering,[a] Urologic Surgery,[b] and Biomedical Engineering,[c] University of Minnesota, Minneapolis, Minnesota 55455, USA

[d]*Center for Engineering in Medicine, Massachusetts General Hospital, Harvard Medical School, Boston, Massachusetts 02114, USA*

ABSTRACT: Protein stability is critical to the outcome of nearly all thermally mediated applications to biomaterials such as thermal therapies (including cryosurgery), burn injury, and biopreservation. As such, it is imperative to understand as much as possible about how a protein loses stability and to what extent we can control this through the thermal environment as well as through chemical or mechanical modification of the protein environment. This review presents an overview of protein stability in terms of denaturation due to temperature alteration (predominantly high and some low) and its modification by use of chemical additives, pH modification as well as modification of the mechanical environment (stress) of the proteins such as collagen. These modifiers are able to change the kinetics of protein denaturation during heating. While pH can affect the activation energy (or activation enthalpy) and the frequency factor (or activation entropy) of the denaturation kinetics, many other chemical and mechanical modifiers only affect the frequency factor (activation entropy). Often, the modification affecting activation entropy appears to be linked to the hydration of the protein. While the heat-induced denaturation of proteins is reasonably well understood, the heat denaturation of structural proteins (e.g., collagen) within whole tissues remains an area of active research. In addition, while some literature exists on protein denaturation during cold temperatures, relatively little is known about the kinetics of protein denaturation during both freezing and drying. Further understanding of this kinetics will have an important impact on applications ranging from preservation of biomaterials and pharmaceutics to cryosurgery. Interestingly, both freezing and drying involve drastic shifts in the hydration of the proteins. It is clear that understanding protein hydration at the molecular, cellular, and tissue level will be important to the future of this evolving area.

KEYWORDS: protein; denaturation; thermal therapy; cryopreservation; burn injury; cryosurgery; biopreservation; stability

IMPORTANCE OF TEMPERATURE

Biomedical applications in which protein stability is important usually involve extremes in temperature excursion from body temperature such as that during thermal therapies, burn injury, and biopreservation. The most dramatic and important protein change during its loss of stability is denaturation and this will be the focus of this review. Burn injury usually affects skin and surface tissues in the extremities causing a significant health care cost in the United States every year. The American Burn Association[1] estimates that more than one million burn injuries occur per year that require medical attention. A small but significant percent of these burn injuries are due to Joule heating during an electrical burn which can affect deeper tissues as well.[2] In thermal therapies (including cryosurgery), elevated or cryogenic temperatures are utilized to locally destroy surface or deep tissues within the body. Two main targets of thermal therapies are cancer and cardiovascular disease. The American Cancer Society statistics list the projected occurrence of various organ cancers in the U.S. population in 2005 as over 200,000 for each of both prostate and breast, over 20,000 male kidney cancers, and close to 20,000 primary liver cancers, and a much higher incidence of colorectal cancer (100,000) that often metastasizes to the liver.[3] In addition, the American Heart Association lists the annual incidence of cardiovascular disease in the U.S. population as 13 million with coronary heart disease and over 2 million with atrial fibrillation.[4]

In biopreservation, there can be overlapping micro- and nanoscale events particularly in the cryothermic regime with thermal therapies and burn injury. In particular, molecular and cellular events during biopreservation are typically important to define the outcome. The various forms of biopreservation include hypothermic storage (above 0°C), cryopreservation (slow freezing to usually below −80°C), vitrification, ultrafast freezing, or glass formation usually in the same temperature regime as cryopreservation, freeze-drying, or drying without freezing. One important issue is that a bioprotective (cryo- or lyo-) agent is often added to the system prior to cooling or drying with the intent to preserve rather than destroy the cell or tissue system. There are many important biotransport challenges with the introduction of the bioprotective agent. In addition, challenges exist in controlling the alteration of the temperature and/or moisture content to achieve the preserved state followed by thaw (or rehydration) and use. Many reviews on the topic of biopreservation are available and some that are from an engineering perspective are listed here: Coger and Toner;[5] Karlsson and Toner;[6] Han and Bischof.[7]

Molecular mechanisms associated with protein denaturation have been suggested for some time to be critical in high-temperature injury, while there is a growing realization of its importance in low temperature and dehydration injury as well. As an example, recent work in one cell type (AT-1 Dunning rodent prostate cancer cell line) has shown the importance of protein denaturation in both extremes of thermal injury.[8] During heating, a recent work has shown that the kinetics associated with protein denaturation overlaps with overall cellular injury kinetics in the same temperature range.[16] It has also been shown that during freezing relatively little protein denaturation occurs at −20°C in AT-1 prostate tumor cells, but massive denaturation occurs at −80°C.[8] It is interesting to note that −20°C is a survivable insult for this cell type while −80°C is not. In order to better understand thermal injury with regard to protein denaturation, it is instructive to understand protein stability and denaturation during both heating and freezing/cooling. As will be seen in this review, it is particularly important to define the temperature range and/or hydration state within which the comparison is being made.

MACROMOLECULAR TRANSITIONS INCLUDING PROTEIN DENATURATION

There are numerous studies that suggest that the mechanism of hyperthermic thermal injury in cells and tissues is related to macromolecular transitions and of these transitions protein denaturation is critical. In simple terms, lipids, proteins, DNA, and RNA are the building blocks of all cells and tissues, and hence their transitions and stability are directly responsible for injury in cells and tissues. Phase diagrams for some individual lipids have been constructed and are available in the literature.[10–12] The gel to liquid crystalline phase transition of pure lipids generally occurs below 37°C. However, no clear evidence of a single gel to liquid phase transition has been demonstrated in mammalian cells which are typically a mixture of many lipid species as reviewed by Lepock.[13] Nevertheless, cell membrane lipids generally go through a melting transition before any alteration of proteins in mammalian cells,[8] and thus are presumed to be important in inducing hyperpermeability of membranes and cell injury during heating.[14] DNA and RNA are also potential macromolecular targets of thermal injury; they are, however, typically only damaged above 85–90°C. This leaves proteins that denature predominantly between 40–80°C (the temperature regime of thermal therapy) as the last and most important macromolecular group to be affected by thermal excursions. When one views the kinetics of thermal injury or denaturation and extracts activation energies as will be discussed later, these are much higher for protein denaturation than for lipid melting or lipid-mediated events (permeability). The activation energy for thermal injury and protein denaturation have in fact been shown to be very similar and have led Westra and Dewey[15] and later others to argue that protein denaturation is in fact the most important event in cell killing due to heat.[13] In fact, the amount of protein denaturation necessary to destroy cells has been quantified for acute and chronic survival. In AT-1 cells, thermal conditions that achieve denaturation of 28% of total cellular protein correlate with almost immediate cell death by PI membrane dye assay (acute membrane integrity assay), whereas denaturation of only 7% of total cellular protein is necessary to destroy the cells as assessed by the ability of the cells to form colonies by the clonogenic assay (survival with reproduction).[16] Similar estimates for other cell types have been reported.[13]

This review will discuss protein stability with regard to both heat and cold freezing. We will then also discuss the measurement and modeling of protein denaturation in various systems and the potential modifiers of this event.

WHAT ARE PROTEINS AND HOW DO THEY DENATURE?

Proteins are complex macromolecules (polypeptides) that play important roles as single molecules (drugs, enzymes), cellular constituents (membrane and cellular organelle components) as well as in tissues such as extracellular matrix components, most notably collagenous tissues. Proteins have complex structures that are important to their function. This structure has numerous levels including primary, secondary, tertiary, and quaternary.[17] The primary structure is associated with the covalent bonds between the atoms making up the protein molecule; the secondary structures involve primarily hydrogen bonding between the atoms (although some disulfide

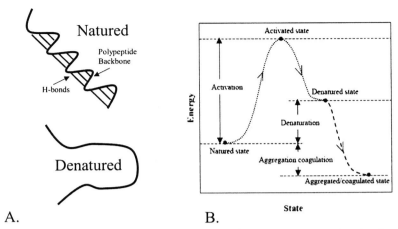

FIGURE 1. Energy states of protein denaturation. (**A**) Hydrogen bonds of a simple alpha helix secondary structure within one protein are shown schematically. During denaturation, these hydrogen bonds break, and the protein backbone unravels. This process is shown energetically for any protein in (**B**) where the natured state is the initial state; there is a period of activation (energy input to the system) followed by denaturation to a energy state above that of the initial natured state, further followed randomly in some cases by coagulation and aggregation to an energy state below that of the initial natured state, which is irreversible. Note that the amount of energy of activation usually is much higher than that of denaturation and aggregation/coagulation; **B** is just a cartoon representation of the relationship between energy and state (i.e., not to scale).

bonding can also occur), thereby creating the well-known alpha helix and beta sheet structures, whereas the ultimate 3D folded structure of a whole globular protein is called the tertiary structure and is important to protein function. Quaternary structure usually involves the conformational fitting of two proteins together associated with specific function (i.e., receptor ligand binding). Function is associated with the native or higher-structured state of the protein as shown for a simple idealized case of an alpha helix in FIGURE 1A. When this structure is changed or altered, the protein is unable to carry out its specific function. This may involve either partial or total unraveling of the protein through changes to the hydrogen bonding which define the higher-order native structure of the protein (FIG. 1A). This process is called denaturation. It can be either partial or total, meaning that the process may not necessarily complete for a specific condition, and it can also be reversible or irreversible. As an example, cold denaturation of proteins is often reversible.[18] Denaturation does not involve breaking the individual covalent bonds between the atoms of the polypeptide backbone of the protein molecule.

In the case of heat denaturation of proteins in the temperature range of thermal therapy (i.e., > 45°C), the process is generally considered irreversible. In single cells, there are groups of proteins that are differentially sensitive to thermal denaturation and hence show different transition temperatures T_m (i.e., temperature at which half the protein denatures). The lowest temperature at which protein denaturation is first detectable is usually called the onset temperature, which usually corresponds to 40–

45°C for mammalian cells and depends on the heating rate. Transitions are usually 10–12°C in width which can extend 5–6°C below T_m.[13] Whole cells usually show 5 to 7 major protein peaks over the temperature range of 40–90°C. These groups separate out according to Lepock's observations in RBCs to be spectrin, membrane cytoskeleton, transmembrane protein groups, and hemoglobin.[13] The term denaturation is often used inappropriately to refer to aggregation, coagulation, and gelation which are related to protein interactions that can be accompanied by denaturation.[19] In brief, denaturation is the process in which a protein is transformed from an ordered ("native") to a less-ordered state due to the rearrangement of hydrogen bonding without any change to covalent bonds (e.g., from α helix to extended β sheet[8]). Aggregation is a general term referring to protein–protein interactions with formation of complexes of higher molecular weights. Coagulation is the random aggregation of already denatured protein molecules and is usually a thermally irreversible process. Finally, gelation is an orderly aggregation of proteins, which may or may not be denatured, forming a three-dimensional network that may be thermally reversible. In this review, we assume that the thermally irreversible loss of protein stability and function is rate limited initially by the denaturation step, which may then be followed by coagulation, aggregation, and/or gelation. The total amount of heat released during denaturation (followed by coagulation, aggregation, and/or gelation) of purified proteins as well as whole cell preparation is usually between 20 and 40 J/g protein[16,20] and 10 and 60 J/g protein for rat tail collagen, depending on hydration.[21] It is important to note that proteins often aggregate during the denaturation process both *in vitro* and in cells. In fact, some studies suggest that these aggregates can contain both native and denatured proteins.[13]

From thermodynamics, denaturation can be viewed as the condition when sufficient energy is transferred to a native (natured–conformational structure) protein such that an alteration in its molecular conformation can take place. This energy usually has two parts—one is an activation energy barrier (kinetic), and the other is enthalpic (total heat absorption or release) (see FIG. 1B). The kinetic (activation energy) barrier determines the temperature and time dependence of the denaturation process. The total enthalpic heat change is calorimetrically measured when the phase transition takes place (i.e., 10–60 J/g protein). As the temperature is raised, it becomes thermodynamically favorable (i.e., sufficient energy to exceed the free-energy barrier of activation) for the protein to denature. The final denatured state can be at a higher or lower total energy than the original state. The final state is at a lower energy than the initial state when coagulation, aggregation, and/or gelation of denatured proteins occur, which is a strongly exothermic process.

Cold and freeze denaturation of proteins is also possible and can be both reversible and irreversible. Privalov and others have carried out work over many years in this area. Privalov has shown that cold temperatures can induce unfolding of proteins (denaturation), but this is usually reversible in nature. The transition temperatures are lower, usually in the 0–10°C range. The total amount of thermal energy change during these transitions, however, is typically less than that for thermal denaturation. Others have shown that protein structural changes occur during freezing and drying in protein solutions[22,23] which has led to studies to try to protect proteins under this stress.[24] Protein denaturation has also recently been measured in whole mammalian cells during freezing.[8] The suggested mechanism for protein denaturation during freezing (and drying) is dehydration because freezing tends to sequester water into growing ice crystals, leaving highly concentrated salts (electrolytes) behind. This can lower the pH, destabilize, and precipitate the proteins. The overall thermodynamics of this process is far less understood compared to those already described for heating. Modeling of the thermodynamic phenomenon or processes and the resulting intense osmotic forces which can change molecular hydration and thus drive lipid and protein transitions during freezing and drying have been examined in the literature.[25–27] There are however, few if any studies that model the kinetics of protein denaturation occurring due to freezing and/or drying conditions.

MEASUREMENT OF PROTEIN DENATURATION

Several techniques, including X-ray crystallography, nuclear magnetic resonance (NMR) spectroscopy, Fourier transform infrared spectroscopy (FTIR), circular

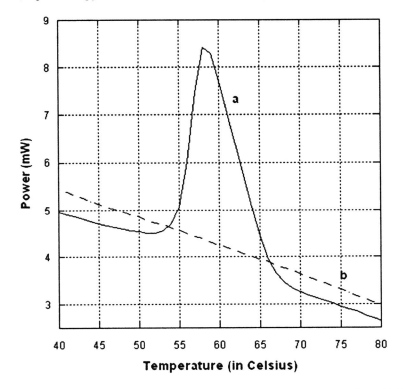

FIGURE 2. Example of how to measure rat tail collagen protein denaturation with DSC (unpublished result). An irreversible denaturation endotherm (**a**) is measured during heating as a function of temperature from 20 to 90°C at a heating rate of 10°C/min. By assignment of an appropriate baseline, the area under the curve can be calculated, which represents the total energy of denaturation. The second curve (**b**) is a rescan after returning to room temperature over the same temperature range (20–90°C) at 10°C/min. The lack of an endotherm in **b** indicates the irreversibility of the denaturation and coagulation/aggregation measured in scan **a**.

dichroism (CD), differential scanning calorimetry (DSC), birefringence, optical opacity, shrinkage, and histology have been used to either directly or indirectly investigate protein structure and structural change. Both X-ray crystallography and NMR spectroscopy can determine the 3D structure of pure proteins; X-ray crystallography, however, requires the protein sample to be in a crystalline state, and NMR spectroscopy can only be used to study small proteins or protein domains (≤20 kDa).[17] Both FTIR and CD have been widely used to study protein secondary structures and their changes during denaturation. DSC, however, is the only technique that can be used to investigate the calorimetric effect associated with protein structural change.

DSC has been used to study the calorimetric events associated with the denaturation of many different isolated pure proteins.[13,28] Considering that protein properties are strongly affected by their biochemical environment (esp. pH),[29,30] several studies investigated the thermally induced protein change in their native environment (i.e., intact cells) using DSC as reviewed in Lepock.[13,28] These studies include the investigation of thermally induced protein denaturation in the erythrocyte membrane, Chinese hamster lung (CHL) V 79 cells, Chinese hamster ovary (CHO) cells, and Wistar rat hepatocytes. Four to seven denaturation peaks, which were thought to be representative of major protein groups, were usually observed in these cell lines. In the case of human erythrocytes, denaturation peaks that were determined *in situ* could be assigned to specific proteins such as spectrin and several other membrane and cytosolic proteins.[31–33] It was further found that a protein/protein group transition around 60°C, rather than the denaturation of spectrin, is the rate-limiting step of thermally induced hemolysis of human erythrocyte.[33] *In situ* thermal denaturation of thermally labile proteins below 50°C has also been implicated as the mechanism of heat shock response and thermotolerance, as discussed in several reviews.[28,34] The DSC data from thermal denaturation of isolated pure proteins have been fit to a first-order kinetic model.[35,36]

An example of how the enthalpy or heat of denaturation for proteins can be measured with DSC is shown in FIGURE 2. Here the total heat absorption is the area between the signal and the baseline. This can be expressed as:

$$\int_0^\infty \partial P dt = \int_{T_{ref}}^{T_{final}} \partial P \frac{dT}{B} = Q_{ex} \qquad (1)$$

Where P is power absorbed (W), t is time (s), T is temperature (°C), B is the constant heating rate (°C/s), and Q_{ex} is the heat of denaturation or enthalpy for the process (J). This approach to measure protein denaturation enthalpies can also be used with some modification to measure total denaturation of proteins within cell systems.[16]

Another device that can be used in the measurement of protein denaturation is FTIR, which can detect alteration of overall protein secondary structure within protein systems. The amide-I and amide-II bands, arising from vibrations of protein polypeptide backbones, have been widely used to determine the protein secondary structures of isolated proteins[38,39] and are diagnostic for the overall protein secondary structure within intact mammalian cells[8,16] and plant systems.[40] FTIR has also been used to monitor thermally induced changes of protein secondary structure in pure proteins[41,42] and other intact cells or organisms.[43] An example of how FTIR

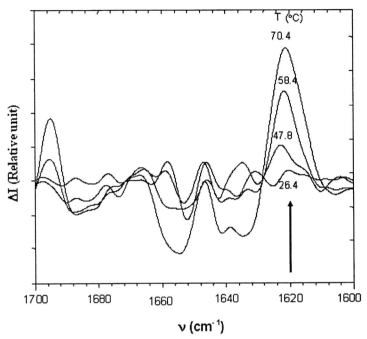

FIGURE 3. Example of how to measure protein denaturation with FTIR. By tracking the area under the growing denaturation (aggregation) band, one can construct a fractional denaturation or kinetic graph as that shown for DSC in FIGURE 5. (Reproduced from Bischof et al.,[8] with permission of Elsevier.)

can be used to measure protein denaturation is shown in FIGURE 3. Here the peak for extended beta sheets centered at 1620 cm^{-1} is shown to grow as the temperature increases. By comparing successive peak areas, one can calculate the kinetics of protein denaturation in whole cells or isolated proteins. This approach has recently been used to show that the kinetics of protein denaturation *in situ* in prostate cancer cells is similar as measured by either DSC or FTIR.[16]

In the case of collagenous tissues, there are further techniques available to assess protein denaturation including shrinkage, and histological changes as reported for tendons in Chen et al.[44] (see FIG. 6) and during "thermal fixation" due to protein coagulation in kidney tissue.[9] Other techniques which measure birefringence change, opacity changes, and other approaches are reviewed in Pearce and Thomsen,[45] Wright and Humphrey,[46] and He and Bischof[47] (see TABLE 4). To date, no measurement of freezing-induced protein denaturation (i.e., collagen) in tissues exists to the authors' knowledge.

KINETIC MODELING OF HEAT PROTEIN DENATURATION

Kinetic models assume that protein denaturation is a rate process. The simplest but most widely used kinetic model is the first order irreversible rate reaction model.

In this model, proteins are assumed either native (N) or denatured (D) as described by the following equation:

$$N \xrightarrow{k} D \tag{2}$$

The heat induced fractional denaturation (F_d) can be derived as follows:

$$F_d = 1 - \exp\left(-\int_0^t k \, dt\right) \tag{3}$$

where the reaction rate k can be calculated using the well-known Arrhenius model[45,47,48]

$$k = A \exp\left(-\frac{\Delta E}{RT}\right) \tag{4}$$

where the activation energy (ΔE) and the pre-exponential factor (A, usually called frequency factor in literature) are the two kinetic parameters that must be determined for a given protein by experiments *a priori*. The reaction rate k can also be calculated using the well-known absolute rate process model[30] as

$$k = \left(\frac{k'}{h'}\right) T \exp\left(-\frac{\Delta g''}{RT}\right) \tag{5}$$

$$\Delta g'' = \Delta h'' - T \Delta s'' \tag{6}$$

where k' and h' are the Boltzmann and Planck's constants, respectively; $\Delta g''$ is the Gibbs free energy of activation that depends on activation enthalpy $\Delta h''$ and entropy $\Delta s''$ (kinetic parameters) that need to be determined for a given protein by experiments *a priori*. The relationship between the parameters of the Arrhenius model and those of the absolute rate theory model for an ith order reaction is as follows:[30,45,47]

$$\Delta E = \Delta h'' - iRT \tag{7}$$

$$A = \left(\frac{k'}{h'}\right) T \exp\left(-\frac{\Delta s''}{R} + i\right) \tag{8}$$

where, 'i' is equal to one for a kinetic process described by equation (2). The difference between ΔE and $\Delta h''$ (i.e., iRT) is less than 0.7 kcal mol^{-1} for a first-order reaction below 80°C. This small difference is negligible for most situations due to the relative high value of the activation energy for protein denaturation (>12 kcal mol^{-1} as reviewed in Ref. 47). Note that the frequency factor in the Arrhenius model for thermal injury or denaturation is usually very large (> $10^{13} s^{-1}$).[47] This term is generally balanced with a much smaller exponential term containing the activation energy that diminishes the rate except within the temperature range where denaturation occurs.

Higher-order kinetic models (i.e., models assuming that one or more intermediate states exist between the native and denatured states) were also adopted in several studies as reviewed in Refs. 47 and 49. It was found that higher-order models could fit experimental results better than the first-order model in those studies. In addition, for collagenous tissue, phenomenological models for the denaturation process have also been developed which fit the data well as reviewed in Wright and Humphrey.[46] However, the first-order kinetic model has been shown to be sufficient for most situations and one or more break points can be introduced to allow a better fit to experimental data using the first-order model if the temperature range is very wide, as argued in several reviews.[45,47,48]

As already noted, protein denaturation has been suggested to be the rate-limiting event in hyperthermic cell death,[13] and a recent work has shown that the measured cell death and protein denaturation can be correlated directly.[16] This study further showed the importance of temperature range in which the studies were undertaken. Another equally important point, which was discussed in He and Bischof[47] is the necessity to clearly appreciate the relationship between cell/tissue survival(s) and cell/tissue injury (Ω) described by the following equation:

$$\Omega = \ln\left(\frac{1}{s}\right) = \int_0^\tau A e^{-\frac{\Delta E}{RT(t)}} dt. \tag{9}$$

If T is constant, then:

$$\frac{\Omega}{\tau} = A e^{-\frac{\Delta E}{RT}}. \tag{10}$$

Taking the natural logarithm of both sides of equation (10) and rearranging the resultant terms give:

$$\ln(\tau) = -\ln\left(\frac{A}{\Omega}\right) + \frac{\Delta E}{RT}. \tag{11}$$

If Ω is equal to 1, then:

$$\ln(\tau) = -\ln(A) + \frac{\Delta E}{RT}. \tag{12}$$

Equation (12) has been used in many studies to extract ΔE and A from experimental data as reviewed in Pearce and Thomsen;[45] Diller et al.;[48] and Cravalho et al.[49] When linear fitting equation (12) to the experimental data of $\ln(\tau)$ vs. T^{-1}, ΔE and A can be found from the slope and intercept of the fitted line (such as shown for cellular injury in FIG. 4 or protein denaturation by time of shrinkage τ_2 in FIG. 6), respectively. Since Ω is equal to 1 in this equation, the survival is implied to be 36.7% according to equation (9). However, in some studies of hyperthermic cell injury, an arbitrary survival criterion s_0 instead of 36.7% was used, and Ω was forced to be 1.

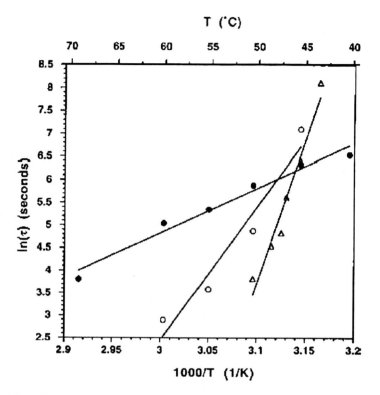

FIGURE 4. Arrhenius plot of Dunning AT-1 prostate cancer cell injury as measured by clonogenics, PI uptake, and calcein leakage. See TABLE 2 for parameter listing. (Reproduced from Bhowmick et al.,[51] with permission of ASME.)

The relationship between the kinetic parameters thus determined and those determined using 36.7% survival can be found using equations (11) and (12) as follows:[47]

$$A_{36.7\%} = A_{s_0} \times \Omega = A \times \ln\left(\frac{1}{s_0}\right), \tag{13}$$

$$E_{36.7\%} = E_{s_0} \tag{14}$$

where subscripts "36.7%" and "s_0" represent the survival used to determine the kinetic parameters (i.e., E and A). Equations (13) and (14) show that the activation energy is not dependent on the survival criterion selected for $\Omega = 1$, while the frequency factor will be different. This conclusion is correct only if the cell killing is indeed governed by first-order kinetics.

Note further that the target temperature and the room temperature under which most biological samples were prepared are different; caution should be taken when extracting kinetic parameters using equation (12) because it assumes that the injury or denaturation induced by the transient heating or cooling between room and target temperature is negligible. This assumption may not be tenable in some situations. For example, He and Bischof [50] have shown that when the target peak temperature was higher than 65°C, the injury percentage of suspended human renal cell carcinoma cells measured by propidium dye uptake was as high as 70% during the transient heating up (at 130°C/min) and cooling down (at 65°/min) between room temperature and 65°C. A general guide on when equation (12) is applicable was given in He and Bischof.[50] If equation (12) is not applicable, a nonlinear method using equation (9) together with an optimization procedure should be employed as has been done by He and Bischof[50] for thermal injury in human renal cell carcinoma cells and He et al.[9] for thermal injury in microwave thermal therapy of normal porcine kidney tissue. Moreover, this nonlinear method is necessary when extracting kinetic parameters from *in vivo* studies using minimally invasive surgical probes, for example, microwave, radiofrequency (RF), laser, or high-intensity focused ultrasound (HIFU) probes,[47,73] where it is very difficult to create a thermal history with a dominant isothermal portion.

KINETIC PARAMETER DISCUSSION

Many studies have been performed to determine the kinetic parameters of the first order kinetic model for many different types of macromolecules, cells and tissue (see TABLE 1–3). It shows that all the data points are located approximately on a straight line. A linear fitting to the data points gives the following relationship between the two parameters ($R^2 = 0.9966$).[47]

$$\ln(A) = 1.591\Delta E - 9.2369 \quad (15)$$

where the units of A and ΔE are s^{-1} and kcal mol^{-1}, respectively. This relationship indicates that the two kinetic parameters are intrinsically connected with each other in a linear fashion. Using equations (6)–(8), the relationship between $\ln(A)$ and ΔE can be derived as follows:[47]

$$\ln(A) = \frac{1}{RT}\Delta E + \ln\left(\frac{k'T}{h'}\right) - \frac{\Delta g}{RT} \quad (16)$$

Comparing equations (15) and (16) gives $T = 43.6°C$ and $\Delta g = 24.4$ kcal mol^{-1}. However, if T changes from 40°C to 80°C, the coefficient before ΔE in equation (15) is from 1.61 to 1.43. The variation is within 10% of 1.591. When the temperature varies between 40°C and 80°C, to keep the same intercept in equations (15) and (16), the Gibbs free energy of activation ($\Delta g''$) must vary between 24.1 and 27.2 kcal/mol. Therefore, besides the small temperature range, another reason for the linear relationship between $\ln(A)$ and ΔE is that the Gibbs free energies of activation ($\Delta g''$) do not vary much despite the large variation in the values of activation enthalpy ($\Delta h''$) and activation entropy ($\Delta s''$), as has been observed for many thermally induced protein denaturation events as reviewed in Johnson et al.[30]

TABLE 1. Activation energy (ΔE) and frequency factor (A) for macromolecules

Molecules	T (°C)	PH	ΔE (kcal mol^{-1})	A (s−1)	Reference
Lipids					
Phospholipid DMPC	23.9		5.4		(Kanehisa and Tsong 1978)[61]
Phospholipid DPPC	41.4		8.7		(Kanehisa and Tsong 1978)[61]
Phospholipid DSPC	54.9		10.6		(Kanehisa and Tsong 1978)[61]
Protein					
Human serum albumin	20	6	24		(Joly 1965)[29]
	70.5		23.9		
Amylase (malt)	60		40.9	6.6×10^{23}	(Eyring and Stearn 1939)[53]
Myoglobin	25	9	42		(Tanford 1968)[72]
Hemoglobin	60.5	5.7	74.9	5.4×10^{45}	(Eyring and Stearn 1939)[53]
Rennin	50		88.8	6.4×10^{57}	(Eyring and Stearn 1939)[53]
Egg albumin	65	5	131.4	2.0×10^{81}	(Eyring and Stearn 1939)[53]
Peroxidase (milk)	70.1		184.6	1.6×10^{114}	(Eyring and Stearn 1939)[53]
Hemolysin	50		197.4	3.9×10^{129}	(Eyring and Stearn 1939)[53]
Invertase (yeast)	50	5.7	51.8	7.6×10^{30}	(Eyring and Stearn 1939)[53]
	50.2	5.2	109.8	5.9×10^{52}	(Eyring and Stearn 1939)[53]
	55	4	85.7	4.9×10^{69}	(Eyring and Stearn 1939)[53]
	55	3	73.7	4.6×10^{45}	(Eyring and Stearn 1939)[53]
RNA/DNA					
RNA			17		(Eigner et al. 1961)[63]
DNA			24		(Eigner et al. 1961)[63]

NOTE: The data from Eyring and Stearn[53] was reported in Johnson et al.[30] as well.

TABLE 2. Activation energy (ΔE) and frequency factor (A) for AT-1 cells

Cells	Assay	ΔE (kcal mol–1)	A (s−1)	T (°C)	Reference
Dunning AT-1 prostate tumor cells (attached)	Calcein	19.4	5.1×10^{10}	40–70	(Bhowmick et al. 2000)[52]
	PI	58.6	3.0×10^{37}	40–70	
	Clonogenics	125.9	1.04×10^{84}	40–50	
Dunning AT-1 prostate tumor cells (suspended)	DSC	26.3	3.8×10^{14}	44–90	(He and Bischof 2005)[16]
	DSC	76.5	2.0×10^{49}	44–61	
	DSC	138.2	1.0×10^{92}	44–53	
	Clonogenics	151.3	1.5×10^{101}	45–52	
	PI	95.7	1.7×10^{62}	45–59	

Once the kinetics of a macromolecular transition have been experimentally determined and then fit to a model such as the Arrhenius, the parameters that govern the process can be extracted and compared. In the case of denaturation these are $\cdot E$, the activation energy, and A the frequency factor. As already mentioned, these two parameters are not entirely independent as shown by equations (15) and (16), and thus we will focus mostly on comparison of activation energies that have been discussed in the literature. Values less than 10 kcal/mol are typically associated with simple diffusion processes, while values in the 10–30 kcal/mol range can involve enzyme-controlled metabolic processes including membrane transport. Values above 30 kcal/mol are usually associated with protein denaturation.[52] A careful review of the literature shows that activation energies for protein denaturation can range from as low as 25 kcal/mol to as high as 200 kcal/mol, depending on the temperature and pH conditions.[47] Lepock and others have argued that protein denaturation usually only occurs for activation energies above 100 kcal/mol, which is a similar range to many thermal injury processes, particularly as measured by clonogenic assays.[13] This argument has been important in identifying protein denaturation as critical to thermal injury processes. However, the wide range of values reported in the literature to identify protein denaturation suggests that further discussion of how these activation energy values are reported (i.e., under what precise conditions measured) is warranted.

In order to assess the assignment of activation energy associated with protein denaturation better and to correlate this with injury within a single cell type, our group has carried out various assays including dye leakage, dye uptake, reproduction measured by clonogenics, and protein denaturation by DSC and FTIR on AT-1 Dunning cells.[16,52] The activation energies for all of the thermal injury assays appeared at first to be quite different with calcein dye leakage at 19.4, PI uptake at 58.6, and clonogenics at 125.9 kcal/mol for AT-1 cells attached on substrate.[52] Later studies[16] using AT-1 cells suspended in media also measured very different activation energies for PI and clonogenics (95.7 vs. 151.3 kcal/mol; see TABLE 2). The increase in activation energy was found to be due to the decrease of working temperature ranges of the viability assays (14°C for PI vs. 7°C for clonogenics; see TABLE 2). Later, DSC measurements of overall protein denaturation within the same cell type also yielded widely varying activation energies depending on the temperature range, for example, 26.3 kcal/mol for 44–90°C ($\Delta T = 46°C$), 76.5 kcal/mol for 44–61°C ($\Delta T = 17°C$), and 138.2 kcal/mol for 44–53°C ($\Delta T = 9°C$) (TABLE 2). Further, we were able to show that the kinetics were similar between all thermal injury assays and protein denaturation in each of the temperature ranges tested as shown in FIGURE 5.[16] This is further quantitative evidence that protein denaturation is linked to thermal injury. However, it may also begin to help explain some of the noted differences in activation energies, especially for thermal injury reported in the literature by putting the measurement into the context of temperature range. As an example, in the case of clonogenics, measurable viability from this assay is usually only up to 50°C, resulting in a temperature range less than 7°C starting from onset temperature (~43°C), which is consistent with very high activation energy (500–1500 kJ/mol) while the activation energies for other assays with detectable viability up to 70°C are usually smaller (100–400 kJ/mol).[16] Clearly, the temperature ranges within which the measurement of thermal injury and protein denaturation are made are critical to the kinetics and therefore may explain some of the activation energy discrepancies in the literature.

TABLE 3. Activation energy (ΔE) and frequency factor (A) for tissues

Tissue	Assay	ΔE (kcal mol^{-1})	A (s^{-1})	T (°C)	Ref.
Skin					
Pig	Histology (necrosis)	150	3.1 10^{98}	40–70	(Henriques 1947)[64]
Rat	Birefringence loss in collagen	73.2	1.6 10^{45}	45–90	(Pearce 1995)[45]
Retina & Cornea & Lens					
Retina	Microscopically round crater-like lesion with whitened edges	150	1.3 10^{99}		(Welch and Polhamus 1984)[65]
Cornea	Complete loss of transparency	23.6	2.9 10^{13}	40–70	(Bhowmick et al. 2000)[51]
Cornea	Optical attenuation	25.3	2.07 10^{15}	60–95	(Kampmeier et al. 2000)[66]
Lens capsule	Calorimetry	205.3	3.85 10^{137}	60–65	(Miles 1993)[67]
Tendon					
Rat tail tendon	Calorimetry	124.3	6.66 10^{79}	57–60	(Miles et al. 1995)[37]
Rat tail tendon	Birefringence loss	88.3	1.9 10^{56}	50–60	(Maitland and Walsh 1997)[68]
Kangaroo tendon	Shrinkage	140.6	1.9 10^{89}	58–62	(Weir 1949)[69]
Capsule					
Joint capsule	Shrinkage	8.1	4 10^{5}	44–60	(Moran et al. 2000)[70]
Joint capsule	Shrinkage	55.8	1.85 10^{32}	60–70	(Moran et al. 2000)[70]
Kidney					
Porcine cortex	Histology (thermal fixation) Histology	95.4	1.48 10^{60}	70–110	(He et al. 2004)[9]
Heart					
Heart muscle	Loss of birefringence	39.2	3.5 10^{22}	70–110	(Han and Pearce 1990)[62]
Rabbit myocardium	Loss of birefringence	30.6	3.1 10^{20}		(Pearce and Thomsen 1995)[45]
Egg					
Egg white	Onset of whiteness	92	3.8 10^{57}	60–90	(Yang et al. 1991)[71]
Egg yolk	Onset of whiteness	93	3.1 10^{56}	60–90	(Yang et al. 1991)[71]

NOTE: In studies that gave the activation enthalpy and entropy, the value of ΔE and A were calculated using equations (7) and (8), and a reference temperature equal to the average of the reported temperature range.

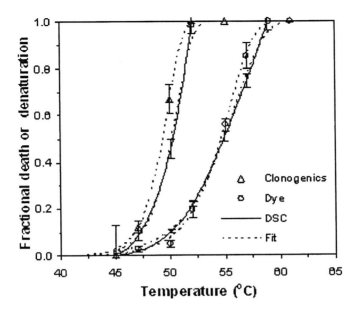

FIGURE 5. The kinetics of protein denaturation within Dunning AT-1 prostate cancer cells matches with the kinetics of cellular injury as measured by clonogenics and PI uptake within the noted temperature ranges. (Reproduced from He et al.,[16] with permission of Springer Science and Business Media.)

MODIFICATION OF KINETICS

Interestingly, it has been shown in several studies that modification of the denaturation process by mechanical and chemical loading can change the kinetics, but not the activation energy of the process. This is shown for chordae tendinae that have either been chemically loaded with glycerol or prestressed with a mechanical load as shown in FIGURE 6.[44] As can be seen, the slope of the process does not change at variable loads although there is a definite shift in kinetics to a higher absolute temperature. This is suggested to be due to a change in frequency factor, or entropy of activation, which is associated with the configurational entropy in the protein molecules themselves not their heat-labile bonds.[44,46]

There are modifiers (sensitizers and protectants) for denaturation in both heat and cold/freeze conditions as shown in TABLE 4. In the case of heat denaturation, pH is known to accentuate heat denaturation in individual proteins.[29,30,53] Other hyperthermic sensitizers in cells include methanol, ethanol, propanol, and butanol. Other agents (thiol-specific oxidative agents) can also sensitize protein to denaturation as reviewed previously.[13] There are also agents such as glycerol, and D_2O, as well as other proteins that can retard or delay protein denaturation in cells. The class of proteins which retard protein denaturation are called heat-shock proteins (HSPs) and are associated with cellular thermotolerance and binding to newly denatured proteins.[13] HSPs can be induced by sublethal heat treatments as reviewed in Lepock[13] and have

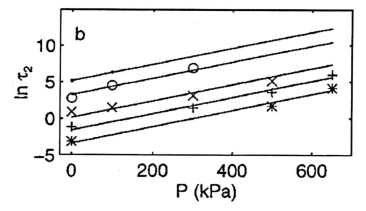

FIGURE 6. The kinetics of chordae tendinae shrinkage (τ_2) vs. mechanical loading. Symbols represent different constant temperatures 65 to 90°C. The graph shows a shift in kinetics but not activation energy (i.e., slope) for different mechanical loading conditions of tendons. Here the time for shrinkage is clearly dependent (i.e., increased) at higher mechanical loads for the same temperature. (Reproduced from Chen et al.,[44] with permission of ASME.)

also been reported for certain cold-shock treatments in some cells.[54] While the presence of HSPs can shift the kinetics of thermal injury during hyperthermia, it has not been determined what if any role HSPs play in cold- or freeze-induced protein denaturation. A recent study documents the kinetics of synthesis of the HSP induction under a variety of conditions (varied temperatures and times) as well as showing the protection afforded during hyperthermia.[55]

Glycerol (and other cryoprotectants) have been proposed to protect against dehydration-induced freezing injury. Arakawa and co-workers have proposed the mechanism whereby protection is afforded by the preferential exclusion hypothesis.[56] Furthermore, Hanafusa has suggested specific protein cryoprotectant interactions that may protect protein from denaturation by manipulation of protein hydration.[57] Crowe and colleagues have suggested the water replacement hypothesis of lyo- or cryoprotectant, e.g., trehalose, for dried membranes.[58,59] In all of these proposed mechanisms, the hydrogen bonding of the glycerol or cryoprotectant with water and with the proteins appears to be important. In collagenous tissues dehydration by cross-linking (or drying) can also lead to increased stability, although the dehydration due to cross-linking appears to be the dominant effect rather than the cross-linking itself.[21] Miles has shown that ethylene glycol (and other protectants) can stabilize collagenous tissues against heat denaturation, most likely by entropic effects that shift the denaturation to higher temperatures.[60] In addition, other studies have shown that increasing the mechanical loading of proteins (collagenous tissues) also shifts the denaturation to higher temperatures as reported in Chen et al.[44] and later reviewed in Wright and Humphrey[46] (see FIG. 6). In all of these studies, the hydration of the protein and the hydrogen bonding within the protein molecule are important to the denaturation process. Interestingly, a thermotolerant state (one that resists protein denaturation) can be induced in yeast and microorganisms by the

TABLE 4. Studies of protein denaturation and modification at the molecular, cellular, and tissue level

Level	Heat	Cold/Freeze	Modifiers
Molecular	DSC (Joly 1965)[29] (Eyring and Stearn 1938)[53] (Johnson et al. 1974)[30]	FTIR (Pikal-Cleland et al. 2000)[23] DSC (Privalov 1990)[18]	PH (Joly 1965)[29] (Eyring and Stearn 1939)[53] (Johnson et al 1974)[30] Cryoprotectants (CPA) (Arakawa et al. 1990)[56] (Hanafusa 1992)[57]
Cellular	DSC (Lepock 2003)[13] DSC and FTIR (He et al. 2004)[16]	FTIR (Bischof et al. 2002)[8]	HSPs (Freeman et al. 1999)[34]
Tissue	Shrinkage (Chen et al. 1998;[44] Wright and Humphrey 2002[46]) DSC (Miles et al. 1995)[37] Birefringence (Pearce and Thomsen 1995)[45] Thermal fixation histology (He et al. 2004)[9]		Chemical loading: glycerol, propanediol, dehydration, X-linking (Miles and Burjanadze 2001;[60] Miles et al. 2005)[21] Mechanical loading (Chen et al. 1998)[44] (Wright and Humphrey 2002)[46]

NOTE: Cryoprotectant groups include various glycols (glycerol, propanediol), sugars (e.g., trehalose, sucrose) and other low molecular weight polymers (e.g., DMSO) (Han et al. 2004)[7]. Compounds which promote heat shock proteins (heat protectant) synthesis include a wide variety of heavy metals, alcohols and other additives (Freeman et al. 1999)[34].

introduction of trehalose and glycerol, as previously reviewed.[13] It is noteworthy that these chemicals are natural cryoprotectants that are suggested to compete with water to hydrogen bond directly with proteins as already discussed above.

SUMMARY AND FUTURE OUTLOOK

This review has presented an overview of protein denaturation from temperature (predominantly high and some low) and its modification by use of chemical additives and pH modification, as well as modification of the mechanical environment (stress) of the proteins such as collagen. These modifiers are able to change the kinetics of the protein denaturation during heating. While pH can change both the activation enthalpy and frequency factor, many chemical and mechanical modifiers reviewed here only change the frequency factor (activation entropy) of the denaturation kinetics. Often, the modification affecting activation entropy appears to be linked to the hydration of the protein. While the heat-induced denaturation of proteins is reasonably well understood, the heat denaturation of structural proteins (i.e., collagen) within whole tissues remains an area of active research. In addition, while a literature exists on reversible protein denaturation during cold temperatures, there is relatively little known about the kinetics of protein denaturation during both freezing and drying. Such understanding will have an important impact on applications ranging from preservation of biomaterials and pharmaceutics to cryosurgery. Interestingly, both freezing and drying involve drastic shifts in the hydration of the proteins. Therefore, understanding the impact of protein hydration on protein stability at the molecular, cellular and tissue level is likely to be increasingly important in this evolving area of research.

ACKNOWLEDGMENTS

The authors acknowledge the mentorship and guidance of Mehmet Toner at Harvard Medical School, who was advisor to both authors at different times and was instrumental in the development of many of the ideas in this work. J.C.B. was supported by a fellowship from the Alexander von Humboldt Foundation as well as a grant from the National Institutes of Health (Grant 2R01CA075284).

REFERENCES

1. AMERICAN BURN ASSOCIATION. 2000. Burn incidence and treatment in the US: 2000 fact sheet. http://www.ameriburn.org/pub/BurnIncidenceFactSheet.htm
2. LEE, R., E. CRAVALHO & J. BURKE. 1992. Electrical Trauma: the Pathophysiology, Manifestations and Clinical Management. Cambridge University Press.
3. AMERICAN CANCER SOCIETY. 2005. Projected Cancer Incidence Table. www.acs.org.
4. AMERICAN HEART ASSOCIATION. 2005. Projected Cardiovascular Disease Statistics. www.americanheartassociation.org.
5. COGER, R. & M. TONER. 1995. Preservation Techniques for Biomaterials. Handbook of Biomedical Engineering, Vol. I. J.D. Bronzoni, Ed.: 1567–1577. CRC Press. Boca Raton, FL.
6. KARLSSON, J. & M. TONER. 2000. Cryopreservation. Principles of Tissue Engineering. J.P. Vacanti & R.P. Langer, Eds.: 293–307. Academic Press. San Diego, CA.
7. HAN, B. & J.C. BISCHOF. 2004. Engineering challenges in tissue preservation. Cell Preservation Technol. **2**: 91–112.
8. BISCHOF, J C., W.F. WOLKERS, N.M. TSVETKOVA, et al. 2002. Lipid and protein changes due to freezing in Dunning AT-1 cells. Cryobiology **45**: 22–32.
9. HE, X., S. MCGEE, J.E. COAD, et al. 2004. Investigation of the thermal and tissue injury behaviour in microwave thermal therapy using a porcine kidney model. Int. J. Hyperthermia **20**: 567–593.
10. CAFFREY, M. 1987. The combined and separate effects of low temperature and freezing on membrane lipid mesomorphic phase behavior: relevance to cryobiology. Biochim. Biophys. Acta **896**: 123–127.

11. QUINN, P.J. 1988. Effects of temperature on cell membranes. Symp. Soc. Exp. Biol. **42:** 237–258.
12. QUINN, P.J. 1989. Membrane lipid phase behavior and lipid-protein interactions. Subcell. Biochem. **14:** 25–95.
13. LEPOCK, J.R. 2003. Cellular effects of hyperthermia: relevance to the minimum dose for thermal damage. Int. J. Hyperthermia **19:** 252–266.
14. LEE, R.C., D. ZHANG & J. HANNIG. 2000. Biophysical injury mechanisms in electrical shock trauma. Annu. Rev. Biomed. Eng. **2:** 477–509.
15. WESTRA, A. & W.C. DEWEY. 1971. Variation in sensitivity to heat shock during the cell cycle of Chinese hamster cells in vitro. Int. J. Radiat. Biol. **19:** 467–477.
16. HE, X., W.F. WOLKERS, J. H. CROWE, et al. 2004. In situ thermal denaturation of proteins in dunning AT-1 prostate cancer cells: implication for hyperthermic cell injury. Ann. Biomed. Eng. **32:** 1384–1398.
17. ALBERTS, B., D. BRAY, A. JOHNSON, et al. 1998. Essential Cell Biology: An Introduction to the Molecular Biology of the Cell. Garland Publishing, Inc. New York.
18. PRIVALOV, P.L. 1990. Cold denaturation of proteins. Crit. Rev. Biochem. Mol. Biology **25:** 281–305.
19. GOSSETT, P., S. RIZVI & R. BAKER. 1984. Symposium: gelation in food protein systems quantitative analysis of gelation in egg protein systems. Food Technology **38:** 67–96.
20. PRIVALOV, P. & N. KHECHINASHVILI. 1974. A thermodynamic approach to the problem of stabilization of globular protein structure: a calorimetric study. J. Mol. Biol. **86:** 665–684.
21. MILES, C.A., N.C. AVERY, V.V. RODIN & A.J. BAILEY. 2005. The increase in denaturation temperature following cross-linking of collagen is caused by dehydration of the fibres. J. Mol. Biol. **346:** 551–556.
22. CARPENTER, J.F., S.J. PRESTRELSKI & T. ARAKAWA. 1993. Separation of freezing- and drying-induced denaturation of lyophilized proteins using stress-specific stabilization. I. Enzyme activity and calorimetric studies. Arch. Biochem. Biophys. **303:** 456–464.
23. PIKAL-CLELAND, K.A., N. RODRIGUEZ-HORNEDO, G.L. AMIDON & J.F. CARPENTER. 2000. Protein denaturation during freezing and thawing in phosphate buffer systems: monomeric and tetrameric beta-galactosidase. Arch. Biochem. Biophys. **384:** 398–406.
24. ANCHORDOQUY, T.J., K.I. IZUTSU, T.W. RANDOLPH & J.F. CARPENTER. 2001. Maintenance of quaternary structure in the frozen state stabilizes lactate dehydrogenase during freeze-drying. Arch. Biochem. Biophys. **390:** 35–41.
25. WOLFE, J., Z. YAN & J.M. POPE. 1994. Hydration forces and membrane stresses: cryobiological implications and a new technique for measurement. Biophys. Chem. **49:** 51–58.
26. WOLFE, J. & G. BRYANT. 1999. Freezing, drying, and/or vitrification of membrane-solute-water systems. Cryobiology **39:** 103–129.
27. PARSEGIAN, V.A., R.P. RAND & D.C. RAU. 2000. Osmotic stress, crowding, preferential hydration, and binding: a comparison of perspectives. Proc. Natl. Acad. Sci. USA **97:** 3987–3992.
28. LEPOCK, J. 1997. Protein denaturation during heat shock. Adv. Mol. Cell Biol. **19:** 223–259.
29. JOLY, M. 1965. A physico-chemical approach to the denaturation of proteins.: 153–170. Academic Press. London.
30. JOHNSON, F.H., H. EYRING & B.J. STONER. 1974. Temperature. In The Theory of Rate Process In Biology and Medicine. John Wiley & Sons. New York.
31. BRANDTS, J., L. REICKSON, K. LYSKO, et al. 1977. Calorimetric studies of the structural transitions of human erythrocyte membrane: the involvement of spectrin in a transition. Biochemistry **16:** 3450–3454.
32. LYSKO, K., R. CARLSON, R. TAVERNA, et al. 1981. Protein involvement in structural transitions of erythrocyte ghosts: use of thermal gel analysis to detect protein aggregation. Biochemistry **20:** 5570–5576.
33. LEPOCK, J.R., H.E. FREY, H. BAYNE & J. MARKUS. 1989. Relationship of hyperthermia-induced hemolysis of human erythrocytes to the thermal denaturation of membrane proteins. Biochim. Biophys. Acta **980:** 191–201.
34. FREEMAN, M., M. BORRELLI, M. MEREDITH & J. LEPOCK. 1999. On the path to the heat shock response: destabilization and formation of partially folded protein intermediates, a consequence of protein thiol modification. Free Radical Biol. Med. **26:** 737–745.

35. ORTEGA, A., J. SANTIAGO-GARCIA, J. MAS-OLIVA & J.R. LEPOCK. 1996. Cholesterol increases the thermal stability of the Ca2+/Mg(2+)-ATPase of cardiac microsomes. Biochim. Biophys. Acta **1283:** 45–50.
36. SENISTERRA, G.A. & J.R. LEPOCK. 2000. Thermal destabilization of transmembrane proteins by local anaesthetics. Int. J. Hyperthermia **16:** 1–17.
37. MILES, C.A., T.V. BURJANADZE & A.J. BAILEY. 1995. The kinetics of the thermal denaturation of collagen in unrestrained rat tail tendon determined by differential scanning calorimetry. J. Mol. Biol. **245:** 437–446.
38. BANDEKAR, J. 1992. Amide modes and protein conformation. Biochim. Biophys. Acta **1120:** 123–143.
39. SUREWICZ, W., H. MANTSCH & D. CHAPMAN. 1993. Determination of protein secondary structure by Fourier transform infrared spectroscopy: a critical assessment. Biochemistry **32:** 389–394.
40. WOLKERS, W., M. ALBERDA, M. KOORNNEEF & F. HOEKSTRA. 1998. Heat stability of proteins in maturation defective mutants of *Arabidopsis thalania*: a FT-IR microspectroscopy study. Plant J. **16:** 133–143.
41. SUREWICZ, W., J. LEDDY & H. MANTSCH. 1990. Structure, stability and receptor interaction of cholera toxin as studied by Fourier-transform infrared spectroscopy. Biochemistry **29:** 8106–8111.
42. ARRONDO, J., J. CASTRESANA, J. VALPUESTA & F. GOÑI. 1994. The structure and thermal denaturation of crystalline and noncrystalline cytochrome oxidase as studied by infrared spectroscopy. Biochemistry **33:** 11650–11655.
43. WOLKERS, W., N. WALKER, F. TABLIN & J. CROWE. 2001. Human platelets loaded with trehalose survive freeze-drying. Cryobiology **42:** 79–87.
44. CHEN, S.S., N.T. WRIGHT & J.D. HUMPHREY. 1998. Heat-induced changes in the mechanics of a collagenous tissue: isothermal, isotonic shrinkage. J. Biomech. Eng. **120:** 382–388.
45. PEARCE, J.A. & S. THOMSEN. 1995. Rate process analysis of thermal damage. *In* Optical and Thermal Response of Laser-Irradiated Tissue. A.J. Welch & M.J.C. van Germert, Eds.: 561–606. Plenum. New York.
46. WRIGHT, N.T. & J.D. HUMPHREY. 2002. Denaturation of collagen via heating: an irreversible rate process. Annu. Rev. Biomed. Eng. **4:** 109–128.
47. HE, X. & J.C. BISCHOF. 2003. Quantification of temperature and injury response in thermal therapy and cryosurgery. Crit. Rev. Biomed. Eng. **31:** 355–422.
48. DILLER, K.R., J.W. VALVANO & J.A. PEARCE. 1999. Bioheat Transfer. *In* The CRC Handbook of Thermal Engineering. F. Kreith, Ed.: 4114–4187. CRC Press. Boca Raton, FL.
49. CRAVALHO, E.G., M. TONER, D.C. GAYLOR & R.C. LEE. 1992. Response of cells to supraphysiological temperatures: experimental measurements and kinetic models. *In* Electrical Trauma: The Pathophysiology, Manifestations and Clinical Management.: 281–300. Cambridge University Press. Cambridge.
50. HE, X. & J.C. BISCHOF. 2005. The kinetics of thermal injury in human renal carcinoma cells. Ann. Biomed. Eng. **33:** 502–510.
51. BHOWMICK, S., J. PEDERSEN & J. BISCHOF. 2000. Investigation of corneal/scleral burning during cataract surgery. ASME Proceedings: Advances in Heat and Mass Transfer in Biotechnology 2000.
52. BHOWMICK, S., D.J. SWANLUND & J.C. BISCHOF. 2000. Suprarphysiological thermal injury in Dunning AT-1 prostate tumor cells. J. Biomech. Eng. **122:** 51–59.
53. EYRING, H. & A.E. STEARN. 1939. The application of the theory of absolute reaction rates to proteins. Chem. Rev. **24:** 253–270.
54. LIU, A., H. BIAN, L. HUANG, *et al.* 1994. Transient cold shock induces the heat shock response upon recovery at 37 degress C in human cells. J. Biol. Chem. **269:** 15710–15717.
55. WANG, S., K.R. DILLER & S.J. AGARWAL. 2003. Kinetics study of endogenous heat shock protein 70 expression. J. Biomech. Eng. **125:** 794–797.
56. ARAKAWA, T., J. CARPENTER, Y. KITA & J. CROWE. 1990. The basis for toxicity of certain cryoprotectants: a hypothesis. Cryobiology **27:** 401.
57. HANAFUSA, N. 1992. The behavior of hydration water of protein with the protectant in the view of 1HNMR. Dev. Biol. Standard. **74:** 241–253.

58. CROWE, J.H., L.M. CROWE & J.F. CARPENTER. 1993. Preserving dry biomaterials: the water replacement hypothesis, part 1. BioPharm. **6:** 28.
59. CROWE, J.H., L.M. CROWE & J.F. CARPENTER. 1993. Preserving dry biomaterials: the water replacement hypothesis, part 2. BioPharm. **6:** 40.
60. MILES, C.A. & T.V. BURJANADZE. 2001. Thermal stability of collagen fibers in ethylene glycol. Biophys. J. **80:** 1480–1486.
61. KANEHISA, M.I. & T.Y. TSONG. 1978. Cluster model of lipid phase transitions with application to passive permeation of molecules and structure relaxations in lipid bilayers. J. Am. Chem. Soc. **100:** 424–432.
62. HAN, A. & J. PEARCE, Eds. 1990. Kinetic model for thermal damage in the myocardium. *In* Advances in Measuring Temperatures in Biomedicine: Thermal Tomography Technique and Bioheat Transfer Models. ASME. New York.
63. EIGNER, J., H. BOEDTKER & G. MICHAELS. 1961. The thermal degradation of nucleic acids. Biochim. Biophys. Acta **51:** 165–168.
64. HENRIQUES, F.J. 1947. Studies of thermal injury, v, the predictability and the significance of thermally induced rate processes leading to irreversible epidermal injury. Arch. Pathol. **43:** 489–502.
65. WELCH, A. & G. POLHAMUS. 1984. Measurement and prediction of thermal injury in the retina of rhesus monkey. IEEE Trans. Biomed. Eng. **BME-31:** 633–644.
66. KAMPMEIER, J., B. RADT, R. BIRNGRUBER & R. BRINKMANN. 2000. Thermal and biomechanical parameters of porcine cornea. Cornea **19:** 355–363.
67. MILES, C.A. 1993. Kinetics of collagen denaturation in mammalian lens capsules studied by differential scanning calorimetry. Int. J. Biol. Macromol. **15:** 265–271.
68. MAITLAND, D. & J. WALSH. 1997. Quantitative measurements of linear birefringence during heating of native collagen. Laser Surg. Med. **20:** 310–318.
69. WEIR, C. 1949. Rate of shrinkage of tendon collagen-heat, entropy, and free energy of activation of the shrinkage of untreated collagen. Effect of acid, salt, pickle, and tannage on the activation of tendon collagen. J. Am. Leather Chemist Assoc. **44:** 108–140.
70. MORAN, K., P. ANDERSON, J. HUTCHESON & S. FLOCK. 2000. Thermally induced shrinkage of joint capsule. Clin. Orthop. Relat. Res. **381:** 248–255.
71. YANG, Y., A. WELCH & H.I. RYLANDER. 1991. Rate process parameters of albumen. Lasers Surg. Med. **11:** 188–190.
72. TANFORD, C. 1968. Protein denaturation. Adv. Protein Chem. **23:** 121–282.
73. MURPHY, D.P. & I.S. GILL. 2001. Energy-based renal tumor ablation: a review. Semin. Urologic Oncol. **2:** 133–140.

The Physics of the Interactions Governing Folding and Association of Proteins

WEIHUA GUO,[a] JOAN-EMMA SHEA,[b] AND R. STEPHEN BERRY[a]

[a]*Department of Chemistry, The University of Chicago, Chicago, Illinois 60637, USA*

[b]*Department of Chemistry, University of California at Santa Barbara, Santa Barbara, California 93106, USA*

ABSTRACT: The review discusses the molecular origins of the forces and free energies that determine several things about proteins, and how experiment and theory reveal this information. The first subject is the stability of the folded, native structures. The second is the range of molecular mechanisms by which proteins find their way to those folded structures in laboratory environments. The third is the much more complex problem of how folding occurs in the cellular environment. This topic includes a discussion of crowding and of the roles of chaperone molecules. The review concludes with a discussion of protein aggregation and fibril formation and of misfolding and therapies associated with it.

KEYWORDS: protein; folding; chaperone molecules; fibril formation

INTRODUCTION

Proteins are fundamental components of all living cells. According to their function, they can be categorized into four main classes: structural proteins (e.g., collagen), immunoglobulins (antibodies), storage proteins (e.g., gluten, albumin), and enzymes (catalytic proteins). In order to carry out these functions, a protein must fold into a particular three-dimensional form (the "native state"), which involves the conformational rearrangement of a linear sequence of amino acids. On the other hand, protein unfolding/misfolding has been associated with incidence of some deadly diseases, such as Alzheimer's disease, type II diabetes, Parkinson's disease, and mad cow disease. In addition, stability of proteins also has other clinical significance. For example, injuries such as burns are associated with a loss of protein function due to denaturation and/or aggregation under thermal stress. One useful question here is what the level of denaturation of vital proteins is for a given temperature history.[1] Understanding the protein folding and aggregation process can provide insight to the molecular basis of these injuries and diseases, leading to development of better treatment methods.

The importance of the protein folding process has motivated endless effort by scientists across the fields of biology, chemistry, physics, computer and medical scienc-

Address for correspondence: R. Steven Berry, Ph.D., Department of Chemistry, The University of Chicago, Chicago, IL 60637. Voice: 773-702-7021; fax: 773-834-4049.
berry@uchicago.edu

es for about 50 years, which has significantly improved our understanding of the process. Nevertheless, protein folding remains one of the most challenging questions in science. In this paper, we review the factors that govern folding and aggregation of proteins and advances in therapeutic approaches to diseases related to protein misfolding and aggregation.

THERMAL STABILITY OF PROTEINS

The thermal stability of a protein is determined by its structure. All naturally occurring proteins are polypeptide chains made of some or all of 20 types of amino acid (residue). Each amino acid has a central carbon atom (C_α) attached to a hydrogen atom and a side-chain group, an amino group (NH_2), and a carboxyl group. During protein synthesis, the amino acids are connected end-to-end into a one-dimensional sequence by the formation of peptide bonds in which the carboxyl group of one amino acid condenses with the amino group of the next and eliminates water (FIG. 1). The only difference between amino acids is the side-chain group attached to the C_α atom. According to the chemical nature of the side chains, the amino acids are usually divided into two types, hydrophobic and polar (including charged and noncharged).

The stability of a protein is determined by the free energy difference between its unfolded state and folded state: $\Delta G = \Delta H - T \Delta S$, where T is temperature, ΔG, ΔH, and ΔS represent changes in free energy, enthalpy, and entropy in the folding process, respectively. More negative ΔG corresponds to higher stability. Various factors determine the folding process and thermal stability of proteins by contributing to different variables in the above formula, including bonded and non-bonded interactions inside the protein, interactions between protein and solvent, etc. The relative contributions of these factors to protein thermal stability vary by protein, and are often intertwined.

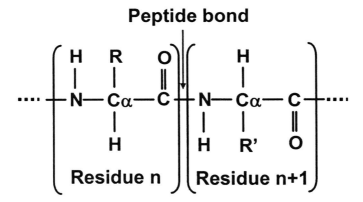

FIGURE 1. The primary structure of a protein. Two amino acids are connected end to end to form a peptide bond.

First, the strength of chemical bonding may affect the thermal stability of proteins. A typical covalent chemical bond has a bonding energy of tens of kcal/mol (e.g., the C—N bond: 35 kcal/mol), while most proteins have a stability of only 5 ~ 20 kcal/mol. Therefore, the protein usually unfolds and breaks many critical physical (non-bonded) interactions before a typical covalent bond can be broken. However, a protein's thermal stability can be strengthened by introducing additional disulfide bridges between two thiol-group-containing amino acids (cysteine or histidine). When the two residues have the correct geometry, the thiol groups can be oxidized and form a $-CH_2-S-S-CH_2-$ bridge in the three-dimensional structure. These bridges strongly restrict the conformational mobility of a protein, thus greatly reducing the number of unfolded conformations (and thus unfolded entropy) and stabilizing the native state. For example, removing either of the two terminal disulfide bonds in bovine pancreatic ribonuclease A by substituting them with alanine residues can decrease its "melting temperature" T_m by 40 degrees.[2] On the other hand, introducing a single disulfide bond into bacteriophage T4 lysozyme can increase its T_m by 4.8, 6.4, or 11.0 degrees, depending on the location of bridges in the chain.[3] Bovine pancreatic trypsin inhibitor (BPTI) is a small protein whose three disulfide links play an important role in the folding process itself.[4] In addition, the α-helices of keratin, a major protein component of hair, are extensively cross-linked by disulfide bonds, providing additional strength of hair.

Other aspects of amino acid sequence also have significant effects on the thermal stability of proteins. For example, glycine is the most flexible among all amino acids, because it has only a single H atom as the side chain. Therefore, in the unfolded state, it may have many conformations, while in native state it only has one conformation. This leads to greater entropy loss than for other proteins in the process of folding and lower stability of proteins containing glycine. On the other hand, proline is the most conformationally restricted amino acid because of its ring structure in the backbone. Including prolines in the sequence significantly decreases the number of available states (entropy) in the unfolded state, thus rendering a more stable native state. However, the superior rigidity can be a double-edged sword: it may disrupt the secondary structure or favorable contacts and/or introduce unfavorable contacts, which could potentially lower the protein's stability. Therefore, its contributions to protein thermal stability must be assessed with caution.

Hydrogen bonding is one of the most prominent interactions in protein structures and plays an essential role in secondary structures (α-helices and β-sheets, etc.). However, its contribution to protein stability has been controversial. It is partly due to the fact that hydrogen bonds can be formed not only between protein residues, but also between protein and water and between water molecules. Numerous theoretical studies have concluded that hydrogen bonds make minor or no contribution to thermodynamic stability of proteins.[5,6] On the other hand, recent experimental studies, contradicting those calculations, have shown that hydrogen bonds contribute significantly to the stability of secondary structures (α-helices and β-sheets, etc.) and the overall stability of the native state. Study of different families of proteins showed that each hydrogen bond provides a stabilization of 1.0 ~ 1.6 kcal/mol.[7]

Electrostatic interactions, including charge–charge, charge–dipole, and dipole–dipole interactions, also contribute significantly to the thermal stability of proteins. Increased number of salt bridges on the protein surface is found in many thermostable or hyperthermostable proteins, compared to their counterparts from meso-

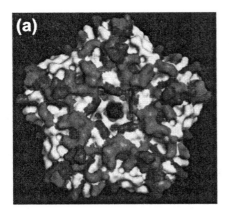
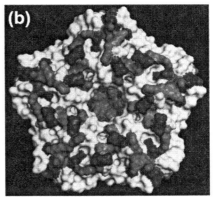

FIGURE 2. Distribution of charged residues on the surface of (a) *Aquifex aeolicus* (PDB code 1HQK) and (b) *Bacillus subtilis* (PDB code 1RVV, chain A-E). Red, blue and green [online version] represent residues with positive charges (Lys, His and Arg), negative charges (Asp and Glu). and neutral residues, respectively. Images generated by PyMOL (Delano Scientific, San Carlos, CA http://www.pymol.org).

philic organisms.[8–11] For example, FIGURE 2 shows the distribution of charged residues on the surface of lumazine synthase from hyperthermostable *Aquifex aeolicus* (T_m = 119.9°C) and its mesophilic counterpart, *Bacillus subtilis* (T_m = 93.3°C). The *Aquifex aeolicus* enzyme has more ion-pairs per subunit, larger accessible surface presented by charged residues, and smaller exposed hydrophobic surface area than other lumazine synthases with similar topology.[11] An α-helix has a dipole moment with positive charge on the N-terminus and negative charge on the C-terminus. Two mutations of a single residue that stabilizes the dipole of helices are found to each increase the T_m of T4 lysozyme by 2 degrees.[3]

Most proteins are functional only in an aqueous environment. Interactions with the solvent contribute to many aspects of stabilization effects. First of all, as a polar molecule, water interacts with charges on the surfaces of proteins and disfavors exposure of hydrophobic side chains. It is entropically costly to have residues with hydrophobic (nonpolar) side chains on the surface of the protein, which leads to ordering of polar water molecules on the nonpolar surface. Therefore, hydrophobic residues tend to point toward the interior of proteins, stick together and form a "hydrophobic core" while exposing polar or charged residues to the solvent. Hydrophobic interactions are the major driving force for protein folding, and major contributors for protein thermal stability. Hydrophobic side chains were shown to provide efficient protection of backbone hydrogen bonds,[12] and guide the folding pathways by decreasing dielectric coefficient of surrounding water and enhancing Coulombic interactions between charged groups.[13] Previous studies have shown that each additional -CH_2-group buried in folding provides 1.3 ± 0.5 kcal/mol in stabilization.[7] Carefully designed mutations to reduce the cavity size in the hydrophobic core successfully increased the T_m, whereas introducing polar residues in the core or loosening the core packing decreased the stability of Archaeal histones.[14] Experiments on T4 lysozyme have also confirmed the correlation between protein thermal stability and reduction of solvent-accessible surface area on folding (an index of hydrophobicity).[3] Mutations on cytochrome *c* that render tighter side-chain packing also increase its thermal stability.[15]

Interactions with solvent contribute to protein stability through other channels as well. For example, water can compete with amino acids for hydrogen bond formation. In addition, water molecules were found to act as lubricant in the final stages of the folding process of the Src SH3 domain.[16] Because the strength of electrostatic interactions is proportional to the reciprocal of dielectric constant, ionic strength and solvent dielectric constant can change the energetics of the folding process and protein stability significantly.[17–19] Trifluoroethanol (TFE), a less polar organic solvent than water, is known to increase the solubility of hydrophobic groups while decreasing that of peptide groups,[20] leading to unfolding or even aggregation of globular proteins.[21,22] On the other hand, peptides that have flexible (i.e., unstable) structures in water such as the Aβ peptide (FIG. 3) can adopt distinct stable secondary structures in different nonpolar environments.[23,24] In addition, solvent acidity (pH value) can also contribute to protein stability by changing the charge state of acidic/basic amino acids.[17]

As shown above, many factors contribute to the thermal stability of proteins collectively. Because protein structures are so complicated, the relative contribution of each factor varies by system: each factor may contribute (positively or negatively) an absolute magnitude more than the overall stability (only 5~20 kcal/mol), and the overall stability is a result of interplay and balance of these factors. For example, mutations designed to stabilize T4 lysozyme by decreasing cavity size in the hydrophobic core were not successful because they introduce unfavorable steric strain.[3] Collagen, a major component of skin, tendons, and other connective tissues, makes use of many factors of stabilization. It has an elongated triple-helical structure (FIG. 4) with a repetitive primary sequence that can be written as: -Gly-Xaa-Yaa-Gly-Xaa-Yaa-Gly-Xaa-Yaa-. Every third residue is occupied by Gly, while Xaa and Yaa are any other amino acids (often proline or lysine). Many of the proline and lysine residues are hydroxylated (i.e., with an additional –OH group), which can form more hydrogen bonds. Glycine is the most flexible residue, while proline is most restricted; the combination of these two forces the helices to have a much smaller radius than a typical α-helix, with hydrogen bonds between adjacent chains

(a) **(b)**

FIGURE 3. Aβ(1-40) peptide adopts different conformations in (a) 40% (by vol.) trifluoroethanol/water (PDB code: 1AML) and (b) aqueous sodium dodecyl sulfate micelles (PDB code: 1BA4). Images generated by PyMOL (Delano Scientific, San Carlos, CA, http://www.pymol.org).

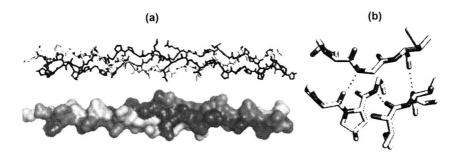

FIGURE 4. (a) Structure of the triple helix of a short collagen model (PDB code: 1BKV) in sticks and surface presentations. Different chains are represented by different colors. (b) Hydrogen bond pattern between the three helices. Side-chain hydrogen atoms are not shown. Color codes are: red for O, green for C and blue for N [online version]. Images generated by PyMOL (Delano Scientific, San Carlos, CA, http://www.pymol.org).

(FIG. 4). A collagen helix is left-handed and has only three residues per turn, while an α helix is right-handed and has 3.6 residues per turn. It is therefore more open than an α helix. The small side chain of glycine allows a compact core with hydrogen bonds between adjacent chains. It was shown that the melting temperature of collagen strongly correlates with the content of hydroxylated proline.[25] Disruption of the Gly-Xaa-Yaa sequence can lead to defective folding of the collagen helices and has been associated with various connective tissue diseases,[25] including osteogenesis imperfecta.

HOW DO PROTEINS FOLD?

The protein-folding process in which the one-dimensional chain folds into its three-dimensional functional form is a massively complex process with highly stochastic nature. Experiments by Anfinsen on the protein ribonuclease in 1957 showed that at least some proteins can be repeatedly unfolded and folded back to the native state without any helper molecules, indicating that folding information is completely encoded in the sequence.[26] Most proteins fold in milliseconds to seconds, indicating that folding does not proceed through completely random search over all available conformations, which would take an astronomical length of time. On the basis of similar arguments, Levinthal proposed in 1968 that protein folding progresses through specific pathways.[27] In the past two decades, advances in experimental techniques and theoretical methods have shifted the paradigm on protein-folding theory from the pathway perspective to a statistical view of the folding landscape.[28,29] This more recent view depicts the folding process as the approach to and then the descent into a funnel-shaped free energy landscape (FIG. 5): the native state lies at the bottom of the funnel with minimal free energy. Folding proceeds through multiple pathways via a trade-off between energy and entropy—as protein moves down the funnel both its energy and entropy decrease, while incomplete cancellation of energy and enthalpy contributions leads to emergence of a free energy barrier. Proteins can get transiently trapped in local free energy minima by roughness of the landscape, but

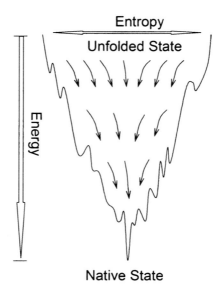

FIGURE 5. The idealized funnel-shaped free energy landscape of proteins. The unfolded state has the highest energy and entropy, while the folded state has the lowest energy and entropy. Folding progresses through multiple pathways; in the extreme, any path that crosses over the rim of the funnel will lead to the minimum.

can eventually access the native state because the barriers are lower than the free energy bias toward the folded state.[30,31] The "folding funnel" model and the original single-pathway concept are essentially polar opposites, at the two ends of a scale. The rich variety and vast number of proteins suggest that real folding processes may well lie along that scale from the extreme of the strict, single-path Levinthal model and the other extreme, the free motion toward and down a broad, open funnel. One of the open challenges in this field now is finding and applying a way to characterize the variety of successful folding paths available to any given protein.

More details from the folding processes were obtained with improved experimental methods.[32] For example, differential scanning calorimetry provides thermodynamic information of the folding process[33,34]; circular dichroism allows us to study the secondary structure of proteins[35]; NMR and hydrogen exchange techniques provide detailed insights into the structure and dynamics of unfolded and partly folded states[36,37]; mass spectroscopy can detect co-existing protein species and kinetic intermediates[38]; rapid mixing and other time-resolved techniques make possible the study of fundamental fast processes and initial stages of folding[32,39,40]; protein engineering enables investigation of the determinants of protein folding and function[41,42]; and single-molecule fluorescence techniques allow us to observe the stochastic process "one protein at a time."[43] Experiments show that folding starts with specific collapse of the chain and formation of compact species for some proteins such as cytochrome c.[44] Larger proteins such as lysozymes follow more complicated mechanisms and may populate one or more intermediates, but whether these intermediates promote (on-pathway) or inhibit folding (off-pathway) is under debate and depend on specific systems.[45–51] On the other hand, folding of small proteins often follows simple two-state kinetics in which only folded and unfolded states are observed without stable detectable intermediates,[32] which may be due to limited sensitivity of experimental techniques.[52] The simplicity of these small systems provided tremendous insight on folding mechanisms, especially about the structure of the rate-limiting step (transition state). Many studies suggest that the transition state involves interactions between a relatively small number of residues, after which condensation of other structural components occurs rapidly (the "nucleation-condensation" mechanism).[53] In addition, many evidences highlight the importance of native topology in guiding the folding process. Proteins with similar native topology but drastically different sequences are found to share similar transition state

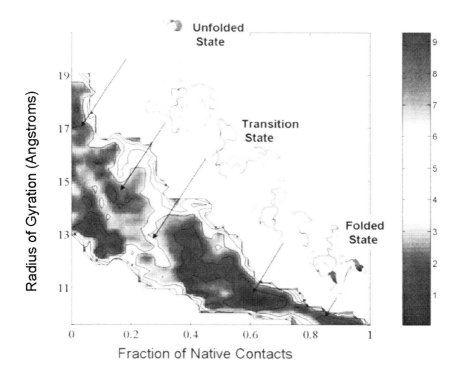

FIGURE 6. The free energy profile of SH3 domain at 343K (near its folding temperature) obtained by computer simulations. Free energy is represented by color bar in units of RT [online version]. Radius of gyration and fraction of native contacts (ρ) measure the size of the protein and similarity to the native state, respectively. Folding starts with collapse of structures, progresses to form the central three-strand β-sheet around $\rho = 0.3$ (the transition state), in good agreement with experimental results. The formation of the second β-sheet comprising terminal strands follows, overcoming a barrier around $\rho = 0.8$ to squeeze out the water molecules in the hydrophobic core, and gets into the native basin.

topology.[54,55] Experimental folding rates of small proteins are highly correlated with the complexity of their folds (measured by "contact order," that is, average number of residues between two spatially interacting residues).[55,56] These measures suggest that there may be relatively simple principles underlying the seemingly overwhelmingly complex problem of protein folding.[54]

Significant success has also been achieved through computer simulations, especially on folding of small proteins.[57–61] Because current computer power is not sufficient to simulate the complete folding process and ensemble of pathways of proteins, a variety of models have been used to simplify the problem.[57,58,60–62] Minimalist models, which utilize simple structural representation of proteins instead of all atomic information, were used to predict mechanisms and timescales of protein folding and to analyze experimental results.[60,62,63] Models treating proteins with all atomic details but solvent molecules with a mean field can also greatly reduce the computing time and generate comparable results to experiments.[61,64] On the other hand, all-atom models including atomic details of both protein and solvent provide not only results directly comparable to experiments, but also details of the folding process beyond the detection limit of current experimental techniques.[58,59,62,65] For example, sampling the free energy surface through all-atom simulations successfully captured the transition state ensemble of the Src SH3 domain with atomic details (FIG. 6),[16,66] but also showed that water molecules function as "lubricant" during the formation of hydrophobic core of the protein and were squeezed out at the final stages of folding (FIG. 7).[16] However, applications of all-atom models are limited to proteins folding at or near the speed limit (microseconds) due to the enormous size of the conformational space. New computing algorithms such as Replica Exchange Molecular Dynamics[67] that enhance sampling efficiency and improvements on distributed computing techniques such as Folding@home[68] are expected to connect the gap in time scales between theory and experiments.

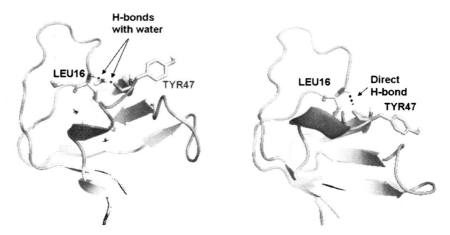

FIGURE 7. Representative structures of Src SH3 before and after crossing the barrier around fraction of native contacts = 0.8. (**a**) Before transition, the hydrogen bonds bridged with water molecules connect LEU16 from the RT-loop and TYR47 from the central β-sheet. (**b**) After transition, a hydrogen bond connects LEU16 and TYR47 directly.

HOW DO PROTEINS FOLD IN THE CELLULAR ENVIRONMENT?

A majority of experimental studies on protein folding have been carried out *in vitro* in dilute solution, where many relevant experimental conditions can be controlled precisely, such as denaturant concentration, ionic strength, pH, etc. However, the interior of cells is a rather crowded environment. It contains not only proteins, but also several other types of macromolecules, such as lipids, sugars, ribosomes, etc. The dry matter can occupy as large as 40% of the cell weight and volume.[69] Crowding also occurs outside the cell; for example, blood contains about 8 percent of proteins by weight. Macromolecular crowding can significantly alter the processes and interactions between macromolecules, including but not limited to protein folding and aggregation. Therefore, more and more efforts, both experimentally and theoretically, have been undertaken to understand these processes *in vivo* by including crowding to mimic the physiological conditions. Recent studies on the effect of macromolecular crowding and interactions were reviewed by Ellis *et al.*,[70,71] and major findings and new results are summarized here.

Crowding is expected to have the following effects on biological systems.[72] Crowding in the cell can slow down the movement of macromolecules by up to 100 times, depending on the size and shape of the molecules. It can also increase the equilibrium constants of macromolecular association by two to three orders of magnitude. When more volume is occupied by other inert substances, less volume is left for the macromolecular reactants, resulting in less flexibility for their distribution (less entropy), while the flexibility (entropy) of the associated state does not decrease as much. Therefore increased crowding results in less overall entropy loss, which leads to more significant decrease in free energy and higher equilibrium constants for the reaction. This can have a significant effect on all processes with a change in excluded volume, such as protein folding/unfolding and aggregation processes. However, this effect does not change monotonically with the concentration of crowding agent, because crowding simultaneously decreases the encountering rate for two macromolecules (a kinetic effect) and increases the free energy bias toward the associated state (a thermodynamic effect). In addition, *in vivo* reactions are affected not only by physical non-specific crowding, but also by other competing reactions, such as specific binding to other molecules,[73] making the processes even more complicated. It was also suggested that in a heterogeneous protein solution, less-stable proteins unfold first and increase crowding, leading to higher thermal stability of more stable proteins, which can be part of the cellular defense system against burn injury.[1,74]

Recent studies have provided significant insight regarding the effect of macromolecular crowding and confinement on folding and aggregation processes of different proteins. High concentration of crowding agents was shown to significantly enhance the self-association of many proteins,[75–79] while the effect can be reversed by the addition of a molecular chaperone.[76,77] On the other hand, when the concentration of crowding agent is low enough to allow some chains to fold rather than aggregate, crowding is also shown to increase the refolding rate of some proteins.[70,80,81] High concentrations of crowding agents were found to induce shifts to more compact structures in the denatured state of the CORE domain of *E. coli* adenylate kinase.[82] Computer simulations have also shown that native state stability, refolding rates, and self-assembly of model proteins are enhanced by crowding and confinement.[83–86] All these results highlight the significance of investigating protein function and stability in environments mimicking *in vivo* conditions to uncover structural and dynamic properties that may not be captured in dilute solution.

Proteins are synthesized in the cell by ribosomes from the N- to the C-terminus. Some proteins start folding before the completion of protein synthesis[87,88]; other proteins may fold in the cytoplasm after the protein is released from the ribosome, while about one-third of newly synthesized proteins only fold to their native states after they are translocated to an organelle or even extracellular compartment where they perform function.[88] Incompletely folded proteins are prone to aggregation in the highly crowded intracellular environment and result in harmful conditions. Therefore, folding processes often require the help of molecular chaperones.[89] Chaperones often work in concert: one type of chaperone would bind to a protein as it is being synthesized on the ribosome, and another would then assist the folding of the protein. The detailed mechanisms by which chaperones operate is controversial (and may be different from chaperone to chaperone). However, a common trait is that they recognize non-native proteins (via a hydrophobic patch that serves as a binding site to exposed hydrophobic residues of unfolded or misfolded proteins).[88–90]

There are two main classes of chaperones. The first class includes small heat-shock proteins (HSPs) which are expressed under conditions of stress[90] such as exposure to supraphysiological temperature during burn injuries. HSPs are very thermally stable proteins that bind tightly to unfolded/misfolded proteins and remove them from circulation (decrease the concentration) until the period of stress is over. It is unclear whether these proteins play a role in actively folding proteins. The second class of chaperones plays a role not only during periods of stress, but also in routine cell maintenance. Many of them are known as chaperonins: they possess a cavity in which a misfolded protein can be encapsulated.

The best understood chaperonin is the GroEL/ES complex,[71] though most results are obtained from studies of their effects on folding *in vitro*. GroEL is composed of 14 subunits, arranged in two rings, stacked back to back. Binding of the misfolded protein occurs at the exposed hydrophobic residues lining the apical domain of one ring. Binding of the protein to GroEL is followed by capping of the GroEL cavity by a GroES co-chaperone and binding of ATP. This process leads to a conformational change in which the hydrophobic residues lining the GroEL interior are mostly buried, leaving the protein encapsulated in a container with weakly hydrophobic walls. The protein remains in the cavity for approximately 10 seconds (the time required for ATP hydrolysis), after which the GroES cap and protein are released. Several cycles of ATP binding and release can occur, and macromolecular crowding is found to increase the capacity of the chaperonin to retain non-native polypeptide throughout the successive reaction cycles.[91] It has been suggested that certain proteins fold inside the cavity in a "passive manner," with the cage offering an "infinite dilution cage," allowing the protein to fold sheltered from other protein, so that aggregation will not occur.[71,89] The "active models" are of two sorts. One (the iterative annealing model) suggests either that folding rates and yields are increased by folding occurring outside the cage through various cycles of ATP-driven binding and unbinding of the protein from the chaperonin,[92] while the other suggests that folding occurs inside the cage through various cycles of stochastic binding and unbinding of the protein from the mildly hydrophobic lining chaperonin wall.[93]

Despite all the efforts by chaperones to help smooth the folding process, the complexity and stochastic nature of the process demands that some proteins are still misfolded after rounds of attempts, especially under cellular stress, such as exposure to high temperature. Secretion of these misfolded proteins into the extracellular environment can lead to accumulation and disturbance of normal cellular function, because the extracellular environment is devoid of molecular chaperones (although one has been discovered recently).[94] (Although extracellular aggregates can be degraded by immune-system pathways including macrophages, this process is often slowed down by disease and aging.) Therefore, eukaryotic systems have developed a remarkable "quality control" process to prevent the secretion of misfolded proteins and give them another chance to either fold in the endoplasmic reticulum under the help of chaperones, or be degraded and recycled by the ubiquitin-proteasome system into short non-aggregating peptides.[95,96] When the external stress or mutations overwhelm the cellular defense system (e.g., insufficient production of chaperones or inefficient "quality control" process), excessive amount of misfolded proteins are produced, which are either degraded and result in lack-of-function diseases such as cystic fibrosis,[95] or they accumulate and form aggregates in the extracellular space, leading to other disorders, such as Alzheimer's disease.[97]

UNFOLDED PROTEINS ARE PRONE TO AGGREGATION

Deposits of protein aggregates on tissues are associated with the pathology of diverse diseases, such as Alzheimer's disease, Parkinson's disease, type II diabetes, and transmissible spongiform encephalopathies (TSEs).[97,98] These aggregates often form fibrils or plaques on the tissue surface. Usually each disease has its own characteristic protein component, for example, Alzheimer's disease fibrils contain mainly amyloid-β peptide (Aβ), while TSE fibrils contain mainly prion proteins. Despite the fact that the sequences and soluble structures of the proteins related to different diseases are significantly different, the fibrils share the following properties: regular fibrillar structure under electron microscope; binding with Congo red and showing bright green fluorescence under polarized light; binding with thioflavin T; and showing β-pleated sheet structure by X-ray diffraction (FIG. 8). Fibrils with these properties are called amyloid fibrils. In addition, common structure has been found for amyloid oligomers of different sequences, implying that different diseases may have a common mechanism of pathogenesis.[99] All these similarities imply that these fibrils may have a similar mechanism of aggregation.

It was not until recently that researchers realized that the amyloid state may be a common state for all proteins, not only those commonly associated with diseases, under certain fine-tuned denaturing conditions. Trifluoroethanol (TFE) is a denaturant capable of partially unfolding proteins while keeping some hydrogen bonds and secondary structures untouched. For example, adding TFE successfully produced amyloid fibrils from soluble proteins such as ribonuclease Sa.[22] A native SH3 domain, although too compact to fit into the fibril density, can also partially unfold and form fibrils.[100] It is even more striking that myoglobin, whose native state contains mainly helical structures, can also be converted into amyloid fibrils rich in β-sheet.[101] Destabilizing mutations can also convert soluble proteins into amyloid fibrils.[98]

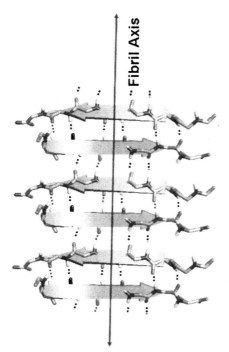

FIGURE 8. An illustration of the β-pleated sheet structure of amyloid fibrils (only backbone atoms are shown). Color codes are: green for C, red for O, gray white for H and blue for N [online version]. The β strands are perpendicular to the fibril axis, while the backbone hydrogen bonds (in black *dashed lines*) between β strands are parallel to fibril propagation.

The fibrilization process usually starts from soluble proteins that are either "naturally unfolded" peptides (unstructured and without a stable native state under normal physiological conditions, such as amyloid-β peptide and islet amyloid polypeptide [IAPP]), or partially unfolded proteins converted from stable native states (e.g., SH3, lysozyme, and myoglobin) under denaturing conditions. Naturally unfolded proteins may need to "partially fold" and form some secondary structures before aggregation.[102] As shown in the above analysis of stabilizing factors, hydrophobic interactions are one of the most important factors that drive the protein-folding process. Under normal conditions, the hydrophobic groups tend to point into the protein interior space and remain shielded from water. However, the partially folded proteins have exposed hydrophobic groups and thus tend to decrease the solvent-accessible surface area through aggregation. In fact, experimental studies on proteins responsible for different diseases under conditions close to physiological have shown that changes in rate of aggregation are positively correlated with changes in residue hydrophobicity and secondary structure propensity, while negatively correlated with changes in total charge.[103] Recent experiments with combinatorial libraries of designed proteins have also shown that sequences with alternating polar and nonpolar residues, which should have high propensity for β structures, form fibrils and are underrepresented in the protein database.[104] Furthermore, proteins with exposed hydrogen bonds on their surfaces tend to be those that form fibrils most readily.[105]

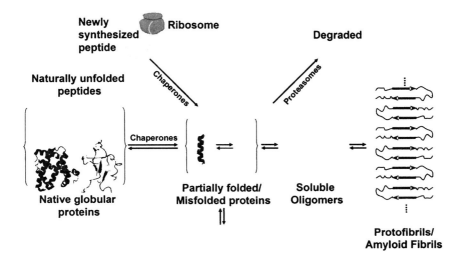

FIGURE 9. Relevant processes in the protein folding/aggregation problem. Partially folded/misfolded proteins can come from nascent peptides, folding of naturally unfolded peptides, and unfolding or misfolding of globular proteins. These partially folded structures have a tendency to aggregate in to disordered form or amyloid fibrils. They can also be degraded by ubiquitin proteasomes into shorter peptides.

In vitro experiments have shown that the fibril formation process generally occurs through slow nucleation-dependent oligomerization followed by fast fibril growth, while the first step can be eliminated through seeding.[98] *In vitro* studies have also revealed the existence of an on-pathway intermediate state, "protofibrils."[98] However, the structure of a fibril at atomic resolution is unavailable even *in vitro*, although there has been some progress in this direction, such as determination of the structure of the PI3 kinase SH3 fibril[100] and Aβ fibrils.[106,107] On the other hand, the detailed mechanism of cytotoxity of amyloids remains unclear. More and more evidence indicates that the oligomers or protofibrils, rather than monomer or fibrils, are cytotoxic.[108-110]

THERAPEUTIC APPROACHES TO PROTEIN MISFOLDING DISEASES

Progress in understanding the pathogenesis of protein-misfolding diseases has motivated researchers to identify molecules to slow down or revert disease progression. Recent advances of these therapeutic approaches were reviewed by Soto[111] and Cohen *et al.*[112] Major findings are summarized here.

Many efforts have been focused on speeding the clearance of misfolded or aggregated proteins, increasing the native state stability or increasing the activation barrier for misfolding or aggregation.[112] Many small molecules such as indoles have been found promising to correct loss-of-function diseases due to mutations, such as cystic fibrosis and α_1-antitrypsin deficiency. These small molecules act as chemical chaperones: they reduce protein misfolding by binding to a protein and stabilizing its native state. Mutations and small molecular inhibitors have also been found to stabilize the native (tetramer) state of transthyretin (TTR), preventing amyloidgenesis and alleviating TTR amyloid disease.[112] Prion disease is associated with aggregates of disease-causing isoform (PrP^{Sc}) converted from the normal host prion protein (PrP^C). Small molecules were found to block prion replication.[113] More efficient clearance of PrP by antibodies are also efficacious in slowing prion disease.[113] In the case of Alzheimer's disease, inhibition of Aβ fibril formation may not be a cure, because recent evidence suggests that soluble oligomers instead of monomer or fibril are toxic.[99,110] Substantial effort has been directed into development of drugs which can decrease Aβ peptide concentration through modulating the Aβ-producing secretases.[114,115] With the increasing number of diseases associated with protein misfolding and aggregation, more emphasis will be put onto therapeutics utilizing new molecular medicine targeting protein interactions.

CONCLUSIONS

A large body of knowledge has been gained toward understanding the protein folding/misfolding processes and diseases related to them. The free energy landscape theory and computer simulations have broadened our view of the mechanisms of these processes. Improved experimental techniques have provided more details about transient processes and intermediate states in the folding and aggregation processes. More emphasis has been directed to studying these processes *in vivo* and in complex environments *in vitro*. Many therapeutic approaches designed on the basis of this knowledge have shown promising in curing the related diseases. In case of thermal burn injuries, better understanding on damages to specific tissue macromolecules *in vivo* is expected to provide more insights on treatment schemes based on protein interactions.

ACKNOWLEDGMENTS

J.-E.S would like to acknowledge the support of the David and Lucille Packard Foundation, the Alfred P. Sloan Foundation, and the National Science Foundation (Career Award No. 0133504).

REFERENCES

1. DESPA, F., D.P. ORGILL, J. NEUWALDER & R.C. LEE. 2005. The relative thermal stability of tissue macromolecules and cellular structure in burn injury. Burns **31:** 568–577.
2. KLINK, T.A., K.J. WOYCECHOWSKY, K.M. TAYLOR & R.T. RAINES. 2000. Contribution of disulfide bonds to the conformational stability and catalytic activity of ribonuclease A. Eur. J. Biochem. **267:** 566–572.
3. MATTHEWS, B.W. 1995. Studies on protein stability with T4 lysozyme. Adv. Protein Chem. **46:** 249–278.
4. FERNANDEZ, A., K.S. KOSTOV & R.S. BERRY. 2000. Coarsely resolved topography along protein folding pathways. J. Chem. Phys. **112:** 5223–5229.

5. HONIG, B. & A.S. YANG. 1995. Free-energy balance in protein-folding. Adv. Protein Chem. **46:** 27–58.
6. SIPPL, M.J., M. ORTNER, M. JARITZ, et al. 1996. Helmholtz free energies of atom pair interactions in proteins. Folding & Design **1:** 289–298.
7. PACE, C., B. SHIRLEY, M. MCNUTT & K. GAJIWALA. 1996. Forces contributing to the conformational stability of proteins. FASEB J. **10:** 75–83.
8. VETRIANI, C., D.L. MAEDER, N. TOLLIDAY, et al. 1998. Protein thermostability above 100°C: a key role for ionic interactions. Proc. Natl. Acad. Sci. USA **95:** 12300–12305.
9. DAS, R. & M. GERSTEIN. 2000. The stability of thermophilic proteins: a study based on comprehensive genome comparison. Functional & Integrative Genomics **1:** 76–88.
10. KARSHIKOFF, A. & R. LADENSTEIN. 2001. Ion pairs and the thermotolerance of proteins from hyperthermophiles: a "traffic rule" for hot roads. Trends Biochem. Sci. **26:** 550–556.
11. ZHANG, X., W. MEINING, M. FISCHER, et al. 2001. X-ray structure analysis and crystallographic refinement of lumazine synthase from the hyperthermophile *Aquifex aeolicus* at 1.6 Å resolution: determinants of thermostability revealed from structural comparisons. J. Mol. Biol. **306:** 1099–1114.
12. FERNANDEZ, A. & R.S. BERRY. 2002. Extent of hydrogen-bond protection in folded proteins: a constraint on packing architectures. Biophys. J. **83:** 2475–2481.
13. DESPA, F., A. FERNANDEZ & R.S. BERRY. 2004. Dielectric modulation of biological water. Phys. Rev. Lett. **93:** 228104.
14. LI, W.-T., R.A. GRAYLING, K. SANDMAN, et al. 1998. Thermodynamic stability of archaeal histones. Biochemistry **37:** 10563–10572.
15. HASEGAWA, J., S. UCHIYAMA, Y. TANIMOTO, et al. 2000. Selected mutations in a mesophilic cytochrome c confer the stability of a thermophilic counterpart. J. Biol. Chem. **275:** 37824–37828.
16. GUO, W.H., S. LAMPOUDI & J.E. SHEA. 2003. Posttransition state desolvation of the hydrophobic core of the src-SH3 protein domain. Biophys. J. **85:** 61–69.
17. KARANTZA, V., A.D. BAXEVANIS, E. FREIRE & E.N. MOUDRIANAKIS. 1995. Thermodynamic studies of the core histones: ionic strength and pH dependence of H2A-H2B dimer stability. Biochemistry **34:** 5988–5996.
18. FERNANDEZ, A., A. COLUBRI & R.S. BERRY. 2002. Three-body correlations in protein folding: the origin of cooperativity. Physica A-Statistical Mechanics and Its Applications. Physica A-**307:** 235–259.
19. DOMINY, B.N., H. MINOUX & C.L.I. BROOKS. 2004. An electrostatic basis for the stability of thermophilic proteins. Proteins: Structure, Function, and Bioinformatics **57:** 128–141.
20. LUO, P. & R.L. BALDWIN. 1997. Mechanism of helix induction by trifluoroethanol: a framework for extrapolating the helix-forming properties of peptides from trifluoroethanol/water mixtures back to water. Biochemistry **36:** 8413–8421.
21. NIELSEN, L., R. KHURANA, A. COATS, et al. 2001. Effect of environmental factors on the kinetics of insulin fibril formation: elucidation of the molecular mechanism. Biochemistry **40:** 6036–6046.
22. SCHMITTSCHMITT, J.P. & J.M. SCHOLTZ. 2003. The role of protein stability, solubility, and net charge in amyloid fibril formation. Protein Sci. **12:** 2374–2378.
23. STICHT, H., P. BAYER, D. WILLBOLD, et al. 1995. Structure of amyloid A4-(1-40)-peptide of Alzheimer's disease. Eur. J. Biochem. **233:** 293–298.
24. COLES, M., W. BICKNELL, A.A. WATSON, et al. 1998. Solution structure of amyloid beta-peptide(1-40) in a water-micelle environment. Is the membrane-spanning domain where we think it is? Biochemistry **37:** 11064–11077.
25. BAUM, J. & B. BRODSKY. 1999. Folding of peptide models of collagen and misfolding in disease. Curr. Opin. Struct. Biol. **9:** 122–128.
26. ANFINSEN, C.B. 1973. Principles that govern folding of protein chains. Science **181:** 223–230.
27. LEVINTHAL, C. 1968. Are there pathways for protein folding? J. Chimie Physique Physico-Chimie Biologique **65:** 44–45.
28. DILL, K.A. & H.S. CHAN. 1997. From Levinthal to pathways to funnels. Nature Struct. Biol. **4:** 10–19.

29. ONUCHIC, J.N., Z. LUTHEYSCHULTEN & P.G. WOLYNES. 1997. Theory of protein folding: the energy landscape perspective. Annu. Rev. Phys. Chem. **48**: 545–600.
30. SHEA, J.E., Y.D. NOCHOMOVITZ, Z.Y. GUO & C.L. BROOKS. 1998. Exploring the space of protein folding Hamiltonians: The balance of forces in a minimalist beta-barrel model. J. Chem. Phys. **109**: 2895–2903.
31. SHEA, J.E., J.N. ONUCHIC & C.L. BROOKS. 1999. Exploring the origins of topological frustration: design of a minimally frustrated model of fragment B of protein A. Proc. Natl. Acad. Sci. USA **96**: 12512–12517.
32. MYERS, J.K. & T.G. OAS. 2002. Mechanisms of fast protein folding. Annu. Rev. Biochem. **71**: 783–815.
33. WEBER, P.C. & F.R. SALEMME. 2003. Applications of calorimetric methods to drug discovery and the study of protein interactions. Curr. Opin. Struct. Biol. **13**: 115–121.
34. BRUYLANTS, G., J. WOUTERS & C. MICHAUX. 2005. Differential scanning calorimetry in life science: thermodynamics, stability, molecular recognition and application in drug design. Curr. Med. Chem. **12**: 2011–2020.
35. KELLY, S.M., T.J. JESS & N.C. PRICE. 2005. How to study proteins by circular dichroism. Biochim. Biophysi. Acta Proteins & Proteomics **1751**: 119–139.
36. ENGLANDER, S.W. 2000. Protein folding intermediates and pathways studied by hydrogen exchange. Annu. Rev. Biophys. Biomol. Struct. **29**: 213–238.
37. DYSON, H.J. & P.E. WRIGHT. 2005. Elucidation of the protein folding landscape by NMR. NMR Biol. Macromol. **394**: 299–321.
38. KONERMANN, L. & D.A. SIMMONS. 2003). Protein-folding kinetics and mechanisms studied by pulse-labeling and mass spectrometry. Mass Spectrometry Rev. **22**: 1–26.
39. PLAXCO, K.W. & C.M. DOBSON. 1996. Time-resolved biophysical methods in the study of protein folding. Curr. Opin. Struct. Biol. **6**: 630–636.
40. EATON, W.A., V. MUNOZ, S.J. HAGEN, *et al.* 2000. Fast kinetics and mechanisms in protein folding. Annu. Rev. Biophys. Biomol. Struct. **29**: 327–359.
41. SAVEN, J.G. 2002. Combinatorial protein design. Curr. Opin. Struct. Biol. **12**: 453–458.
42. VENTURA, S. & L. SERRANO. 2004. Designing proteins from the inside out. Proteins: Structure Function and Bioinformatics **56**: 1–10.
43. SCHULER, B. 2005. Single-molecule fluorescence spectroscopy of protein folding. Chemphyschem. **6**: 1206–1220.
44. SHASTRY, M.C.R., J.M. SAUDER & H. RODER. 1998. Kinetic and structural analysis of submillisecond folding events in cytochrome c. Acc. Chem. Res. **31**: 717–725.
45. KIM, P.S. & R.L. BALDWIN. 1990. Intermediates in the folding reactions of small proteins. Annu. Rev. Biochem. **59**: 631–660.
46. RADFORD, S.E., C.M. DOBSON & P.A. EVANS. 1992. The folding of hen lysozyme involves partially structured intermediates and multiple pathways. Nature **358**: 302–307.
47. PRIVALOV, P.L. 1996. Intermediate states in protein folding. J. Mol. Biol. **258**: 707–725.
48. RODER, H. & W. COLON. 1997. Kinetic role of early intermediates in protein folding. Curr. Opin. Struct. Biol. **7**: 15–28.
49. WAGNER, C. & T. KIEFHABER.1999. Intermediates can accelerate protein folding. Proc. Natl. Acad. Sci. USA **96**: 6716–6721.
50. KUWATA, K., R. SHASTRY, H. CHENG, *et al.* 2001. Structural and kinetic characterization of early folding events in [beta]-lactoglobulin. Nature Struct. Biol. **8**: 151–155.
51. CAPALDI, A.P., C. KLEANTHOUS & S.E. RADFORD. 2002. Im7 folding mechanism: misfolding on a path to the native state. Nature Struct. Mol. Biol. **9**: 209–216.
52. SANCHEZ, I.E. & T. KIEFHABER. 2003. Evidence for sequential barriers and obligatory intermediates in apparent two-state protein folding. J. Biol. Chem. **325**: 367-376.
53. FERSHT, A.R. 1997. Nucleation mechanisms in protein folding. Curr. Opin. Struct. Biol. **7**: 3–9.
54. BAKER, D. 2000. A surprising simplicity to protein folding. Nature **405**: 39–42.
55. FERSHT, A.R. 2000. Transition-state structure as a unifying basis in protein-folding mechanisms: contact order, chain topology, stability, and the extended nucleus mechanism. Proc. Natl. Acad. Sci. USA **97**: 1525–1529.

56. PLAXCO, K.W., K.T. SIMONS & D. BAKER. 1998. Contact order, transition state placement and the refolding rates of single domain proteins. J. Mol. Biol. **277:** 985–994.
57. DINNER, A.R., A. SALI, L.J. SMITH, et al. 2000. Understanding protein folding via free energy surfaces from theory and experiment. Trends Biochem. Sci. **25:** 331–339.
58. MIRNY, L. & E. SHAKHNOVICH. 2001. Protein folding theory: from lattice to all-atom models. Annu. Rev. Biophysi. Biomol. Struct. **30:** 361–396.
59. DAGGETT, V. & A. FERSHT. 2003. The present view of the mechanism of protein folding. Nature Rev. Molec. Cell Biology **4:** 497–502.
60. CHEUNG, M.S., L.L. CHAVEZ & J.N. ONUCHIC. 2004. The energy landscape for protein folding and possible connections to function. Polymer **45:** 547–555.
61. SNOW, C.D., E.J. SORIN, Y.M. RHEE & V.S. PANDE. 2005. How well can simulation predict protein folding kinetics and thermodynamics? Annu. Rev. Biophys. Biomol. Struct. **34:** 43–69.
62. SHEA, J.E., M. FRIEDEL & A. BAUMKETNER. 2005. Simulations of protein folding. Rev. Comp. Chem. **22:** chapt. 3.
63. THIRUMALAI, D. & D.K. KLIMOV. 1999. Deciphering the timescales and mechanisms of protein folding using minimal off-lattice models. Curr. Opin. Struct. Biol. **9:** 197–207.
64. CUI, B., M.-Y. SHEN & K.F. FREED. 2003. Folding and misfolding of the papillomavirus E6 interacting peptide E6ap. Proc. Natl. Acad. Sci. USA **100:** 7087–7092.
65. SHEA, J.E. & C.L. BROOKS. 2001. From folding surfaces to folding proteins: a review and assessment of simulation studies of protein folding and unfolding. Annu. Rev. Phys. Chem. **52:** 499–535.
66. DING, F., W.H. GUO, N.V. DOKHOLYAN, et al. 2005. Reconstruction of the src-SH3 protein domain transition state ensemble using multiscale molecular dynamics simulations. J. Mol. Biol. **350:** 1035–1050.
67. MITSUTAKE, A., Y. SUGITA & Y. OKAMOTO. 2001. Generalized-ensemble algorithms for molecular simulations of biopolymers. Biopolymers **60:** 96–123.
68. PANDE, V.S., I. BAKER, J. CHAPMAN, et al. 2003. Atomistic protein folding simulations on the submillisecond time scale using worldwide distributed computing. Biopolymers **68:** 91–109.
69. ZIMMERMAN, S.B. & S.O. TRACH. 1991. Estimation of macromolecule concentrations and excluded volume effects for the cytoplasm of *Escherichia coli*. J. Mol. Biol. **222:** 599–620.
70. ELLIS, R.J. 2001. Macromolecular crowding: an important but neglected aspect of the intracellular environment. Curr. Opin. Struct. Biol. **11:** 114–119.
71. ELLIS, R.J. 2001. Molecular chaperones: inside and outside the Anfinsen cage. Curr. Biol. **11:** R1038–R1040.
72. MINTON, A.P. 2005. Influence of macromolecular crowding upon the stability and state of association of proteins: predictions and observations. J. Pharm. Sci. **94:** 1668–1675.
73. MINTON, A.P. 1995. Confinement as a determinant of macromolecular structure and reactivity. II. Effects of weakly attractive interactions between confined macrosolutes and confining structures. Biophys. J. **68:** 1311–1322.
74. DESPA, F., D.P. ORGILL & R.C. LEE. 2005. Effects of crowding on the thermal stability of heterogeneous protein solutions. Ann. Biomed. Engineer. **33:** 1125–1131.
75. RIVAS, G., J.A. FERNANDEZ & A.P. MINTON. 1999. Direct observation of the self-association of dilute proteins in the presence of inert macromolecules at high concentration via tracer sedimentation equilibrium: theory, experiment, and biological significance. Biochemistry **38:** 9379–9388.
76. VAN DEN BERG, B., C.M. DOBSON & R.J. ELLIS. 1999. Effects of macromolecular crowding on protein folding and aggregation. EMBO J. **18:** 6927–6933.
77. LI, J., S. ZHANG & C.C. WANG. 2001. Effects of macromolecular crowding on the refolding of glucose-6-phosphate dehydrogenase and protein disulfide isomerase. J. Biol. Chem. **276:** 34396–34401.
78. RIVAS, G., J.A. FERNANDEZ & A.P. MINTON. 2001. Direct observation of the enhancement of non-cooperative protein self-assembly by macromolecular crowding. Proc. Natl. Acad. Sci. USA **98:** 3150–3155.

79. VAN DEN BERG, B., R.J. ELLIS, R. WAIN & C.M. DOBSON. 2000. Macromolecular crowding perturbs protein refolding kinetics: implications for folding inside the cell. EMBO Journal **19:** 3870–3875.
80. QU, Y. X. & D.W. BOLEN. 2002. Efficacy of macromolecular crowding in forcing proteins to fold. Biophys. Chem. **101:** 155–165.
81. TOKURIKI, N., M. KINJO, S. NEGI, et al. 2004. Protein folding by the effects of macromolecular crowding. Protein Sci. **13:** 125–133.
82. ITTAH, V., E. KAHANA, D. AMIR & E. HAAS. 2004. Applications of time-resolved resonance energy transfer measurements in studies of the molecular crowding effect. J. Mol. Recog. **17:** 448–455.
83. KLIMOV, D.K., D. NEWFIELD & D. THIRUMALAI. 2002. Simulations of beta-hairpin folding confined to spherical pores using distributed computing. Proc. Natl. Acad. Sci. USA **99:** 8019–8024.
84. FRIEDEL, M. & J.E. SHEA. 2004. Self-assembly of peptides into a beta-barrel motif. J. Chem. Phys.**120:** 5809–5823.
85. CHEUNG, M.S., D. KLIMOV & D. THIRUMALAI. 2005. Molecular crowding enhances native state stability and refolding rates of globular proteins. Proc. Natl. Acad. Sci. USA **102:** 4753–4758.
86. GRIFFIN, M.A., M. FRIEDEL & J.E. SHEA. 2005. Effects of frustration, confinement, and surface interactions on the dimerization of an off-lattice beta-barrel protein. J. Chem. Phys. **123:** 174–707.
87. HARDESTY, B. & G. KRAMER. 2001. Folding of a nascent peptide on the ribosome. Prog. Nucl. Acid Res. Mol. Biol. **66:** 41–66.
88. DEUERLING, E. & B. BUKAU. 2004. Chaperone-assisted folding of newly synthesized proteins in the cytosol. Crit. Rev. Biochem. Molec. Biol. **39:** 261–277.
89. YOUNG, J.C., V.R. AGASHE, K. SIEGERS & F.U. HARTL. 2004. Pathways of chaperone-mediated protein folding in the cytosol. Nat. Rev. Mol. Cell Biol. **5:** 781–791.
90. TREWEEK, T.M., A.M. MORRIS & J.A. CARVER. 2003. Intracellular protein unfolding and aggregation: the role of small heat-shock chaperone proteins. Aust. J. Chem. **56:** 357–367.
91. MARTIN, J. & F.U. HARTL. 1997. The effect of macromolecular crowding on chaperonin-mediated protein folding. Proc. Natl. Acad. Sci. USA **94:** 1107–1112.
92. THIRUMALAI, D. & G.H. LORIMER. 2001. Chaperonin-mediated protein folding. Annu. Rev. Biophys. Biomolec. Struct. **30:** 245–269.
93. JEWETT, A.I., A. BAUMKETNER & J.-E. SHEA. 2004. Accelerated folding in the weak hydrophobic environment of a chaperonin cavity: creation of an alternate fast folding pathway. Proc. Natl. Acad. Sci. USA **101:** 13192–13197.
94. WILSON, M.R. & S.B. EASTERBROOK SMITH. 2000. Clusterin is a secreted mammalian chaperone. Trends Biochem. Sci. **25:** 95–98.
95. SHERMAN, M.Y. & A.L. GOLDBERG. 2001. Cellular defenses against unfolded proteins: a cell biologist thinks about neurodegenerative diseases. Neuron **29:** 15–32.
96. DOBSON, C.M. 2003. Protein folding and misfolding. Nature **426:** 884–890.
97. STEFANI, M. & C.M. DOBSON. 2003. Protein aggregation and aggregate toxicity: new insights into protein folding, misfolding diseases and biological evolution. J. Mol. Med. **81:** 678–699.
98. ROCHET, J.-C. & P.T.J. LANSBURY. 2000. Amyloid fibrillogenesis: themes and variations. Curr. Opin. Struct. Biol. **10:** 60–68.
99. KAYED, R. 2003. Common structure of soluble amyloid oligomers implies common mechanism of pathogenesis. Science **300:** 486–489.
100. JIMENEZ, J.L. 1999. Cryo-electron microscopy of an SH3 amyloid fibril and model of the molecular packing. EMBO J. **18:** 815–821.
101. FANDRICH, M., M.A. FLETCHER & C.M. DOBSON. 2001. Amyloid fibrils from muscle myoglobin. Nature **410:** 165–166.
102. KAYED, R., J. BERNHAGEN, N. GREENFIELD, et al. 1999. Conformational transitions of islet amyloid polypeptide (IAPP) in amyloid formation in vitro. J. Mol. Biol. **287:** 781–796.
103. CHITI, F., M. STEFANI, N. TADDEI, et al. 2003. Rationalization of mutational effects on protein aggregation rates. Nature **424:** 805–808.

104. BROOME, B.M. & M.H. HECHT. 2000. Nature disfavours sequences of alternating polar and non-polar amino acids: implications for amyloidogenesis. J. Mol. Biol. **296:** 961–968.
105. FERNANDEZ, A., J. KARDOS, L.R. SCOTT, *et al.* 2003. Structural defects and the diagnosis of amyloidogenic propensity. Proc. Natl. Acad. Sci. USA **100:** 6446–6451.
106. PETKOVA, A.T., Y. ISHII, J.J. BALBACH, *et al.* 2002. A structural model for Alzheimer's beta-amyloid fibrils based on experimental constraints from solid state NMR. Proc. Natl. Acad. Sci. USA **99:** 16742–16747.
107. LUHRS, T., C. RITTER, M. ADRIAN, *et al.* 2005. 3D structure of Alzheimer's amyloid-{beta}(1-42) fibrils. Proc. Natl. Acad. Sci. USA **102:** 17342–17347.
108. BHATIA, R., H. LIN & R. LAL. 2000. Fresh and globular amyloid {beta} protein (1-42) induces rapid cellular degeneration: evidence for A{beta}P channel-mediated cellular toxicity. FASEB J. **14:** 1233–1243.
109. CONWAY, K.A., S.-J. LEE, J.-C. ROCHET, *et al.* 2000. Acceleration of oligomerization, not fibrillization, is a shared property of both alpha-synuclein mutations linked to early-onset Parkinson's disease: implications for pathogenesis and therapy. Proc. Natl. Acad. Sci. USA **97:** 571–576.
110. WALSH, D.M. 2002. Naturally secreted oligomers of amyloid [beta] protein potently inhibit hippocampal long-term potentiation in vivo. Nature **416:** 535–539.
111. SOTO, C. 2003. Unfolding the role of protein misfolding in neurodegenerative diseases. Nat. Rev. Neurosci. **4:** 49–60.
112. COHEN, F.E. & J.W. KELLY. 2003. Therapeutic approaches to protein-misfolding diseases. Nature **426:** 905–909.
113. CASHMAN, N.R. & B. CAUGHEY. 2004. Prion diseases: close to effective therapy? Nat. Rev. Drug Disc. **3:** 874–884.
114. VASSAR, R. 2002. [beta]-Secretase (BACE) as a drug target for Alzheimer's disease. Advan. Drug Delivery Rev. **54:** 1589–1602.
115. WOLFE, M.S. 2002. [gamma]-Secretase as a target for Alzheimer's disease. Curr. Topics Med. Chem. **2:** 371–383.

Molecular Crowding Effects on Protein Stability

FLORIN DESPA,[a] DENNIS P. ORGILL,[b] AND RAPHAEL C. LEE[a]

[a]*Department of Surgery, The University of Chicago, Chicago, Illinois 60637, USA*

[b]*Department of Surgery, Brigham and Woman's Hospital, Harvard Medical School, Boston, Massachusetts 02115, USA*

ABSTRACT: The volume fraction occupied by the dry matter of the cell can be as large as 40%, of which more than half (~60%) are proteins. Thus, cellular proteins and protein assemblies occupy a large volume that can have a profound effect on their own native-state stabilities and on their unfolding/refolding rates. In addition, macromolecular crowding can change the properties of a significant fraction of the water in the cell. We review features of the molecular crowding effect which are relevant for describing the microscopic mechanism of thermal injuries.

KEYWORDS: crowding effects; protein denaturation; thermal injury

INTRODUCTION

To a large extent cells are made of proteins, which constitute more than half (~60%) of the dry weight of the cell.[1] Proteins determine the structure of the cell and, more importantly, they represent the physical apparatus which performs designed functions in the cell. Specific proteins, such as actin and myosin, are organized in large macromolecular arrays (e.g., cytoskeleton fibers) and play the essential role in shaping the cell. Besides proteins, the interior of cells contains several other kinds of macromolecules like lipids, sugars, and nucleic acids. Because no single macromolecular species may be present at high concentration, but all species taken together occupy a significant fraction of the volume of the medium, such media are referred to as "crowded." The volume fraction occupied by the dry matter of the cell (FIG. 1) can be as large as $\varphi = 0.4$. The large volume occupied by these crowding agents can have profound effect on the native state stability and unfolding/refolding rates of cellular proteins. Molecular crowding is considered as a source of nonspecific interactions between cellular proteins. Steric repulsion is the most common of all interactions between macromolecules and is always present in crowded environments, independent of the magnitude of the general electrostatic and hydrophobic interactions.

Because molecules are mutually impenetrable, the presence of a significant volume fraction of macromolecules in the medium is a source of constraints on the

Address for correspondence: Dr. Raphael Lee, Department of Surgery, MC 6035, University of Chicago, Chicago 60637, IL. Voice: 773-702-6302; fax: 773-702-1634.
r-lee@uchicago.edu

placement of an additional macromolecule. These constraints depend upon the relative sizes, shapes, and concentrations of all macromolecules in that environment. Volume may be excluded also by the surfaces of "immobile" structures, that is, membranes and large macromolecular assemblies (FIG. 2). Excluded volume effect, as described by Minton and others,[2–6] can predict many of the aspects of molecular crowding *in vivo*, but other physical factors need to be considered. For instance, a

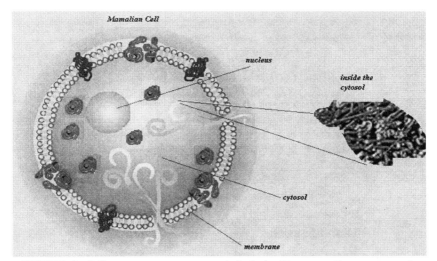

FIGURE 1. Cell compartments are crowded. Actin filaments, ribosomes, membrane structures and other macromolecular assemblies occupy a volume fraction which can be as large as $\varphi = 0.4$.

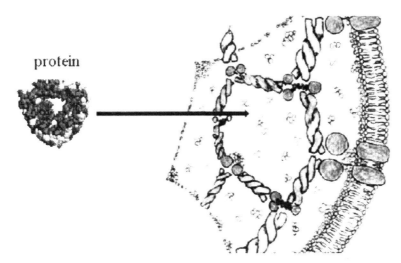

FIGURE 2. The volume of certain cellular compartments, though comparable with protein dimensions, is excluded by membranes and cytoskeleton filaments.

significant outcome of molecular crowding is exerted via diffusion effects on the process of aggregation of unfolded proteins.[7] In addition, interfacial water molecules within a few hydration layers are also a sensor of the cell crowding.[8,9] The physical properties of confined water differ considerably from those corresponding to bulk water and affect protein–protein interactions.[8]

In this chapter we will review features of the molecular crowding effect that are relevant for describing the microscopic mechanism of thermal injuries. For details regarding various mathematical formulations of the crowding effects, the reader is directed to the original papers[2–7] in the reference list.

NONSPECIFIC INTERACTION AND ENTROPIC EFFECTS ON PROTEIN STABILITY

Proteins are made from an assortment of 20 very different amino acids, each with a distinct chemical personality. This leads to specific interactions among the amino acids (FIG. 3), which are important for the primary and secondary structure, as well

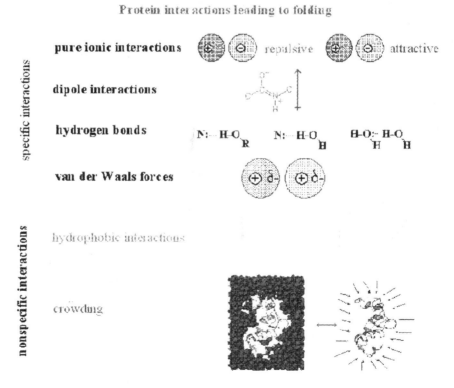

FIGURE 3. The diversity of the chemical constituents of the protein and the properties of the surrounding media lead to both specific and nonspecific interactions during folding.

as nonspecific interactions. A nonspecific interaction does not depend strongly upon details of the primary, secondary, or tertiary structure(s) of the interacting molecules, but rather upon global properties of the molecules, such as polarity and macromolecular shape, or/and properties of the surrounding environment. Hydrophobic interactions between molecules are promoted by structuring effects of water. Molecular crowding is a source of nonspecific interactions.

Because of the complexity of these interactions between the protein of interest and crowding agents (all the other cellular components), it is difficult to predict the net energetic effect of macromolecular crowding on the protein dynamics.[6] Small molecules (water, amino acids, etc.) alter protein dynamics by short-range site–site excluded interactions typically over distances of a few angstroms. In contrast, the range of macromolecular excluded volume interactions can be on the order of tens of angstroms, which is given by the actual size of globular proteins. By modeling the crowding particles as hard spheres, Cheung, Klimov, and Thirumalai[6] predicted the changes in the folding of two-state folders (proteins that can be characterized by two states, folded or unfolded, and have no other intermediate states) by using entropic arguments. They assume that proteins would prefer to be localized in a region that is free of the macromolecular objects. The probability to find such a region decreases at a high fractional volume occupancy $f > 0$. Therefore, at high values of f, there is an increased probability that a protein is in its compact form (folded in its native state). If the protein is compact at large f values, then the entropy change

$$\Delta S = S(f > 0) - S(0) < 0$$

because conformations involving unfolded states are suppressed. Thus, the stability of the native state of a protein is predicted to increase as f increases.

MOLECULAR CROWDING INCREASES THERMAL STABILITY OF CELLULAR PROTEINS

Proteins and protein assemblies optimally perform their function when they are in specific three-dimensional conformations. High temperatures alter these conformations and often lead to irreversible processes (denaturation) which affect the cell viability and trigger cell death. Because the functional structure of each protein and organelle is unique, so is its vulnerability to denaturation at high temperatures. Characteristic vulnerability to thermal denaturation of each cellular component can efficiently be characterized by two main thermodynamic parameters, the melting temperature (T_m) and denaturation enthalpy (ΔH_m).[10] T_m represents the temperature at which half of the proteins are denaturated and is the enthalpy of unfolding at this temperature. T_m and ΔH_m are obtained routinely by calorimetric measurements of proteins in dilute solutions.[11] In the crowded environment of a living cell, the work required for a protein to unfold is much greater than that required for unfolding in a dilute solution. Crowding increases usually the value of the melting temperature of a protein.[12,13] Evidence for the increase of T_m due to crowding can be obtained simply by calorimetric measurements of proteins incubated with surfactants. For example, the melting temperature of *actin* increases by approximately 5°C in the presence of 100 mg/mL PEG-6000, a nonionic surfactant.[14] Minton derived a correction for the melting temperature of the protein in solution to account for volume exclusion effects,[12]

$$\Delta T_m \cong 2.303 \frac{R}{\Delta H_m} \Delta \log K. \quad (1)$$

where R is the gas constant, $R \approx 8.315\ JK^{-1}\ mol^{-1}$. K represents the equilibrium reaction constant defined as the ratio between the unfolding (k_u) and folding (k_f) rates of the protein. The above equation states that any isothermal variation of K changes the temperature at which half of the proteins are denaturated. TABLE 1 displays the values of the thermodynamic parameters T_m and DH_m, as well as the expected values T_m^*, $T_m^* = T_m + \Delta T_m$, for the melting temperatures of these biomolecules *in situ*. We can see that crowding effects can substantially increase the thermal stability of the cellular components. However, the amount of unfolded protein increases dramatically at supraphysiological temperatures. We have shown that,[7] at temperatures above 60°C, tissue proteins are most likely denatured, with probabilities approaching unity.

We can infer from TABLE 1 that the lipid bilayer and membrane-bound ATPases are the proteins most predisposed to thermal denaturation. Therefore, the alteration of the plasma membrane is likely to be the most significant cause of the tissue necrosis. This hypothesis correlates with the observation that edema is considered to be the first evidence of thermal injury in tissue. This edema is likely due to early disturbances in the cell membrane or cell membrane ion pumps (NKP). In many cases it appears that these cells can recover from this injury (i.e., most first-degree burns heal). Temperatures above the first-degree burn threshold lead to irreversible dam-

TABLE 1. The values of the thermodynamic parameters T_m and ΔH_m as determined in calorimetric experiments and the expected values T_m^* corrected for crowding effects

	$\Delta H_m(kJ)$	$T_m(°C)$	$T_m^*(°C)$
Lipid bilayer	290	41.6	45.6
Spectrin	197.19	66	72.8
NKP	490	54.5	57
PMCP	224	47.4	52.8
SRCP	411	60	63.1
DNA	314	55.5	59.5
RNA	326	58.4	62.3
Histone	259.83	47.2	51.8
Cytochrome c	338.9	60	63.8
ATP synthase e	539.74	57.5	59.8
F actin	782.41	67	68.7
Myosin	355.64	53.7	57.2
Tubulin	627.6	55.8	57.8
ApoCaM	332	60	63.9
Collagen	289	58	62.4

age to the cell membrane or other macromolecules and yield a critical injury. *F actin* seems to be a very stable protein at elevated temperatures, as one can deduce from TABLE 1. This protein has a low probability of unfolding in the temperature range corresponding to a second degree burn and is damaged extensively only at higher temperatures, that is, in a third-degree burn.[7] Cells contain also other very thermally stable proteins, as for example, heat-shock proteins (*Hsps*). *Hsps* are assumed to act as molecular chaperones to assist in refolding denatured proteins. *Hsp25* and *Hsp27* have a midpoint transition temperature of $T_m = 69.9°C$,[15] which is higher than that corresponding to *F actin*, for example. Crowding effects inherently enhance the stability of these proteins, too. One can predict that *Hsps* can exist in functional form of 80% even above 75°C. However, above 45°C, the cell membrane breakdown is so extensive that it is improbable that *Hsps* occur in high enough concentration to be an effective protector against cell disruption.[7]

MOLECULAR CROWDING PLACES GEOMETRICAL RESTRICTIONS ON THE UNFOLDING OF PROTEINS

Let r_g be a measure of the compactness of the protein structure, i.e., the radius of gyration of a protein (FIG. 4). This r_g expands during unfolding excluding volume to other surrounding proteins. If the increase Δr of the radius of gyration is in the range of the protein interspace, defined as the mean distance between proteins in solution [$d = (\frac{3}{4}\pi n)^{-\frac{1}{3}}$, where n is the protein concentration], the subsequent confinement would provide stability for adjacent proteins which are in compact, native states. The self-stabilization effect develops progressively during protein unfolding. The results are also relevant for describing the stability of proteins in tightly packed fibers and membranes.[16]

To understand this effect in a more quantitative way, we can write the apparent equilibrium constant K of the protein in a crowded environment as[7]

$$\frac{K}{K_0} = \left(1 + \frac{\Delta r}{r_N} \frac{1}{1 - f^{-1/3}}\right)^3 \qquad (2)$$

where K_0 stands for the equilibrium constant of proteins in the ideal case of a dilute solution, r_N is the radial size of the molecule in the native form and f represents the fractional volume occupancy of the protein,

$$f^{-1/3} = \left(\frac{3}{4\pi N_p} \frac{V_{tot}}{r_N^3}\right)^{1/3} \cong \frac{d}{r_N} \cdot N_p$$

is the total number of proteins in the particular environment. It is not difficult to observe that any conformational change that increases the volume of a protein

$$\frac{4\pi}{3}\left(r_N^3 \to \frac{4\pi}{3}(r_N + \Delta r)^3\right)$$

changes the protein interspace d. Consequently, d is related to the evolution of the density distribution of the population in the unfolded state of the protein P_U,

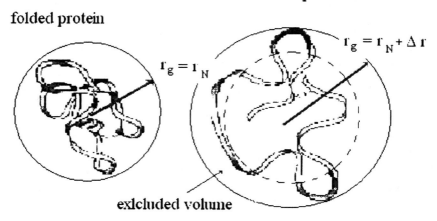

FIGURE 4. The compactness of the protein structure in the folded state is different from that in the unfolded state. r_g expands during unfolding, excluding volume to other surrounding proteins.

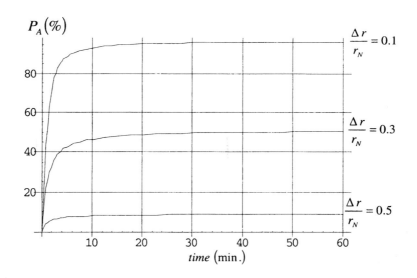

FIGURE 5. Denaturation of *adenylate kinase* in time for a temperature history corresponding to a muscle electrical shock injury of 10 kV, 1s hand-to hand contact at the distal forearm location.[7] The self-stabilization occurs owing to steric effects (Δr increases).

$P_U = 1 - P_N$, where P_N is the density distribution of the population in the native state. The above observation leads to the conclusion that the protein unfolding process is progressively inhibited by the conformational changes of proteins that increase their coresponding volumes and shrink the characteristic interspace (d) between proteins.[7] In FIGURE 5 one can observe that, for example, the probability of distribution of *adenylate kinase* in the denaturated state (P_A) is much lower in a solution in which proteins have a finite volume occupancy (i.e., $f = 20\%$) than in an ideal case of a dilute solution ($f \to 0$). The self-stabilization occurs because of steric effects (Δr increases) induced by the unfolding of a fraction of proteins in solution. As the volume available per unfolded protein is larger than that corresponding to a native protein, this will impose geometrical constraints (volume exclusion) on the proteins in native states, as described above.

A more general approach will describe the unfolding of a protein species i in a crowd formed by M various other protein species. This shows that the volume exclusion effect leads to a decrease of the equilibrium constant K_i of the protein species i in a mixture of different protein kinds J.[7]

UNFOLDING OF THE MOST THERMOLABILE PROTEINS IN A CELL INCREASES THE STABILITY OF THE OTHER CELLULAR PROTEINS

The vulnerability to denaturation at high temperatures of various cellular proteins is different.[10] Thus, by increasing the temperature over the physiological level, proteins with low midpoint transitions will unfold first. The excluded volume theory tells us that the unfolding of these proteins can provide extra stability for the other proteins in the cell, having presumably higher melting temperatures. Obviously, the stabilization of proteins with the highest melting point transitions (e.g., proteins making up the cytoskeleton, such as *actin* and *myosin*) is a result of the excluded volume yielded by the unfolding of all the other protein species. The stability of the most thermolabile proteins in the cell are also affected by molecular crowding, that is, these proteins have a higher stability in a cell than in a dilute solution. However, it is unlikely that their dynamics can be influenced at any extent by the unfolding of the proteins with high melting temperatures.[7]

DIFFUSION OF THE PROTEINS MODIFIES THE NET EFFECTS OF MOLECULAR CROWDING

The net outcome of the steric effects on individual proteins can be modified by diffusional motion of the molecules.[4] This is because the driving force in the diffusion process of particles requires the existence of a gradient of concentration, which makes the crowd to disperse in time. Inherent local fluctuations in the particle density may also alter the steric effects. In addition, the likely presence of direct intermolecular interactions might even reverse the excluded volume effects discussed above. In this context we recall that protein unfolding leads to an exposure of their hydrophobic core residues to water. Unfolded protein may stick together to minimize the area of hydrophobic exposure. Therefore, the tendency of aggregation in-

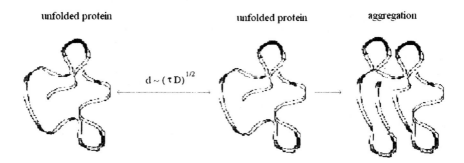

FIGURE 6. If the distance between two unfolded proteins is comparable with the diffusion length d, then the probability of irreversible aggregation of unfolded proteins increase.

creases with the increase of the population in the unfolded state. However, the rate of aggregation is limited by the diffusion of the unfolded proteins (FIG. 6). Inherently, crowding in homogeneous solutions of unfolded proteins leads to a rapid irreversible aggregation of those proteins.

The rate of aggregation can be approximated by the inverse of the diffusion time τ_i, $k_{a,i} = 1/\tau_i$. τ_i for an unfolded molecule of a protein species i relates to its corresponding diffusion coefficient D_i by $\tau_i = d'^2/4D_i$. Here, D_i is an effective diffusion coefficient which includes a correction due to the restriction on the movement of the molecules in a crowded environment.[7]

As discussed above, the aggregation applies only to unfolded proteins. Under such circumstances, the translational diffusion length d depends on density distribution of the population in the unfolded state of the protein (P_U), and on the fractional volume occupancy of the protein (f).[7] In this way, the rate of protein aggregation is directly related to the extent of crowding.

QUANTITATIVE ESTIMATION OF CROWDING EFFECTS ON THERMAL DENATURATION OF PROTEINS

As shown above, crowding can substantially affect the transition of a protein between its native (N) and unfolded (U) states via volume exclusion effects. Also, crowding influences considerably the aggregation (A) of unfolded proteins. To examine the details, one can study the protein transition

$$N \underset{k_{f,i}}{\overset{k_{u,i}}{\leftrightarrow}} U \overset{k_{a,i}}{\to} A . \qquad (3)$$

as described by Despa, Orgill and Lee.[7] $k_{u,i}$ and $k_{f,i}$ are the unfolding/folding rate coefficients and $k_{a,i}$ denotes the rate constant for the irreversible aggregation. $k_{u,i}$ is a function of the melting temperature $T_{m,i}$ and the enthalpy of denaturation $\Delta H_{un,i}$

$$k_u = A\exp\left(-\frac{\Delta H_{un}}{RT}\left(1 - \frac{T}{T_m}\right)\right). \tag{4}$$

A is a constant which determines the time scale of the unfolding process. This depends, among others, on the coupling of the protein with the solvent.[17–20] The backward rate is simply $k_{f,i} = k_{u,i}/K_i$. $T_{m,i}$ and $\Delta H_{un,i}$ are derived by calorimetric measurements of dilute protein solutions. Corrections for crowding effects are incorporated in K_i (see Eq. 2) via volume exclusion and in $k_{a,i}$ by rescaling the translational diffusion length d (see above). This is a model for an experiment in which temperature is changed with time according to a "temperature history" $T(t)$.[7] The approach yields $P_{N,i}$, $P_{U,i}$ and $P_{A,i}$, representing the distribution density of the population in the native, unfolded and aggregated state of the protein species i. A suggestive result is presented in FIGURE 7. Here, one can observe the effects of crowding in a mixture of proteins (*adenylate kinase, creatine kinase, ATP synthase e* and *cytochrome c*) with different thermal stabilities. Steric effects brought by the unfolding of thermolabile proteins enhance the stability of those proteins in the mixture which have higher melting points. The top curve represents the denaturation of cytochrome c in a homogeneous solution, while the bottom curve describes the course of denaturation of this protein in a mixture with the other three proteins. A low probability for aggregation means, implicitly, an increased stability in the native state.

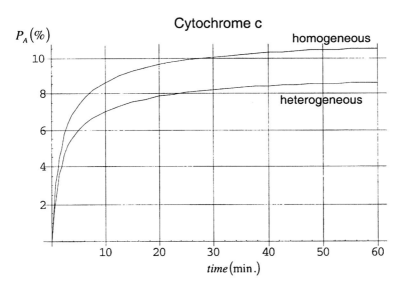

FIGURE 7. Steric effects brought by the unfolding of thermolabile proteins enhance the stability of those proteins in the mixture that have higher melting points. The time–temperature course is the same as the one used in FIGURE 4. The propensity of unfolded proteins to aggregate in a homogeneous solution is higher than in a mixture with other three proteins.

MOLECULAR CROWDING AFFECTS THE BEHAVIOR OF PROTEINS VIA A WATER EFFECT

For a mean protein mass in the cell of ~50 kDa, at an assumed protein concentration of 300 mg/mL, the fraction of interfacial water would be about 30% (two layers of interfacial water) to 70% (four layers) of the total water in the cell. Under circumstances in which macromolecular crowding can change the properties of a significant fraction of the water in the cell, crowding effects could exert a strong influence on the behavior of smaller solutes as well as on larger macromolecules.

Recent *in vivo* expriments[21] showed that the interfacial water component increases from 23.5% of total water in the control sample to 25% in yeast heat-shocked at 315K and to 30% in yeast cells at 323K. The heat shock, which causes some proteins to become unfolded in the cell and, therefore, increases molecular crowding, will change also the hydration of the cell components. An alteration of the properties of the interfacial water could have a direct influence on a range of cellular functions and properties.

Physical properties of water confined in microscopic environments differ from the properties corresponding to bulk water.[8] For example, it was shown recently[8] that water molecules under hydrophobic confinement move about an order of magnitude slower than those in the bulk, and that the dielectric constant of this water layer is significantly reduced. Water's high dielectric constant is the reason why it is a good solvent for ions: it screens their electrical charges and so prevents them from aggregating. But in the vicinity of hydrophobic residues in a protein chain, the reduction in dielectric constant means that charged residues will interact much more strongly, potentially helping to fix the protein's folds in place.

Another example in which the behavior of biomolecules is altered via a water effect is the interaction between hydrophobes. The hydrophobic interaction—the apparent attraction between hydrophobic species in water—is considered a key factor in maintaining the correct folded conformation of a protein molecule and also the main cause of protein aggregation. This attraction is thought to result, in a way that is still imperfectly understood, from changes in the arrangement of hydrogen bonds between water molecules surrounding a hydrophobe.[8,22] This gives rise to a local polarization of the interfacial water, which is shown to be strong enough to induce long-range attraction between hydrophobic molecules.

SUMMARY

Proteins are three-dimensional structures with conformations dictated by the characteristic amino acid sequences. At body temperature, proteins are in native conformations that allow them to perform their designated functions. At supraphysiological temperatures, proteins are driven towards unfolded conformations. *Steric effects increase the stability of the proteins which are in compact, native states. As each type of protein has its own thermal stability, the unfolding of the most thermolabile proteins will increase the stability of the other proteins.* In unfolded conformations, many proteins tend to form stable insoluble aggregates. Aggregation is a major source of irreversibility of protein unfolding when the temperature returns to normal. The tendency of aggregation increases with the increase of the population in the unfolded state and the rate of aggregation is limited by the diffusion of the unfolded proteins. *The net outcome of the steric effects on individual proteins can be modified by diffusional motion of the molecules.*

Understanding the regulation of these processes may lead to clinical strategies for limiting the devastating development of the injury after the thermal insult of the tissue stopped. Within the computational complexity theory, protein dynamics is defined rigorously as *NP* hard[a] and, so far, we can simulate exactly the *in vitro* structural dynamics only for a very limited pool of proteins. However, the volume of data sets is often so big that the efficiency of manipulation and extraction of useful information becomes problematic. Bringing the inherent solvent effects as well as crowding into play increases the complexity of the problem.

Despite difficulties, studying proteins can advance by a profitable utilization of recent statistical mechanical treatments of potential energy surfaces[17–20] conjoined with experimental observations on the energetics and stability of proteins.[11] Here, we have reviewed several quantitative analyses of the kinetic stability of cellular components confronted with the destabilizing effect of irreversible alteration which are relevant for describing the microscopic mechanism of thermal injuries.

REFERENCES

1. ALBERTS, B., D. BRAY, J. LEWIS, *et al.* 1983. *In* Molecular Biology of the Cell.: 111. Garland. New York & London.
2. ELLIS, J.R. 2001. Macromolecular crowding: an important but neglected aspect of the intracellular environment. Curr. Opin. Struct. Biol. **11**: 114–119.
3. BURG, M.C. 2000. Macromolecular crowding as a cell volume sensor. Cell. Physiol. Biochem. **10**: 251–256.
4. VERKMAN, A.S. 2002. Solute and macromolecule diffusion in cellular aqueous compartments. Trends Biochem. Sci. **27**: 27–33.
5. HALL, D.& A.P. MINTON. 2003. Macromolecular crowding: qualitative and semiquantitative successes, quantitative challenges. Biochim. Biophys. Acta **1649**: 127–139.
6. CHEUNG, M.S., D. KLIMOV & D. THIRUMALAI. 2005. Molecular crowding enhances native state stability and refolding rates of globular proteins, Proc. Natl. Acad. Sci. USA **102**: 4753–4758.
7. DESPA, F., D.P. ORGILL & R.C. LEE, 2005. Effects of crowding on the thermal stability of heterogeneous protein solutions. Ann. Biomed. Eng. **33**: 1125-1131 .
8. DESPA, F., A. FERNANDEZ & R.S. BERRY. 2004. Dielectric modulation of biological water. Phys. Rev. Lett. **93**: 228104
9. FORD, R.C., S.V. RUFFLE, A.J. RAMIREZ-CUESTA, *et al.* 2004. Inelastic incoherent neutron scattering measurements of intact cells and tissues and detection of interfacial water. J. Am. Chem. Soc. **126**: 4682–4688.
10. DESPA, F., D.P. ORGILL, J. NEWALDER, & R.C. LEE. 2005. The relative thermal stability of tissue macromolecules and cellular structure in burn injury. Burns **31**: 568–577.
11. MAKHATADZE, G.I. & P.L. PRIVALOV. 1995. Energetics of protein structure. Adv. Protein Chem. **47**: 307–425.
12. MINTON, A.P. 2000. Effect of a concentrated "inert" macromolecular cosolute on the stability of a globular protein with respect to denaturation by heat and by chaotropes: a statistical-thermodynamic model. Biophys. J. **78**: 101–109.

---

[a]This is when the number of spatial configurations available for a system increases exponentially with its size. In such a case, it becomes difficult to infer dynamics from the system's characteristic potential surface.

13. ZHANG, S. & C. WANG. 2001. Effects of macromolecular crowding on the refolding of glucose-6phosphate dehydrogenase and protein disulfide isomerase. J. Biol. Chem. **37:** 34396–34401.
14. TELLAM, R.L., M.J. SCULLEY, L.V. NICHOL & P.R. WILLS. 1983. Influence of poly(ethylene glycol) 6000 on the properties of skeletal-muscle actin. Biochem. J. **213:** 651–659.
15. DUDICH, I.V,, V.P. ZAV'YALOV, W. PFEIL, et al. 1995. Dimer structure as a minimum cooperative subunit of small heat-shock proteins. Biochim. Biophys. Acta. **11253:** 163–168.
16. YANNAS, I.V., J.F. BURKE, P.L. GORDON, et al. 1980. Design of an artificial skin. II. Control of chemical composition. J. Biomed. Mater. Res. **12:** 7–32.
17. DESPA, F. & R.S. BERRY. 2001. Inter-basin dynamics on multidimensional potential surfaces: escape rates on complex basin surfaces. J. Chem. Phys. **115:** 8274–9278.
18. DESPA, F. & R.S. BERRY. 2003. Inter-basin dynamics on multidimensional potential surfaces: kinetic traps, Eur. Phys. **24:** 203–206.
19. DESPA, F., A. FERNÀNDEZ, R.S. BERRY, et al. 2003. Inter-basin motion approach to dynamics of conformationally constrained peptides. J. Chem. Phys. **118:** 5673–5682.
20. DESPA, F., D.J. WALES & BERRY RS. 2005. Archetypal energy landscapes: dynamical diagnosis. Escape rates on complex basin surfaces. J. Chem. Phys. **122:** 24–103.
21. R.C. FORD, S.V. RUFFLE, I. MICHALARIAS, et al. 2004. Neutron scattering measurements of intact cells show changes after heat shock consistent with an increase in molecular crowding. J. Mol. Recogn. **17:** 505–511.
22. CHANDLER, D. 2002. Two faces of water. Nature **417:** 491.

Mechanical Cell Injury

KENNETH A. BARBEE

School of Biomedical Engineering, Science & Health Systems, Drexel University, Philadelphia, Pennsylvania 19104, USA

ABSTRACT: The tissues of the body are continually subjected to mechanical stimulation by external forces, such as gravity, and internally generated forces, such as the pumping of blood or muscle contraction. Within a physiological range, the forces elicit adaptive responses acutely (to rapidly alter function) and chronically (to remodel tissue structure to optimize load-bearing capabilities). When the forces exceed certain thresholds, injury results. To understand the mechanisms of mechanical injury at the cellular level, we must analyze the structural response of the cell to various modes of deformation and examine the biological consequences of the structural alterations caused by the trauma. This chapter reviews the mechanics of cell membrane deformation and failure. Evidence for the strain-rate–dependent, transient disruption of cell membranes, or mechanoporation, is presented for a variety of cell types. The complex interactions between the structural damage and the biological sequelae are illustrated using clinically relevant forms of cell injury. Finally, novel therapeutic approaches targeting membrane integrity are described.

KEYWORDS: neurotrauma; angioplasty; mechanoporation; poloxamer

INTRODUCTION

Mechanical injury of cells can produce a wide range of responses depending on the tissue as well as the severity and type of damage to the cell. Injury may elicit reparative processes, cause transient or persistent dysfunction, or initiate progressive degenerative changes that propagate from the site of injury to surrounding tissue. Mechanical injury can, of course, cause cell death by a variety of mechanisms from immediate cell lysis to delayed apoptosis. Important, clinically relevant instances of cell injury include traumatic brain injury (TBI)[1] and blood cell trauma[2,3] caused by flow through artificial valves and other blood-contacting devices, including left ventricular–assist devices and artificial hearts. Surgical procedures may also cause cellular trauma, as in angioplasty, in which distention of an occluded blood vessel can injure the vascular smooth muscle cells,[4] the reaction to which may contribute to the re-stenosis response. Cell injury may also become an important issue in tissue engineering when engineered constructs are placed in load-bearing tissues in the body.[5]

Address for correspondence: Kenneth A. Barbee, School of Biomedical Engineering, Science & Health Systems, Drexel University, 3141 Chestnut Street, Philadelphia, PA 19104. Voice: 215-895-1335; fax: 215-895-4983.
 barbee@drexel.edu

Before proceeding, it is important to define what we mean by "injury." From an engineering mechanics perspective, structural failure and injury may be thought to be synonymous. Broken bones and ruptured blood vessels are examples of injury for which this definition appears to be wholly appropriate. From a biological perspective, dysfunction of a cell or tissue resulting from some external insult may be the more common usage. Cell death is often a measure of injury. While structural failure of a cell, or lysis, may be equated with cell death, the correspondence between structural and functional failure is not always so clear cut. Often, it is a question of scale. Dysfunction at the organ or tissue level may be indicative of structural failure (or death) at the cellular level. Similarly, cellular dysfunction may be indicative of structural failure of subcellular structures. At the tissue level, structural failure may lead to secondary functional failure in a temporally segregated, cause-and-effect manner. For example, structural failure of an artery leads to dysfunction of the tissue normally supplied with oxygen and nutrients by that vessel. At the cellular and molecular scale, structural and functional failure are difficult to distinguish, and primary and secondary mechanisms of injury are intermingled spatially and temporally. The purpose of this chapter is to define mechanical injury at the cellular level focusing primarily on membrane damage. First, we will describe the structural properties and failure criteria for cell membranes. We will then discuss experimental approaches to producing and characterizing both the mechanics and the functional consequences of cell injury. Finally, some novel approaches to therapeutic intervention will be outlined. Examples from the clinically relevant applications noted above will be used to illustrate various aspects of the subject.

MEMBRANE MECHANICS

The plasma membrane forms the boundary between intracellular and extracellular compartments. As such, it mediates mass transport into and out of the cell, maintains membrane potential, serves as a biochemical transduction interface, and provides mechanical coupling between the cell and its environment. The cytoskeleton provides mechanical structure to the cytoplasm. It transmits tension throughout the cell and generates the forces responsible for cell motility and intracellular trafficking of membrane-bound vesicles. There are a great deal of data available on the mechanical properties and failure criteria for these structures; however, most of this work has been performed in the context of normal physiological function. Thus, large deformations at high strain rates were not usually considered. This section will review the experimental data on the mechanical properties of cellular membranes and relate them to the response to traumatic loading conditions.

The plasma membrane is a composite structure consisting of a two-dimensional "fluid mosaic" with a fibrous protein "backbone." The term, "fluid mosaic," was coined by Singer and Nicholson[6] to indicate the two-dimensional (2-D) fluid nature of the bilayer of amphiphilic lipids and the inclusion of a wide variety of membrane-associated proteins. The membrane proteins include ion channels, receptors, and signaling molecules. They can be freely diffusing (in 2-D) in the lipid bilayer or linked to the cortical cytoskeleton, to the extracellular matrix, or to both. (The distinction between the fibrous protein backbone of the membrane and the cytoskeleton is arbitrary. For the purposes of this discussion, the membrane-associated "backbone" or cortical cytoskeleton will be considered a part of the composite of the membrane, while the cytoskeleton will refer to the fibrous proteins that traverse the cytoplasm.)

A great deal is known about the mechanical properties of cell membranes. The data come mostly from experiments on two relatively ideal systems. Artificial lipid bilayer vesicles have been used to determine the properties of the fluid lipid bilayer component of the membrane without the influence of the integral membrane proteins or the fibrous backbone. The red blood cell provides a lipid bilayer membrane with the protein backbone but with no internal cytoskeleton.

There are a number of excellent reviews on the mechanics of lipid bilayers and cell membranes.[7–9] I refer the reader to these works for a comprehensive treatment of the subject. I shall highlight the basic concepts and discuss the aspects of the subject that are most relevant to injury. As its name implies, the lipid bilayer is two molecules thick, and, thus, can be considered a continuum is two dimensions only. The primary modes of deformation are in the plane of the membrane. The bending modulus for cell membranes is very small compared to the other moduli such that bending stresses can be safely neglected except when the radius of curvature approaches molecular dimensions. We will also not concern ourselves with changes in thickness since the compressibility in the thickness direction is sufficiently small that the deformations of interest may be considered to occur without change in thickness. That leaves us with membrane tensions and in-plane shear stresses. It is convenient to decompose the two principal membrane tensions, T_1 and T_2 (FIG. 1), into an isotropic tension, $\bar{T}$, and a deviator, or shear resultant, which is equivalent to the maximum surface shear stress, $\bar{T}_s$, which acts at a 45-degree angle with respect to the principal axes. The corresponding deformations can be represented in terms of the principal stretch ratios, λ_1 and λ_2 (FIG. 1). They are the area dilatation

$$\alpha = \lambda_1 \lambda_2 - 1$$

and the shear strain at constant area ($\lambda_1 \lambda_2 = 1$)

$$\varepsilon_s = \frac{1}{4}(\bar{\lambda}^2 - \bar{\lambda}^{-2})$$

where $\bar{\lambda} = \lambda_1 = 1/\lambda_2$ is the stretch ratio at constant area.

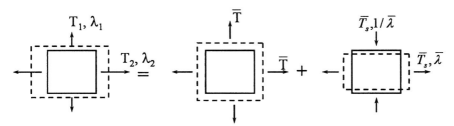

FIGURE 1. Modes of membrane deformation. The two-dimension state of membrane tension, indicated by the principal tensions, T_1 and T_2 (and the corresponding stretch ratios, λ_1 and λ_2), can be represented by the superposition of an isotropic membrane tension, $\bar{T}$, and a pure shear component, $\bar{T}_s$. $\bar{T}$ and $\bar{T}_s$, are related to the area dilatation and the in-plane shear strain, respectively.

The fractional area expansion, α, in a membrane is related to the isotropic tension by the area expansion modulus, K_α:

$$\overline{T} = K_\alpha \alpha$$

The relationship between $\overline{T}$ and α for lipid vesicles and cell membranes is typically linear to the point of rupture. In other words, in dilatation, the membranes behave like brittle materials. Physically, because the membrane is a bilayer, an area dilatation means an increase in the area occupied by the individual lipid molecules in the membrane. The cohesion of the membrane is due to the separation of the hydrophobic chains that form the core of the bilayer from the surrounding aqueous environment. The energy associated with exposing the hydrophobic interior to water provides the elastic restoring force.[10] When the area per molecules reaches a critical point, it becomes energetically favorable to form a hole or tear. For red cell membranes, the critical area dilatation for rupture, α_c, is 2–4%.[11]

The composition of the lipid components of the membrane, particularly cholesterol, can have a large effect on both K_α and α_c. Varying the cholesterol content from 0 to 50% can cause a three-fold increase in K_α and reduce α_c by 20-30%.[12] Thus, when we turn our attention to injury, the cell type and its membrane composition will have to be considered.

The in-plane shear resultant is related to the shear strain by the elastic shear modulus, μ. For in-plane shear deformations (or elongation at constant area), the lipid bilayer behaves as a fluid. Thus, the elastic shear modulus is essentially zero, and the membrane can deform indefinitely under low stress without failure. When the protein backbone is added, the membrane can support a static shear stress. The elastic shear modulus for red blood cells is on the order of 10^{-2} dyn/cm.[13]

For the analysis of traumatic loading conditions, the response to deformation rates greater than quasi-static must be considered. It was stated above that the lipid bilayer behaves as a fluid for in-plane shear deformations. This behavior is characterized by a coefficient of viscosity. Analogous to the elastic moduli, two viscosity coefficients (or viscous moduli) relate the surface stresses to the rate of deformation corresponding to the two modes of deformation described for the elastic case. They are defined as follows:

$$\overline{T} = \kappa \frac{\partial \alpha}{\partial t}$$

$$T_s = 2\eta \frac{\partial \ln \overline{\lambda}}{\partial t}$$

The coefficient, κ, represents the resistance to the rate of area dilatation or contraction, $\partial \alpha / \partial t$. The coefficient, η, is the surface shear viscosity. The expression for the shear rate of deformation ($\partial \ln \overline{\lambda} / \partial t$, which could also be written, $(1/\overline{\lambda}) \partial \overline{\lambda} / \partial t$) is derived from an Eulerian description of the deformation, in which the rate of deformation at any point in time is referenced to the current, deformed geometry rather than the initial, undeformed geometry. Details can be found in Evans and Skalak[7] (pp. 212–208). The simplest expression combining the viscous and elastic behavior is a linear combination:

$$\overline{T} = K_\alpha \alpha + \kappa \frac{\partial \alpha}{\partial t}$$

$$T_s = \frac{\mu}{2}(\overline{\lambda}^2 - \overline{\lambda}^{-2}) + 2\eta \frac{\partial \ln \overline{\lambda}}{\partial t}$$

These expressions are equivalent to the Kelvin-Voigt linear viscoelastic model. The relaxation time constants associated with both modes of deformation are given by:

$$\tau_\alpha = \kappa/K_\alpha \text{ and } \tau_\lambda = \eta/\mu.$$

If such a viscoelastic membrane were held under tension, then suddenly released, the elastic recoil tending to restore the initial unloaded state will be resisted by the viscosity of the material. The relaxation time constant gives the characteristic time to recover the unloaded shape. When loads are applied to a membrane over time periods much greater than the relaxation time constant, the viscosity contributes little to the response and the material behaves elastically. For loads applied over periods of time that are very short compared to the times constant (or with frequencies much greater than $1/\tau$), the material also behaves elastically but with a higher apparent stiffness because of the contribution of the viscosity to the resistance to deformation. At intermediate loading rates, the material behavior is strongly rate-dependent.

The relaxation time constant for shear or elongation, τ_α, is in the range of 10 ms to 1 second for red cell membranes.[9] Other cell membranes with different compositions may vary somewhat from this range, but in general we should expect the response to loading rates associated with trauma to be loading rate–sensitive.

The viscous area dilatation modulus, κ, has been estimated to be on the order of 10^{-7} dyn-s/cm.[9] The elastic area dilatation modulus, K_α, has been determined for cell membranes and lipid bilayers and is the range of 10^2–10^4 dyn/cm.[11] The relaxation time constant for area dilatation, τ_α, is then on the order of 10^{-9} seconds or less. Thus, for loading rates that could reasonably occur even during traumatic injury, it appears that this mode of deformation can be considered to be elastic.

The experimental conditions for producing isotropic membrane tension and measuring area dilatation achieve a uniform dilatation over the whole membrane in a closed system (either a cell or a vesicle). The red blood cell is unusual in the relative smoothness of its plasma membrane. However, most cells have more irregular surfaces with bumps and invaginations in the plasma membrane. Furthermore, as discussed below, the mechanical loading of cells residing in tissue or even cultured on a flat substrate cannot be expected to produce uniform loading of the membrane. In regions of area dilatation, the mechanism of stress relaxation will be the recruitment of lipid from adjacent regions with lower tension. This process will be governed by the surface shear viscosity. Thus, loading rates associated with trauma can be expected to produce area dilatation focally while lower loading rate would not.

EVIDENCE FOR MECHANICAL INJURY OF THE PLASMA MEMBRANE

Although the maintenance of plasma membrane integrity is generally accepted to be a necessary condition for cell viability, the partial or transient loss of membrane integrity is seldom considered as a causative factor in cellular pathology. More often membrane damage is considered either as an all-or-none acute response to some injurious stimulus causing cell lysis and immediate death or as the final common event in necrotic cell death.

The idea that transient survivable membrane disruptions, or pores, may be formed by mechanical deformation, or mechanoporation, and that such pores may play important roles in the development of various pathologies is not universally accepted. In the field of neural injury, for example, Smith *et al.* find no evidence of mechanoporation in their cell-culture model of neural trauma[14,15] while other groups using different models have demonstrated membrane damage both *in vivo*[16,17] and *in vitro*.[18–21] The most straightforward and familiar example of mechanoporation is the formation of red cell ghosts by osmotic swelling.[22,23] Carefully controlled exposure to hyposmotic media causes swelling of the cell until membrane pores form, allowing hemoglobin to escape. The osmotic gradient thus is relieved and the plasma membrane spontaneously reseals. This procedure demonstrates two important phenomena: (1) that membrane pores large enough to allow diffusion of macromolecules can be formed without lysing the cell and (2) that the membrane is capable of resealing. In fact, these properties are exploited in a number of procedures used for transfecting cells including sonication,[24] "scrape-loading,"[24,25] and syringe loading.[26]

The field of neural trauma provides an obvious motivation for understanding the response of cells and tissue to mechanical loading. An isolated tissue model for neural injury provides an excellent example of the strain-rate–dependence of membrane damage suggested by the theory presented above. Galbraith *et al.*[27] used the same

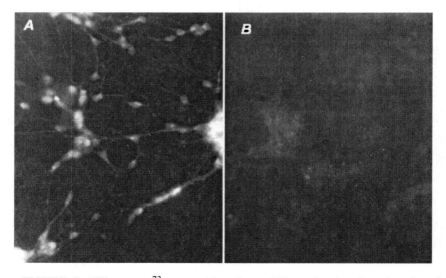

FIGURE 2. NT2 neurons[73] were subjected to a 200-ms duration shear impulse of 50 dyn/cm^2 with a 50-ms rise time in the presence of fluorescently conjugated 3000 MW dextran. The tracer was excluded from unsheared controls (A). Injured cells exhibited diffuse fluorescence throughout the cytoplasm, including neuritic processes. Retention of the tracer after rinsing 5 minute post injury indicates membrane resealing.

isolated squid giant axon preparation used in the electrophysiologic studies of Hodgkins and Huxley,[28] but with a force transducer connected to one end of the axon and a displacement actuator attached to the other.[27] Quasi-static elongation to 50% strain caused no change in membrane potential while dynamic stretches of 5–20% produced membrane depolarizations whose magnitude and duration were dependent on the applied strain. The depolarization response was unaffected by the application of tetrodotoxin (TTX) and tetraethylammonium (TEA), blockers of sodium and potassium channels, respectively, suggesting a nonspecific, or leak, pathway due to mechanical trauma to the membrane. More recent *in vitro* models have shown the loading-rate–dependence of injury in a variety of neuronal cell types[19–21,29–30] as well as in non-neuronal cells.[31] In some of these studies, the creation of nonspecific membrane pores was demonstrated directly using macromolecular tracer molecules.[19–21,31] One strategy is to injure the cells in the presence of a fluorescently labeled molecule that normally does not cross the cell membrane (FIG. 2). Fluorescent labeling of the cytoplasm indicates both that the membrane became permeable to the tracer and that the membrane resealed to trap the tracer within the cell.

EXPERIMENTAL MODELS OF CELL INJURY

Some of the techniques used to apply mechanical loads to cells and characterize the stresses and deformations have been described above. The osmotic swelling experiment produces uniform isotropic tension that can be easily calculated, but measurement of changes in diameter are difficult because of the small size of the cells.[32] Micropipette aspiration techniques overcame that limitation to provide accurate assessment of the area expansion modulus and the critical area dilatation for rupture.[11] The technique is applicable to cells in suspension; however, besides blood cells, most cells are anchorage-dependent, and their morphology and structure depend strongly on their association with a solid support. Creating the dynamic mechanical loading conditions associated with trauma in cell culture models that reasonably reproduce the normal structure of the cell presents some experimental challenges. Furthermore, the desire to monitor the biological responses to the mechanical insult in real time has lead to the development of novel cell-culture models of injury. The two most common approaches are hydrodynamic shear stress and elastic substrate stretch. In the hydrodynamic category, spinning disk and cone-and-plate devices, in which flow is driven by direct displacement of the disk or cone, provide the best control over the temporal characteristics of the stimulus in the range of loading rates relevant to injury. LaPlaca *et al.* used the spinning parallel disk configuration to injure cultured NT2 neurons.[33] Shear stress magnitude and the onset rate were varied by using different combinations of disk speed and the height of the gap between the parallel disks. Blackman designed a cone-and-plate device that has the advantage that the shear stress is uniform across the plate surface (shear stress varies radially in the spinning disk device). The cone is driven by a microstepper motor that allows arbitrary and independent control of the stress magnitude and the loading rate.[31]

Fluid flow over cells cultured on a flat surface does present some difficulties in the analysis and interpretation of results. Although the nominal shear stress at the culture surface may be spatially uniform (as in the cone-and-plate system), the surface topography of the cells themselves influences the local flow field, causing steep gradients in the shear stress acting on the cell surface.[34] The variation in shear stress may be on the order of ± 35% of the average shear stress for a monolayer of fairly uniform and regularly spaced cells (such as endothelial cells).[34] The peak shear stress on the surface of an isolated spherical cell will be on the order of four times the average shear stress.[35] In the case of a spherical cell, like the soma of a neuron, the resultant drag and torque have to be balanced by the adhesive forces between the cell and the substrate. Stress concentrations at the points of attachment or focal stretching of neurites where they meet the soma may be more likely to cause injury than the surface shear stress *per se*. In these models, cell adhesion to the substrate is a critical factor both for promoting the proper differentiated cell structure and to withstand forces great enough to cause injury without detaching the cells from their substrate.

The other major approach to producing injury *in vitro* is to grow cells on a compliant substrate that can be deformed. Devices that apply uniaxial[36,37] and biaxial strain fields[38–40] have been developed to study a wide range of physiological responses to mechanical stimuli. Again, to study the response to traumatic injury requires special considerations. For injury, it is necessary to be able to achieve large deformations at high strain rate. Two types of devices have been used for this purpose. One involves applying pressure (or vacuum) to a circularly clamped membrane. A thin membrane will be deformed into a spherical cap.[38] The radial strain is nearly uniform over the entire surface while the circumferential strain gradually decreases toward the clamped periphery.[41] The central two-thirds of the area has a nearly equibiaxial (isotropic) strain field. A popular commercially available device[42] originally employed a thick membrane (a plate, in engineering terms). Because of the bending stresses that could be born by the plate during deformation and the edge clamping condition, this device produced a small (sometimes negative) isotropic strain in the center of the membrane. Towards the periphery, there was a very large radial strain and a vanishing circumferential strain such that the strain field was uniaxial.[42] Besides the nonuniformity of the strain field in this model, anisotropic strain fields in general require special care when interpreting results because the orientation of the cell or cell process with respect to the principal strain directions must be considered. For example, long, thin cell processes tend to have actin filaments oriented longitudinally, terminating in focal adhesions sites near the ends of the process. Such a process oriented in the direction of a uniaxial strain field will experience the substrate strain, whereas the same process oriented transverse to the strain direction will experience little or no deformation. The membrane inflation techniques have other drawbacks as well. The devices are capable of high strain rates,[29] but controlling the strain rate is difficult, though possible, using pneumatic controls.[43] Finally, the movement of the culture surface out of the plane of focus prevents microscopic measurements during the deformation.

A method for producing uniform, isotropic membrane strains was introduced by Hung *et al.*[39] The technique involves pressing a circularly clamped membrane down over a cylindrical form like a drum head. Schaffer *et al.* used the same basic principle, but modified the design to produce larger cyclic strains using a cam-driven displacement.[20] Geddes and Cargill created a voice coil–actuated device capable of the large deformations and high strain rates necessary for the study of trauma.[30] The uniform, isotropic strain field produced by these devices is attractive because one need not be concerned about the orientation of cells with respect to any principle strain axis.

In all of the stretching devices, if the cells remain attached to the substrate, then the average cell strain (at least on the surface adjacent to the substrate) should approximate the substrate strain. The substrate strain is readily verified experimentally by tracking the displacements of fiducial markers.[44] Quantification of the subcellular distribution of strain is more difficult. Depending on the distribution of focal adhesions and the morphology of the cell, the cell strain can be non-uniform and anisotropic even for uniform, isotropic substrate strains.[44] Long, thin cell processes or spindle-shaped cells can be expected to deform uniaxially with little area dilatation, while broad, flat processes or cells well spread in two dimensions can be expected to experience a more biaxial strain with significant area dilatation. On the basis of the discussion above on the role of area dilatation in membrane failure, one would expect greater injury to the well-spread cell. This is illustrated in the stretch injury experiment shown in FIGURE 3.

CONSEQUENCES OF MECHANICAL INJURY

At the cellular level, the structural damage caused by trauma has immediate biological consequences. The maintenance of membrane integrity is of critical importance for the barrier function of the plasma membrane. The formation of pores large enough for macromolecules to pass through will also allow ions to freely diffuse down their concentration gradients. This will cause depolarization of the membrane, affecting voltage-dependent ion channels and the signaling processes that depend upon them (e.g., a depolarized axon cannot pass an action potential). The pores will also allow molecules normally contained within the cytoplasm to be released to the extracellular space where they can act on the cell that released them or neighboring cells. The biological consequences depend upon the cell types involved, the extracellular environment, and the severity of injury. Two examples are given below to illustrate.

Neural Injury

The response of neural tissue to injury involves a complex sequence of events following the initial traumatic event, which typically lasts a few hundred milliseconds or less. It is useful to divide the response to mechanical trauma into three phases. The first phase consists of the initial physical damage of cellular structures, especially the plasma membrane, and the immediate consequences of this damage, which includes membrane depolarization[27] and elevation of intracellular calcium.[19,29] The injury at this stage is not lethal; however, the metabolic challenge of restoring ion homeostasis and various repair mechanisms as well as the initiation of signaling processes can lead to subsequent death of the injured neuron.[19,21] Hence, in the second phase of the injury response, the injured cells may recover to some degree, or they may proceed to die within 24 hours of the initial injury. Cell death during this period is typically characterized by necrosis, or unregulated death involving loss of membrane integrity and release of intracellular contents to the interstitial space. *In vivo,* the survival or death of injured neurons may depend upon several factors in addition to the severity of the mechanical damage to the cell itself. If blood flow is also compromised, the energy supply to meet the metabolic demands of the injured cell will be impaired. The release of excitatory amino acids (EAA) by nearby injured or dying cells can contribute to the calcium overload thought to be a major factor in cell death.[45] The third stage of the response to injury (>24 hours) is characterized primarily by secondary degeneration. Cell death during this period is typically via apoptosis or programmed cell death. The causal link between cell death and the initial mechanical insult at this stage is much less clear. Often cells remote from the sites of early cell death (Phase 2) enter apoptosis even though these remote sites may have experienced minimal tissue deformation during the traumatic event. Thus, their death would seem to be a secondary reaction to the death of the initially injured cells. Though this temporal pattern of cell death (early necrosis, later apoptosis) is often observed,[46] it is not clear whether the initially injured cells are able to recover enough to avoid the early necrosis only to subsequently enter apoptosis or whether the cells entering apoptosis were initially uninjured and driven into apoptosis solely be secondary mechanisms (e.g., EAA stimulation, the presence of reactive oxygen species (ROS), or inflammatory responses).

Within the injured brain, at each time and length scale, there are structural and functional changes that are inextricably linked. The initial membrane damage causes membrane depolarization and elevation of intracellular calcium. These initial phenomena illustrate the intimate interaction between mechanical and biological events. For example, membrane depolarization itself can cause calcium entry into the cell, so whether mechanical trauma leads to calcium influx directly through membrane disruptions or indirectly by first causing depolarization is unclear. At later times, further disruption of the cytoskeleton and axonal swelling observed *in vivo*, if unchecked, can lead to secondary disruption of the plasma membrane.[1] These structural alterations are not necessarily the direct consequence of the initial insult *per se*, but are likely due to the elevated calcium concentration within the cell. Calcium activated proteases (such as calpain)[47] enzymatically degrade the cytoskeletal structure, disrupting axonal transport. The combination of the accumulation of proteins and organelles with the weakening of the membrane leads to focal axonal swelling. During the same period, a variety of biochemical signaling pathways are activated, such as the mitogen-activated protein kinases (MAP kinases) involved in determining cell survival or apoptosis.[48,49] Later events such as secondary degeneration are even less tightly coupled to the initial mechanical trauma, but it may be possible to construct a sequence of cause-and-effect relationships that will allow key points to be targeted for therapeutic intervention. However, because of the interactions between structural or mechanical events and the biological response, particularly in the acute phase, it is essential to develop a complete understanding of the structural changes due to injury and the ensuing biochemical reactions.

Angioplasty

Coronary artery disease remains the leading cause of death in America.[50] It is characterized by the narrowing of the lumen of the coronary artery and its branches due to the intimal accumulation of lipids, infiltration of macrophages, migration, and proliferation of smooth muscle cells from the media and the deposition of fibrous connective tissue. Severely occluded vessels can be treated with bypass surgery in which the diseased segment is excised and replaced with either a vein or arterial graft. An important and widely used alternative to bypass surgery is percutaneous transluminal coronary angioplasty (PTCA). In this procedure, a catheter with a balloon at the end is threaded through the narrowed lumen with the balloon spanning the lesion. Inflation of the balloon compresses the plaque, stretches the vessel wall, and, thus, expands the lumen. The potential for PTCA to eliminate the need for bypass surgery in all but the most complex cases has not been realized because of the phenomenon of restenosis, in which the vessel, whose lumen was initially successfully expanded, reoccludes within 6 months of the procedure in approximately one-third of patients.[50] Despite advances in the understanding of the processes involved in this phenomenon, efforts to develop therapeutic interventions have been unsuccessful in reducing the clinically observed rate of restenosis.

In vivo experimental models of restenosis have provided information about some of the molecular mediators of the process. The two immediate consequences of the balloon inflation are the destruction of the endothelium and the injury of the medial smooth muscle cells. The endothelium normally provides a non-thrombogenic blood-contacting surface and a permeability barrier between the flowing blood and the vessel wall. In addition, the endothelium serves as a signal transduction interface regulating vascular smooth muscle (VSM) behavior in response to blood-borne agonists and hemodynamic stresses. Destruction of the endothelium by the balloon catheter removes the control over growth that endothelial cells normally exert on the VSM and exposes to the blood the thrombogenic subendothelial connective tissue. The VSM cells (VSMs) are then exposed to serum elements normally shielded from

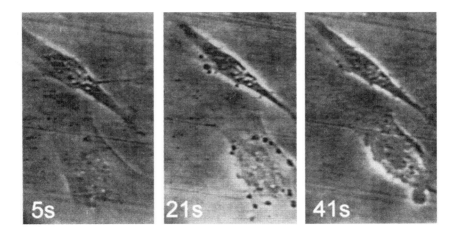

FIGURE 3. Vascular smooth muscle (VSM) cells were cultured on thin polyurethane membranes.[44] A rapid 15% biaxial strain impulse (applied at $t = 0$) caused the formation of membrane "blebs" indicative of local cytoskeletal degradation and swelling. The lower cell is well spread in two dimensions while the upper cell is more spindle-shaped. Cell strain analysis[44] suggests that the upper cell would experience simple elongation with little area dilatation, while the lower cell would experience an area dilatation similar to the substrate (30%). Blebs form in both cells in less than 10 seconds after the stretch. In the more spread cell *(lower)*, the blebs are more numerous, and they grow with time. The cell eventually becomes partially detached and retracts. During the same period, the blebs shrink on the spindle-shaped cell *(upper)* as it appears to be recovering its pre-injury morphology.

them by the endothelium as well as growth and chemoattractant factors released by activated platelets adherent to the denuded region of the vessel wall.

The role of direct injury to the VSMs in the restenosis process has received relatively less attention. However, a series of studies by Reidy and colleagues clearly demonstrate that the VSM response to stretch is at least as important as the endothelial/platelet effects described above. They showed that endothelial denudation alone (i.e., without vessel distention) can elicit VSM proliferation; however, they also found that if the endothelium re-populated the denuded area within 7 days, no intimal lesions formed.[51] Furthermore, the proliferation of VSMs was much less than when the denudation was combined with distention of the vessel.

The most compelling result in support of the hypothesis that VSM injury *per se* is an important stimulus for the restenosis response comes from an experiment in which vessels were distended by fluid pressure without contacting the endothelium.[4] In this model, there was minimal disruption of the endothelium, and endothelial integrity was reestablished within 3 days—well before the time required for intimal lesion formation in the denudation without stretch experiment.[51] Furthermore, the proliferative response of VSMs was much greater than in the denudation without stretch and comparable to the proliferative response to combined denudation and stretch.

Further support for the concept that an intrinsic response to mechanical injury by the VSM is an important component of the restenosis phenomenon comes from the following findings: after vessel distention, VSMs express and release basic fibroblast growth factor (bFGF) and express the bFGF receptor;[52] bFGF is a more potent mitogen than PDGF;[53] and bFGF is released by mechanically injured cells.[54–57] This last point is significant: bFGF is a cytoplasmic protein with no known release mechanism; it lacks the classic signal peptide for direct secretion.[58] This has led to the speculation that it is released by injured or dead cells with damaged membranes as a wound-healing mechanism.[59] The importance of bFGF in restenosis has been

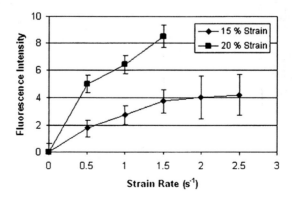

FIGURE 4. Strain-rate dependence of VSM cell injury. VSM cells were cultured on thin silicon membranes and subjected to biaxial strain impulses in the presence of fluorescently conjugated 3000 MW dextran using a device similar to that of Cargill.[30] Membrane damage was quantified by the fluorescence intensity of the tracer trapped in the cells (see FIGURE 2). The injury is strongly strain-rate dependent over the range $0–1.5 \text{ s}^{-1}$.

demonstrated in animal models in which binding of bFGF by antibodies,[60] heparin,[61] or heparin mimics[62] inhibits the restenosis response.

The mechanism of release of bFGF is not well understood. That its release is a result of damage to the cell membrane resulting from mechanical injury was suggested by the work on restenosis discussed above and an *in vitro* study of the response of VSMs to stretch.[63] Cheng showed membrane damage and FGF release with cyclically loaded cells. The device used was capable of a maximum 1-Hz sinusoidal strain wave form. They could not detect membrane damage from a single stretch, perhaps because the strain rate was not sufficient. We have used a stretching device with the same membrane deformation mechanics, but with an actuator capable of producing high strain rates. We found that injury severity was strongly dependent on the strain and strain rate (FIG. 4).

While the cyclic loading regime used by Cheng may not produce a single large stretch injury, as in angioplasty, their data may suggest that cyclic mechanical stimuli in the physiological range induces continual sub-lethal injuries of the plasma membrane. Release of bFGF has also been described in other tissues exposed to mechanical deformation and, thus, may represent a common wound-healing response mechanism.[55–57,59]

MEMBRANE REPAIR AS A THERAPEUTIC STRATEGY

As described above, membrane disruption can initiate a sequence of pathologic events leading to tissue dysfunction and death. If, in these scenarios, physical damage to the cell membrane is the precipitating event, it is natural to ask whether repairing the membrane damage can stave off the ensuing pathology. Therapeutic approaches to membrane repair for other types of membrane damage (e.g., electroporation) are reviewed elsewhere in this volume. However, relatively little work has been done with regard to membrane damage due to mechanical trauma. One approach is the use of surfactants such as poloxamer 188 (P188) to restore membrane integrity. This approach has been shown to protect cells from thermal,[64] electrical,[65] and radiation[66] insults. P188 is a tri-block co-ploymer with a hydrophic polypropylene oxide chain flanked by hydrophilic polyethylene oxide chains. The mechanism by which P188 promotes the resealing of membrane pores is not completely understood. It has been shown to insert into damaged membranes such that the hydrophobic stretch spans the lipophilic core of the plasma membrane.[67]

Borgens and colleagues have used P188[68] and polyethylene glycol (PEG)[69] in a spinal cord injury model and have shown partial recovery of function. Restoration of action potential conduction suggests that the treatment indeed restored membrane continuity along the white matter tracts in the spinal cord. The mechanism of action of the P188 is presumed to be via insertion into the membranes as described in other systems. The PEG is presumed to act as a fusegen as it does in *in vitro* cell systems. While these treatments did not restore function completely, the finding that waiting 6–8 hours after the injury to apply the treatment provided the same degree of recovery is encouraging for the potential clinical application of this technique.

In an *in vitro* model of neural injury, treatment with P188 at 10–15 minutes post injury restored membrane integrity as assessed by LDH release acutely and restored cell viability at 24 hours to control values compared to approximately 60% viability in untreated injured cells.[21] As described in the previous section, the response to injury in neural tissue is complex, and cell death may be due to necrosis or apoptosis. Necrosis is characterized by the breakdown of the cell membrane, so it is reasonable to expect strategies to repair the cell membrane to reduce necrotic cell death. Indeed, we found that P188 treatment virtually eliminated necrosis. However, in this model, the majority of the dead cells at 24 hours post injury were found to be undergoing apoptosis. P188 treatment also significantly reduced the rate of apoptosis.[70] The mechanisms by which mechanical injury leads to apoptosis are not completely understood, but it is clear that multiple signaling pathways are involved. That P188 treatment provided profound neuroprotection suggests that the membrane damage is the primary precipitating event and that repairing the membrane can forestall the signaling cascades leading to apoptosis.

It is possible that the mechanism of action of P188 goes beyond simply restoring membrane integrity. In a model of excitotoxic neural cell death without mechanical trauma, Marks et al. found that P188 significantly inhibited necrosis.[71] They showed that the P188 was inserting into the cell membranes and that it was capable of arresting molecular transport across the membrane. In addition, they found that the presence of P188 in the membrane appeared to protect the membrane lipids from oxidative damage. The generation of reactive oxygen species is also a feature of traumatic neural injury that can lead to further membrane damage or apoptosis. This model differs from the direct injury of the membrane due to mechanical trauma in that the loss of membrane integrity is a rather late consequence of cellular processes triggered by biochemical stimulation.

It may seem surprising that a cell that has progressed towards necrosis to the point that the membrane is breaking down could still be rescued by repairing the membrane. However, in addition to necrosis induced by excitotoxicity in neurons, ischemic injury to cardiac myocytes can also be rescued from death by membrane repair. In an *in vitro* model of ischemia, cardiac myocytes were subjected to hypoxic conditions to the point that viability was reduced to a small fraction of the controls.[72] In a novel approach to membrane resealing, immunoliposomes targeted to the intracellular protein myosin were added to the cells. Binding only where breaches in the membrane exposed the myosin, the liposomes plugged the pore and eventually fused with the membrane. Thus, it appears that at multiple stages of the injury process, targeting the restoration of membrane integrity is a viable approach to the treatment of cell injury.

ACKNOWLEDGMENTS

The author gratefully acknowledges support of his work on cell injury from the Centers for Disease Control (CDC R49/CCR316574-01-3), the National Science Foundation (NSF BES-9984276), and the Whitaker Foundation.

REFERENCES

1. MAXWELL, W.L., J.T. POVLISHOCK & D.L. GRAHAM. 1997. A mechanistic analysis of nondisruptive axonal injury: a review. J. Neurotrauma **14:** 419–440. Erratum in J. Neurotrauma **14:** 755.

2. BLACKSHEAR, P.L., JR. 1972. Mechanical hemolysis in flowing blood. *In* Biomechanics: Its Foundations and Objectives. Y.C. Fung, N. Perrone, and M. Anliker, Eds. :501–528. Prentice-Hall. Englewood Cliffs, NJ.
3. DEWITZ, T.S., T.C. HUNG, R.R. MARTIN & L.V. MCINTIRE. 1977. Mechanical trauma in leukocytes. J. Lab. Clin. Med. **90:** 728–736.
4. CLOWES, A.W., M.M. CLOWES & M.A. REIDY. 1986. Kinetics of cellular proliferation after arterial injury. III. Endothelial and smooth muscle growth in chronically denuded vessels. Lab. Invest. **54:** 295–303.
5. LIU, S.Q., M.M. MOORE & C. YAP. 2000. Prevention of mechanical stretch-induced endothelial and smooth muscle cell injury in experimental vein grafts. J. Biomech. Eng. **122:** 31–38.
6. SINGER, S.J. & G.L. NICOLSON. 1972. The fluid mosaic model of the structure of cell membranes. Science **175:** 720–731.
7. EVANS, E.A. & R. SKALAK. 1979. Mechanics and thermodynamics of biomembranes: part 1. CRC Crit. Rev. Bioeng. **3:** 181–330.
8. EVANS, E.A. & R. SKALAK. 1979. Mechanics and thermodynamics of biomembranes: part 2. CRC Crit. Rev. Bioeng. **3:** 331–418.
9. EVANS, E.A. & R.M. HOCHMUTH. 1978. Mechano-chemical properties of membrane. Curr. Top. Membr. Transp. **10:** 1–64.
10. BOAL. D. 2002. Mechanics of the Cell. Cambridge University Press. Cambridge, UK.
11. EVANS, E.A., R. WAUGH & L. MELNIK. 1976. Elastic area compressibility modulus of red cell membrane. Biophys. J. **16:** 585–595.
12. NEEDHAM, D. & R.S. NUNN. 1990. Elastic deformation and failure of lipid bilayer membranes containing cholesterol. Biophys. J. **58:** 997–1009.
13. HOCHMUTH, R.M., N. MOHANDAS & P.L. BLACKSHEAR, JR. 1973. Measurement of the elastic modulus for red cell membrane using a fluid mechanical technique. Biophys. J. **13:** 747–762.
14. SMITH, D.H., J.A. WOLF, T.A. LUSARDI, *et al.* 1999. High tolerance and delayed elastic response of cultured axons to dynamic stretch injury. J. Neurosci. **19:** 4263–4269.
15. WOLF, J.A., P.K. STYS, T. LUSARDI, *et al.* 2001.Traumatic axonal injury induces calcium influx modulated by tetrodotoxin-sensitive sodium channels. J. Neurosci. **21:** 1923–1930.
16. PETTUS, E.H., C.W. CHRISTMAN, M.L. GIEBEL & J.T. POVLISHOCK. 1994. Traumatically induced altered membrane permeability: its relationship to traumatically induced reactive axonal change. J. Neurotrauma **11:** 507–522.
17. PETTUS, E.H. & J.T. POVLISHOCK. 1996. Characterization of a distinct set of intra-axonal ultrastructural changes associated with traumatically induced alteration in axolemmal permeability. Brain Res. **722:** 1–11.
18. ELLIS, E.F., K.A. MCKINNEY, S.L. WILLOUGHBY & J.T. POVLISHOCK. 1995. A new model for rapid stretch-induced injury of cells in culture: characterization of the model using astrocytes. J. Neurotrauma **12:** 325–339.
19. LAPLACA, M.C., V.M. LEE & L.E. THIBAULT. 1997. An in vitro model of traumatic neuronal injury: loading rate-dependent changes in acute cytosolic calcium and lactate dehydrogenase release. J. Neurotrauma **14:** 355–368.
20. GEDDES, D.M., R.S. CARGILL II & M.C. LAPLACA. 2003. Mechanical stretch to neurons results in a strain rate and magnitude-dependent increase in plasma membrane permeability. J. Neurotrauma **20:** 1039–1949.
21. SERBEST, G., J. HORWITZ & K. BARBEE. 2005.The effect of poloxamer-188 on neuronal cell recovery from mechanical injury. Neurotrauma **22:** 119–132.
22. LIEBER, M.R. & T.L. STECK. 1982. A description of the holes in human erythrocyte membrane ghosts. J. Biol. Chem. **257:** 11651–11659.
23. LIEBER, M.R. & T.L. STECK. 1982. Dynamics of the holes in human erythrocyte membrane ghosts. J. Biol. Chem. **257:** 11660–11666
24. FECHHEIMER, M., J.F. BOYLAN, S. PARKER, *et al.* 1987. Transfection of mammalian cells with plasmid DNA by scrape loading and sonication loading. Proc. Natl. Acad. Sci. USA **84:** 8463–8467.
25. MCNEIL, P.L., R.F. MURPHY, F. LANNI & D.L. TAYLOR. 1984. A method for incorporating macromolecules into adherent cells. J. Cell. Biol. **98:** 1556–1564.

26. CLARKE, M.S. & P.L. MCNEIL. 1992. Syringe loading introduces macromolecules into living mammalian cell cytosol. J. Cell. Sci. **102:** 533–541.
27. GALBRAITH, J.A., L.E. THIBAULT & D.R. MATTESON. 1993. Mechanically induced depolarization in the squid giant axon to simple elongation. J. Biomech. Eng. **115:** 13–22.
28. HODGKIN, A.L., A.F. HUXLEY & B. KATZ. 1952. Measurement of current-voltage relations in the membrane of the giant axon of Loligo. J. Physiol. **116:** 424–448.
29. CARGILL, R.S., II & L.E. THIBAULT. 1996. Acute alterations in $[Ca^{2+}]_i$ in NG108-15 cells subjected to high strain rate deformation and chemical hypoxia: an in vitro model for neural trauma. J. Neurotrauma **13:** 395–407.
30. GEDDES, D.M. & R.S. CARGILL II. 2001. An in vitro model of neural trauma: device characterization and calcium response to mechanical stretch. J. Biomech. Eng. **123:** 247–255.
31. BLACKMAN, B.R., K.A. BARBEE & L.E. THIBAULT. 2000. In vitro cell shearing device to investigate the dynamic response of cells in a controlled hydrodynamic environment. Ann. Biomed. Eng. **28:** 363–372.
32. RAND, R.P. & A.C. BURTON. 1963. Area and volume changes in hemolysis of single erythrocytes. J. Cell Comp. Physiol. **61:** 245–253.
33. LAPLACA, M. & L. THIBAULT. 1997. An in vitro traumatic injury model to examine the response of neurons to a hydrodynamically induced deformation. Ann Biomed. Eng. **25,** 665–677.
34. BARBEE, K.A., T. MUNDEL, R. LAL & P.F. DAVIES. 1995. Subcellular distribution of shear stress at the surface of flow-aligned and nonaligned endothelial monolayers. Am. J. Physiol. **268:** H1765–772.
35. O'NEILL, M.E. 1968. A sphere in contact with a plane wall in a slow linear shear flow. Chem. Eng. Sci. **23:** 1293–1298.
36. LEUNG, D.Y., S. GLAGOV & M.B. MATHEWS. 1977. A new in vitro system for studying cell response to mechanical stimulation: different effects of cyclic stretching and agitation on smooth muscle cell biosynthesis. Exp. Cell Res. **109:** 285–298
37. DARTSCH, P.C. & E. BETZ. 1989. Response of cultured endothelial cells to mechanical stimulation. Basic Res. Cardiol. **84:** 268–281.
38. WINSTON, F.K., E.J. MACARAK, S.F. GORFIEN & L.E. THIBAULT. 1989. A system to reproduce and quantify the biomechanical environment of the cell. J. Appl. Physiol. **67:** 397–405.
39. HUNG, C.T. & J.L. WILLIAMS. 1994. A method for inducing equi-biaxial and uniform strains in elastomeric membranes used as cell substrates. J. Biomech. **27:** 227–232.
40. SCHAFFER, J.L., M. RIZEN, G.J. L'ITALIEN, *et al.* 1994. Device for the application of a dynamic biaxially uniform and isotropic strain to a flexible cell culture membrane. J. Orthop. Res. **12:** 709–719.
41. WILLIAMS, J.L., J.H. CHEN & D.M. BELLOLI. 1992. Strain fields on cell stressing devices employing clamped circular elastic diaphragms as substrates. J. Biomech. Eng. **114:** 377–384.
42. GILBERT, J.A., WEINHOLD, P.S., BANES, A.J., *et al.* 1994. Strain profiles for circular cell culture plates containing flexible surfaces employed to mechanically deform cells in vitro. J Biomech. **27:** 1169–77.
43. MORRISON, B., 3RD, D.F. MEANEY & T.K. MCINTOSH. 1998. Mechanical characterization of an in vitro device designed to quantitatively injure living brain tissue. Ann. Biomed. Eng. **26:** 381–390.
44. BARBEE, K.A., E.J. MACARAK & L.E. THIBAULT. 1994. Strain measurements in cultured vascular smooth muscle cells subjected to mechanical deformation. Ann. Biomed. Eng. **22:** 14–22.
45. CHOI, D.W. 1987. Ionic dependence of glutamate neurotoxicity. J. Neurosci. **7:** 369–379
46. RAGHUPATHI, R., D.L. GRAHAM & T.K. MCINTOSH. 2000. Apoptosis after traumatic brain injury. J. Neurotrauma **17:** 927–938.
47. SAATMAN, K.E., B. ABAI, A. GROSVENOR, *et al.* 2003. Traumatic axonal injury results in biphasic calpain activation and retrograde transport impairment in mice. J. Cereb. Blood Flow Metab. **23:** 34–42.

48. XIA, Z., M. DICKENS, J. RAINGEAUD, et al. 1995. Opposing effects of ERK and JNK-p38 MAP kinases on apoptosis. Science **270:** 1326–1323.
49. OZAWA, H., S. SHIODA, K. DOHI, et al. 1999. Delayed neuronal cell death in the rat hippocampus is mediated by the mitogen-activated protein kinase signal transduction pathway. Neurosci. Lett. **262:** 57–60.
50. AMERICAN HEART ASSOCIATION. 2000. 2001 Heart and Stroke Statistical Update. American Heart Association. Dallas, Texas.
51. FINGERLE, J., Y.P. AU, A.W. CLOWES & M.A. REIDY. 1990. Intimal lesion formation in rat carotid arteries after endothelial denudation in absence of medial injury. Arteriosclerosis **10:** 1082–1087.
52. LINDNER, V. & M.A. REIDY. 1993. Expression of basic fibroblast growth factor and its receptor by smooth muscle cells and endothelium in injured rat arteries: an en face study. Circ. Res. **73:** 589–595.
53. JAWIEN, A., D.F. BOWEN-POPE, V. LINDNER, et al. 1992. Platelet-derived growth factor promotes smooth muscle migration and intimal thickening in a rat model of balloon angioplasty. J. Clin. Invest. **89:** 507–511.
54. KLAGSBRUN, M. & E.R. EDELMAN. 1989. Biological and biochemical properties of fibroblast growth factors:: implications for the pathogenesis of atherosclerosis. Arteriosclerosis **9:** 269–278.
55. CLARKE, M.S., R.W. CALDWELL, H. CHIAO, et al. 1995 Contraction-induced cell wounding and release of fibroblast growth factor in heart. Circ. Res. **76:** 927–934.
56. CLARKE, M.S. & D.L. FEEBACK. 1996. Mechanical load induces sarcoplasmic wounding and FGF release in differentiated human skeletal muscle cultures. FASEB J. **10:** 502–509.
57. CLARKE, M.S., R. KHAKEE & P.L. MCNEIL. 1993. Loss of cytoplasmic basic fibroblast growth factor from physiologically wounded myofibers of normal and dystrophic muscle. J. Cell Sci. **106:** 121–133.
58. ABRAHAM, J.A., A. MERGIA, J.L. WHANG, et al. 1986. Nucleotide sequence of a bovine clone encoding the angiogenic protein, basic fibroblast growth factor. Science **233:** 545–548.
59. MCNEIL, P.L. & R.A. STEINHARDT. 1997. Loss, restoration, and maintenance of plasma membrane integrity. J. Cell Biol. **137:** 1–4.
60. LINDNER, V. & M.A. REIDY. 1991. Proliferation of smooth muscle cells after vascular injury is inhibited by an antibody against basic fibroblast growth factor. Proc. Natl. Acad. Sci. USA **88:** 3739–3743.
61. CLOWES, A.W. & M.J. KARNOWSKY. 1977. Suppression by heparin of smooth muscle cell proliferation in injured arteries. Nature **265:** 625–626.
62. HERRMANN, H.C., S.S. OKADA, E. HOZAKOWSKA, et al. 1993. Inhibition of smooth muscle cell proliferation and experimental angioplasty restenosis by beta-cyclodextrin tetradecasulfate. Arterioscler. Thromb. **13:** 924–931.
63. CHENG, G.C., W.H. BRIGGS, D.S. GERSON, et al. 1997. Mechanical strain tightly controls fibroblast growth factor-2 release from cultured human vascular smooth muscle cells. Circ. Res. **80:** 28–36.
64. MERCHANT, F.A., W.H. HOLMES, M. CAPELLI-SCHELLPFEFFER, et al. 1998. Poloxamer 188 enhances functional recovery of lethally heat-shocked fibroblasts. J. Surg. Res. **74:** 131–140.
65. LEE, R.C., L.P. RIVER, F.S. PAN, et al. 1992. Surfactant-induced sealing of electropermeabilized skeletal muscle membranes in vivo. Proc. Natl. Acad. Sci. USA **89:** 4524–4528.
66. HANNIG, J., D. ZHANG, D.J. CANADAY, et al. 2000. Surfactant sealing of membranes permeabilized by ionizing radiation. Radiat. Res. **154:** 171–177.
67. MASKARINEC, S.A., J. HANNIG, R.C. LEE, et al. 2002. Direct observation of poloxamer 188 insertion into lipid monolayers. Biophys. J. **82:** 1453–1459.
68. BORGENS, R.B., D. BOHNERT, B. DUERSTOCK, et al. 2004. Subcutaneous tri-block copolymer produces recovery from spinal cord injury. J. Neurosci. Res. **76:** 141–154.
69. BORGENS, R.B., R. SHI & D. BOHNERT. 2002. Behavioral recovery from spinal cord injury following delayed application of polyethylene glycol. J. Exp. Biol. **205:** 1–12.

70. SERBEST, G., J. HORWITZ, M. JOST & K. BARBEE. 2005. Mechanisms of cell death and neuroprotection by poloxamer 188 after mechanical trauma. FASEB J. In press.
71. MARKS, J.D., C. PAN, T. BUSHELL, et al. 2001. Amphiphilic, tri-block copolymers provide potent, membrane-targeted neuroprotection. FASEB J. **15:** 1107–1109.
72. KHAW, B.A., V.P. TORCHILIN, I. VURAL, et al. 1995. Plug and seal: prevention of hypoxic cardiocyte death by sealing membrane lesions with antimyosin-liposomes. Nat. Med. **1:** 1195–1198.
73. PLEASURE, S.J., C. PAGE & V.M.Y. LEE. 1992. Pure, postmitotic, polarized human neurons derived from NTera 2 cells provide a system for expressing exogenous proteins in terminally differentiated neurons. J. Neurosci. **12:** 1802–1815.

Cell Injury by Electric Forces

RAPHAEL C. LEE

Electrical Trauma Research Program, Department of Surgery, Organismal Biology and Anatomy (Biomechanics), The University of Chicago, Chicago, Illinois 60637, USA

ABSTRACT: The molecular architecture of biological systems is heavily influenced by the highly polar interactions of water. Thus, macromolecules such as proteins that are highly water soluble must be electrically polar. Energy generation methods needed to support cell metabolic processes depend on compartmentalizing mobile ions and thus require electrical ion transport barriers such as membranes. One consequence of these biological design constraints is vulnerability to injury by electrical forces. Suprahysiological electric forces cause damage to cells and tissues by disrupting cell membranes and altering the conformation of biomolecules. In addition, prolonged passage of electrical current leads to damage by thermal mechanisms. This review will focus on the non-thermal effects.

KEYWORDS: cell injury; electrical current; nonthermal electrical damage; lipid bilayer; electroporation

INTRODUCTION

The passage of injurious magnitudes of electrical current through tissue is capable of causing injury through one or multiple distinct biophysical energy transduction mechanisms. These mechanisms include the direct action of direct electrical forces on proteins, membranes, and other biomolecular structures, as well as the indirect action mediated by the generation of heat. When considering the effects of electrical current on the entire body, adding to this complexity are the multiple modes of frequency-dependent tissue–current interactions and the variation in current density along the path through the body, as well as variations in body size, body position, and use of protective gear. The dominant mode of injury for any particular electrical trauma victim and how it manifests depends on several different factors. Given this, it is not surprising that survivors of electrical shock and lightning injuries have widely variable patterns of physiological problems. The purpose of this chapter is to review the fundamental biophysical pathways and mechanisms of electrical trauma injury. The focus will be on basic nonthermal modes of electrical damage modes at the cellular level.

Basically, electric forces cause cell damage by altering molecular and organelle structures. While these forces can alter all tissue components, it is the bilayer lipid structure and the protein conformation in membranes of cells that have the greatest vulnerability of all cellular constituents. The most important function of membrane is to provide a diffusion barrier against free ion exchange. Under normal healthy conditions, transmembrane ion traffic takes place through a tightly regulated system of pumps and channels. Most of the metabolic energy used in mammalian cells is invested in maintaining the electrochemical potential across the cell membrane. Disruption of the plasma membrane leads rapidly to metabolic exhaustion. Thus, the importance of the structural integrity of the lipid bilayer to cell viability is apparent.

ELECTRICAL FORCES ACTING ON THE CELL MEMBRANE

Because cells are highly compartmentalized structures with membranes that restrict and control transmembrane solute transport, the current imposed by an extracellular constant DC electric field will attempt to pass around a cell in the field rather than pass through it.[1] As long as the cell membrane is intact, the intracellular current density will be less than the current density outside the cell.[2] This means that the electric field inside the cell will be less than the extracellular field. This further indicates that the voltage difference imposed across the entire cell by the field is mostly dropped across the membrane. The magnitude of the induced V_m (FIG. 1) across the membrane depends on a variety of factors, such as the intra- and extracellular medium conductivity, cell shape and size, the external electric field strength E, as well as how the electric field vector orients with respect to the point of interest on the cell membrane.[2,3] While this is very useful in allowing biological cells to detect small electrical fields in its environment, it also renders the cell membrane vulnerable to damage by an electric field.

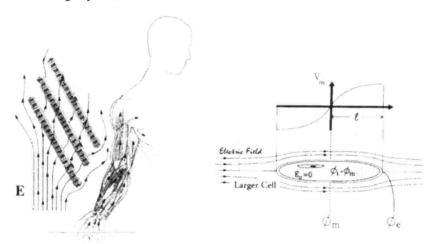

FIGURE 1. Graphic illustration of the induced transmembrane potential (ΔV_m) distribution in a cylindrical cell, like a skeletal muscle fiber, induced by an external DC electric field E_o in a physiologically conducting bath. If the cell is really small, the field lines do not penetrate the cell because the pathway of least resistance for the current is around the cell. The largest induced (ΔV_m) is at the points of the cell that project longest in the direction of the field.

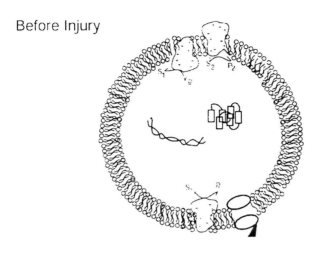

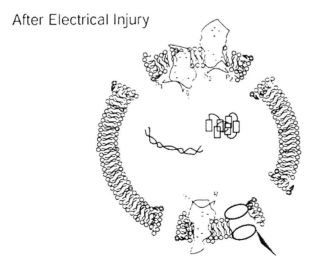

FIGURE 2. Diagram illustrating the membrane alterations following high field exposure. Membrane breakdown by electroporation leads to altered structural integrity of the membrane, increased membrane conductivity, decreased transmembrane potential, and increased membrane water content

MEMBRANE ELECTROPORATION

The natural transmembrane potential difference of a cell, typically 50–70 mV in magnitude, originates from the difference in ionic strengths of the cell's intra- and extracellular fluids. When the transmembrane potential of a cell is in excess of 200–300 mV, structural rearrangement or breakdown of the membrane occurs. This event is called electroporation; it is mediated by water penetration into the cell membrane which is coincident with reorganization of lipids in a lipid bilayer and electroconformational denaturation of membrane proteins.[4–8] The threshold transmembrane potential for induction of membrane electroporation is remarkably similar across cell types and ranges from 250 to 350 mV.[9,10] Several authors have developed reasonably accurate models based on chemical energies of membrane–water interactions to explain the experimentally observed values of V_m required for electroporation.[11]

Membrane electroporation can be either transient or stable, depending on the magnitude and the duration of the imposed transmembrane potential.[12,13] Because physical properties of the membrane are extremely temperature dependent, the membrane temperature is also important. The field-induced generation and growth of pores result in increased membrane electrical conductance, short-circuiting the transmembrane potential. The kinetics of electroporation of cell membranes is dependent on the voltage applied across the membrane. Characteristic time constants for electroporation are in the 0.01–0.10 ms range at normal mammalian body temperatures.[14–17] This reduces the effort of water leaving the membrane pores and increases the probability of spontaneous bilayer sealing. If sealing occurs, sealing kinetics can be orders of magnitude slower than the field relaxation because pore closure requires rearrangement of the membrane fragments and pushing water out of electropores so that the defect may close. These are energy-requiring and time-consuming processes.[1]

Typical DC field strengths required for electroporation are in the range of 1 kV cm^{-1} for most cell types that have a characteristic length scale of 10–40 μm. On the other hand, specialized cells that are designed to communicate via small electric fields are typically much larger. These are nerve and striated muscle (skeletal and cardiac) cells. Due to their relatively long lengths, skeletal muscle cells are up to 8 cm long in large animals, and nerve cells range upwards of 2 m long; these cells have much lower electroporation thresholds. The magnitude of the transmembrane potential induced by an externally applied electric field scales with the dimensions of the cell in the direction of the applied field. Therefore, muscle and nerve cell membranes can be damaged with electrical fields as small as 60 V cm^{-1}. Skeletal muscle cells within the human upper extremity can reach up to several centimeters in length, and peripheral nerve cells are substantially longer. These dimensions are enormous compared to cells of other tissues that have characteristic dimensions in the range of 10–100 μm. Therefore, the strength of the electric field required to electroporate skeletal muscle and peripheral nerve cells is typically 100–1000-fold less than that required to electroporate smaller cell types. Skeletal muscle and nerve cell membranes can be damaged with electrical fields as small as 60 V cm^{-1}.

The distribution of imposed field-induced electropore formation in a large skeletal muscle cell placed aligned parallel to an applied field is well studied.[18,19] Expanding from previous mathematical models based on the traditional "Cable" distributed RC circuit model,[2] and including the fact that the membrane charging time of about 1 μs

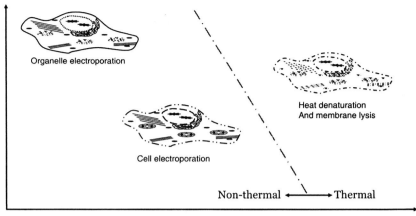

FIGURE 3. Cells are compartmentalized by membranes. Because the time required to change the induced transmembrane potential of a cell or organelle scales with size, then it is possible to gain some selectivity on which structure is electroporated by controlling the duration of the field pulse. Ultrashort pulses of large amplitude can damage intracellular organelles without major damage to the plasma membrane. If the pulse width is long enough, heat-mediated damage will dominate the field effect.

is very short compared to a 1-ms field duration, they concluded that large supraphysiological V_m imposed by the field at the ends of the cell rapidly creates pores, thereby effectively preventing a damaging increase in V_m in these areas.

Electroporation will lead to cell death if the membrane is not quickly sealed. If the membrane is permeabilized, the work required in maintaining the transmembrane concentration differences increases proportionately. The conductance of electroporated cell membranes can increase by several orders of magnitude. When ATP-fueled protein ionic pumps are not able to keep pace with the diffusion of ions through the membrane defects, metabolic energy exhaustion results. If the membrane is not sealed, the cell will progress to biochemical arrest and then to necrosis. Thus, in discussing tissue injury resulting from electrical shock, the principal focus is directed at the kinetics of cell membrane injury and the reversibility of that process.[20–22]

ELECTROCHEMICAL DENATURATION OF MEMBRANE PROTEINS

The composition of cell membranes consists of 40–60% proteins. These membrane proteins function as ion channels, transporters, and signal receptors. These proteins are acted upon by the electric field within the membrane. Under natural conditions, the membrane electric field ranges between 10^6 and 10^7 V/m. Each peptide unit of the typical protein represents an electric dipole moment of about 3.5 Debye (D). For each α-helical structure, which is oriented perpendicular to the cell membrane, many small peptide dipoles are aligned to form a larger dipole on the order of 120 D.[23] An electric field-induced membrane potential will strongly affect the orientations of these electric dipoles. Thus, the intramembrane electric field has a powerful allosteric effect on membrane proteins.

For voltage-dependent membrane proteins, many charged side groups are movable in response to changes in membrane potential. This gating action allows them to function as voltage sensors. One example is the ion channel. Segments of the four inter-repeat domains that line the channel consist of repeated motifs of positively charged amino acid residue followed by two hydrophobic residues. They have been suggested to be a voltage sensor in the proteins' gating. These voltage sensors are obviously susceptible to an applied intensive electric field.

Tsong and Teissie examined the effects of a strong field pulse exposure on the membrane protein Na/K ATPase in erythrocytes.[24] After using a microsecond pulsed intense electric field to shock red blood cell membranes, they found that in a low ionic medium at least 35% of membrane pores induced by the shock pulse were linked to channels in denatured Na/K ATPase in the cell membrane. They attributed this ionic leakage to the electroporation of the Na/K ATPase.

In a series of experiments performed to study the effects of field-induced large transmembrane potentials on ion channel proteins in skeletal muscle fibers, both the open channel currents from the voltage-dependent Na^+ channels and the delayed rectifier K^+ channels were reduced by an intense electric shock.[25,26] An electric shock by a single 4-ms, −500-mV pulse may decrease about 20% of the Na^+ channel currents and 30% of the delayed rectifier K^+ channels currents. The channel conductance of the delayed rectifier K^+ channels were also substantially reduced by the above electric shock. The decrement varies from 10 to 40% depending on the type of cells.

SUMMARY AND CONCLUSIONS

The continuing development of uses for electricity mandates a better understanding of the potential harmful effects of direct electrical contact on biological systems. The molecular structure of biological systems can be severely altered by high-energy commercial-frequency electrical power. The mechanisms of damage include cell membrane electroporation, Joule heating, electroconformational protein denaturation, and others. Effective diagnosis and treatment for electrical injury patients require understanding and addressing each of the direct and indirect modes of tissue–field interaction.

ACKNOWLEDGMENTS

This work has been partly supported by National Institutes of Health Grants R01 GM61101 and R01 GM64757, and by the Electric Power Research Institute.

REFERENCES

1. WEAVER, J.C. 1993. Electroporation: A general phenomenon for manipulating cells and tissue. J. Cell. Biochem. **51:** 426–435.
2. GAYLOR, D.G., A. PRAKAH-ASANTE & R.C. LEE. 1988. Significance of cell size and tissue structure in electrical trauma. J. Theor. Biol. **133:** 223–237.

3. LEE, R.C. & M.S. KOLODNEY. 1987. Electrical injury mechanisms: electrical breakdown of cell membranes. Plast. Reconstr. Surg. **80**: 672–679.
4. KINOSITA, K. & T.Y. TSONG. 1977. Voltage-induced pore formation and hemolysis of human erythrocytes. Biochim. Biophys. Acta **554**: 479–497.
5. NEUMANN, E., A. SPRAFKE, E. BOLDT & H. WOLFF. 1992. Biophysical considerations of membrane electroporation. Plast. Reconstr. Surg. **86**: 77–90.
6. TEISSIE, J., N. EYNARD, B. GABRIEL & P. ROLS. 1999. Electropermeabilization of cell membranes. Adv. Drug Deliv. Rev. **35**: 3–19.
7. CHANG, D.C., B.M. CHASSY, J.A. SAUNDERS & A.E. SOWERS, EDS. 1992. Guide to Electroporation and Electrofusion. Academic Press, Inc. New York.
8. CHIZMADZHEV, A.Y., V.B. ARAKELYAN & V.F. PASTUSHENKO. 1979. Electric breakdown of bilayer membranes: III. Analysis of possible mechanisms of defect origin. Bioelectrochem. Bioenerget. **6**: 63–70.
9. GLASER, R.W., S.L. LEIKIN, L.V. CHERNOMORDIK, et al. 1988. Reversible electrical breakdown of lipid bilayers: formation and evolution of pores. Biochim. Biophys. Acta **940**: 275–287.
10. GOWRISHANKAR,T.R., W. CHEN & R.C. LEE. 1998. Non-linear microscale alterations in membrane transport by electropermeabilization. Ann. N.Y. Acad. Sci. **858**: 205–216.
11. WEAVER, J.C. & Y.A. CHIZMADZHEV. 1996. Theory of electroporation: a review. Bioelectrochem. Bioenerget. **41**: 135–160.
12. BIER, M., S.M. HAMMER, D.J. CANADAY & R.C. LEE. 1999. Kinetics of sealing for transient electropores in isolated mammalian skeletal muscle cells. Bioelectromagnetics **20**: 194–201.
13. GABRIEL, B. & J. TEISSIE. 1997. Direct observation in the millisecond time range of fluorescent molecule asymmetrical interaction with the electropermeabilized cell membrane. Biophys. J. **73**: 2630–2637.
14. BIER, M., S.M. HAMMER, D.J. CANADAY & R.C. LEE. 1999. Kinetics of sealing for transient electropores in isolated mammalian skeletal muscle cells. Bioelectromagnetics **20**: 194–201.
15. BIER, M., W. CHEN, T.R. GOWRISHANKAR, et al. 2002. Resealing dynamics of a cell membrane after electroporation. Phys. Rev. E **66**: 62905.
16. GABRIEL, B. & J. TEISSIE. 1998. Mammalian cell electropermeabilization as revealed by millisecond imaging of fluorescence changes of ethidium bromide in interaction with the membrane. Bioelectrochem. Bioenerget. **47**: 113–118.
17. HIBINO, M., M. SHIGEMORI, H. ITOH, et al. 1991. Membrane conductance of an electroporated cell analyzed by submicrosecond imaging of transmembrane potential. Biophys. J. **59**: 209–220.
18. DEBRUIN, K.A. & W. KRASSOWSKA. 1999. Modeling electroporation in a single cell. I. Effects of field strength and rest potential. Biophys. J. **77**: 1213–1224.
19. DEBRUIN, K.A. & W. KRASSOWSKA. 1999. Modeling electroporation in a single cell. I. Effects of ionic concentrations. Biophys. J. **77**: 1225–1233.
20. LEE, R.C., D.C. GAYLOR, K. PRAKAH-ASANTE, et al. 1988. Role of cell membrane rupture in the pathogenesis of electrical trauma. J. Surg. Res. **44**: 709–719.
21. BHATT, D.L., D.C. GAYLOR & R.C. LEE. 1990. Rhabdomyolysis due to pulsed electric fields. Plast. Reconstr. Surg. **86**: 1–11.
22. BLOCK, T.A., J.N. AARSVOLD, K.L. MATTHEWS II, et al. 1995. Nonthermally mediated muscle injury and necrosis in electrical trauma. J. Burn Care Rehabil. **16**: 581–588.
23. HOL, W.G., P.T. VAN DUIJNEN & H.J. BERENDSEN. 1978. The alpha-helix dipole and the properties of proteins. Nature **273**: 443–446.
24. TEISSIE, J. & T.Y. TSONG. 1980. Evidence of voltage-induced channel opening in Na/K ATPase of human erythrocyte membrane. J. Membr. Biol. **55** : 133–140.
25. CHEN, W. & R.C. LEE. 1994. Altered ion channel conductance and ionic selectivity induced by large imposed membrane potential pulse. Biophys. J. **67**: 603–612.
26. CHEN, W., Y. HAN, Y. CHEN & D. ASTUMIAN. 1998. Electric field-induced functional reductions in the K+ channels mainly resulted from supramembrane potential-mediated electroconformational changes. Biophys. J. **75** : 196–206.

Electroconformational Denaturation of Membrane Proteins

WEI CHEN

Department of Physics, University of South Florida, Tampa, Florida 33620, USA

ABSTRACT: Because of high electrical impedance of cell membrane, when living cells are exposed to an external electric field, the field-induced voltage drops will mainly occur on the cell membrane. In addition to Joule heating damage and electroporation of the cell membrane, the electric field–induced supraphysiological transmembrane potential may inevitably damage the membrane proteins, especially the voltage-dependent membrane proteins. That is because the charged particles in the amino acid of the membrane proteins and, in particular, the voltage-sensors in the voltage-dependent membrane proteins are vulnerable to the membrane potential. An intensive, brief electric shock may induce electroconformational damage or denaturation in the membrane proteins. As a result, the cell functions are significantly reduced. This electric field–induced denaturation in the membrane proteins strongly suggests a new underlying mechanism involved in electrical injury.

KEYWORDS: membrane proteins; electrical injury; electroporation; denaturation; channel conductance

INTRODUCTION

Electrical injury is a frequent trauma in industrialized countries. A better understanding of the underlying mechanisms involved in electrical injury will significantly improve capabilities for patient management and the development of therapeutic treatment. This is a common goal of medical doctors, engineers, and scientists.

The electrical impedance of cell membrane is about six to eight orders of magnitude higher than that of electrolytes in cytoplasmic and extracellular fluids. When living cells are exposed to an external electric field with a frequency equivalent to that found in power lines, cell membranes suffer major voltage drops induced by this electric field. Because cell dimensions are in the order of hundreds or thousands of times larger than the thickness of the cell membrane, an applied electric field with a field strength in tens or hundreds of volts per centimeter (V/cm) may generate millions of V/cm field strength in the cell membrane.[1,2] In all probability, such super-high field strength will inevitably induce damage to the membrane phospholipid bilayer and its embedded membrane proteins.

Joule heating damage has long been accepted as a mechanism involved in electrical injury. Recently, electroporation has been postulated as an important mechanism

---

Address for correspondence: Wei Chen, Department of Physics, University of South Florida, Tampa, FL 33620. Voice: 813-974-5038; fax:813-974-5813.
wchen@cas.usf.edu

involved in electrical injury.[3-5] Electropores are thought to be formed in the lipid bilayer resulting from the extremely high induced supraphysiological transmembrane potential in skeletal muscle and nerve.[6,7] These microlesions in the cell membranes can cause leakage of the cellular metabolic substrates and consequent ionic exchange across the cell membranes.[8,9]

In addition to the phospholipid bilayer, approximately 40% of the cell membrane, by weight, consists of proteins functioning as ionic channels, transporters, and signal receptors. The most basic subunits, the amino acids of the membrane protein molecules, carry electrical charges. In general, each peptide unit represents an electric dipole moment of about 3.5 Debye (D). In an α-helical structure, which is a major functional structure of the membrane proteins, many small peptide dipoles are aligned almost perfectly to form larger dipoles with an order of 120 D across the cell membrane.[10-12] These electric dipoles are strongly affected by an external electrical field. In addition, some charged elements in the voltage-dependent membrane proteins are moveable in response to physiological membrane potential changes. They are often expressed as gating currents in channel proteins,[13] as intramembrane charge movement currents in other voltage-dependent membrane proteins, such as electrogenic pumps,[14,15] and dihyropuridine receptors or the voltage-sensors in skeletal muscle fibers.[16-19] The external electric field-induced supramembrane potential exerts extraordinary force on these moveable charged particles in the membrane proteins, which may cause proteins' conformational damages and denaturation.

In the clinic, we often see electrically injured patients, whose skin and muscles around the current pathway show no sign of damage. However, the functions of the injured limbs or organs show significant reductions. These observations imply that in addition to membrane electroporation, proteins, especially the membrane proteins, may suffer some structural and functional damages during the electric injury.

SUPRAPHYSIOLOGICAL, PULSED TRANSMEMBRANE POTENTIAL-INDUCED FUNCTIONAL REDUCTIONS IN MEMBRANE PROTEINS

Electromediated Leakages in the Na/K ATPases

Pioneer work on the electrical shock–induced conformational changes in membrane proteins was done by Tsong and Teissie on the Na/K ATPase using erythrocyte cells.[20,21] After being shocked by an intensive electrical field in a time range of microseconds, the red cell membrane was perforated.[22] They found that in a low ionic medium at least 35% of the pores were related to the leakage through the Na/K ATPase in the cell membrane. That is because the shock-induced increase in the membrane conductance could be partially blocked by a specific inhibitor, ouabain, or by a specific cross-linking reagent, Cu-phenanthroline, of the ATPase. They attributed this ionic leakage to the electroporation on the Na/K ATPase in cell membranes. However, the proteins' functional changes were not confronted.

Shock Effects on the K^+ Channel Conductance

Abramov *et al.*[23] performed *in vivo* studies on changes in the functions of rat sensory nerves and skeletal muscles after an intensive electrical shock. They found that

4-ms shock pulses of 500 V/cm can significantly reduce the magnitude of the action potential and increase the latency period of the action potential. They predicted that these alterations in nerve functions may be due to some conformational damages in the membrane proteins.

In order to study the electrical shock–induced damages in the membrane proteins, we have developed a modified double Vaseline gap-voltage technique to directly monitor functional changes in membrane proteins.[24] Using this technique, we have investigated several membrane proteins, including ion channel proteins, pump molecules, and dihydroporidine receptors in frog skeletal muscle fibers.

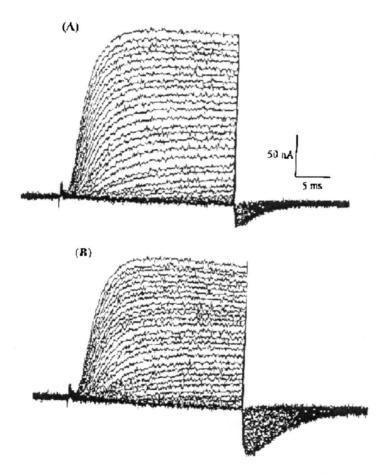

FIGURE 1. Shock field-induced changes in the delayed rectifier K$^+$ channel currents. *Upper* and *lower panels* show the channel currents responding to the same stimulation pulse sequence (25 ms, holding the membrane from −65 to 0 mV) recorded before and after shock by a 4-ms, −500-mV pulse, respectively. (From Chen.[24] Reproduced by permission.)

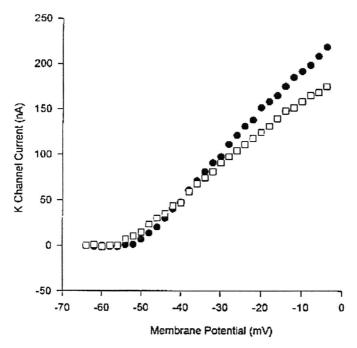

FIGURE 2. The K^+ channel current as a function of the membrane potential. *Dark cycles* and *open squares* represent the channel currents recorded before and after the electric shock, respectively. (From Chen et al.[25] Reproduced by permission.)

For the delayed rectifier K^+ channels, in order to elicit the channel currents, a stimulation pulse sequence was employed which consisted of 30 consecutive pulses holding the membrane potentials in a range from −65 mV to 0 mV. The evoked delayed rectifier K^+ channel currents were resolved and are shown in the upper panel of FIGURE 1, where the Na^+ channels were blocked by tetrodotoxin (TTX).[24]

After a shock by a 4-ms pulse with a supramembrane potential of −500 mV, the same stimulation sequence was applied again to the cell membrane; the recorded K^+ channel currents are shown in the lower panel of FIGURE 1. By comparing the pre- and post-shocked channel currents, one can easily notice that the K^+ channel currents were reduced by a single brief electric shock. For this fiber, the maximum value of the pre-shock K^+ channel currents, about 220 nA, was reduced to around 175 nA, a reduction of approximately 20%.

The K^+ channel currents in the upper and lower panels are plotted as functions of the membrane potential shown in FIGURE 2, represented by dark circles and open squares, respectively. The slope of the post-shocked I-V curve that represents the channel conductance is lower than that of the pre-shocked curve, indicating a reduction in the K^+ channel conductance. For this fiber, the pre- and post-shock K^+ channel conductance is 4.4 μS and 3.5 μS, respectively. Approximately 20% of the channel conductance was reduced by a single 4-ms, −500-mV pulsed shock.

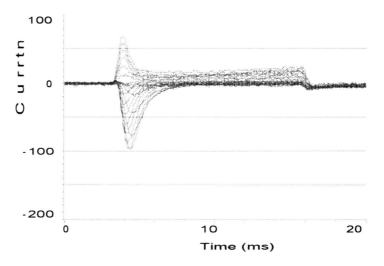

FIGURE 3. Na channel currents recorded from frog skeletal muscle fibers. The membrane holding potential was −90 mV. A sequence of twenty 10-ms stimulation pulses held the membrane at a potential ranging from −55 mV to 40 mV with a 6-mV step. *Upper panel*: control; *lower panel*: after shock by two 4-ms, −450-mV pulses. (From Chen *et al.*[25] Reproduced by permission.)

Supraphysiological Membrane Potential-Induced Reduction in the Na^+ Channel Conductance

Using the same method, the electric shock pulse–induced denaturation of the Na channels was studied.[26] A sequence of 17 pulses of 10-ms duration stimulation that held the membrane potential at a value ranging from −56 mV to 40 mV was applied to the cells. The magnitude step between two consecutive pulses through the sequence was 6 mV. The evoked Na channel currents were resolved and are shown in the upper panel of FIGURE 3. After the cell membrane was shocked by two 4-ms, −450-mV supraphysiological membrane potentials with a time interval of separation of 5 s, the fiber was relaxed until the holding current was observed to have fully recovered. The same stimulation pulse sequence was again applied to the cell membrane. The recorded post-shock Na channel currents are shown in the lower panel of FIGURE 3.

By comparing the Na channel currents recorded before and after the electrical shock, we can see that the two 4-ms, −450-mV shock pulses significantly reduce the peak value of the Na channel currents. For this fiber, the peak values of the Na channel current were 170 nA and 95 nA before and after the electric shock, respectively. This corresponds to a 45% reduction in the Na channel currents after the electrical shock.

The peak values of the pre- and post-shock Na channel currents shown in FIGURE 3 were plotted as functions of the membrane potential of the stimulation pulses, shown in FIGURE 4. The open squares represent the peak values of the Na channel currents recorded before the electrical shock, and the upward pointing triangles represent the channel currents after application of the electrical shock.

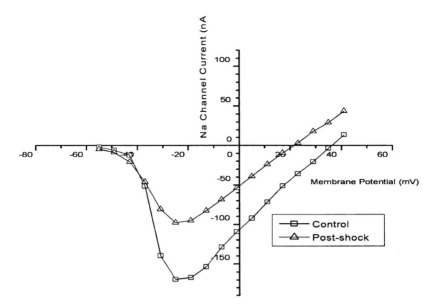

FIGURE 4. The Na channel currents as a function of the membrane potentials. The squares represent the Na channel currents as a control; the upward pointing triangles represent the Na channel currents after shock by two 4-ms, −450-mV pulses. (From Chen et al.[25] Reproduced by permission.)

Channel conductance can be represented by the slope of the I-V curves. For the control, the Na channel conductance, or the slope of the pre-shocked Na channel I-V curve is 3.11 μS. After two 4-ms duration, −450-mV pulsed shocks, the slope of the I-V curve, or the channel conductance, was reduced to 2.26 μS, about 70% of that before the electric shock. A statistical analysis of nine experiments shows that after two 4-ms, −450-mV pulsed shocks, the mean decrease of the channel conductance is 27%.

Potential Threshold for Membrane Electroporation versus the Threshold for Damaging the Channel Proteins

Experimental results from our and other laboratories showed that the field-induced leakage current could be measured when the cell membrane is shocked by a 4-ms supramembrane potential pulse in a range between −250 mV and −300 mV.[9,27] Our results showed that the membrane potential threshold for damaging the voltage-gated channel proteins is from −450 mV to −500 mV, which is higher than that for damaging the phospholipid bilayer. The detailed reason for this discrepancy is unknown. One explanation may relate to the structure of the phospholipid bilayer as a supramolecular assembly held only by hydrophobic forces without chemical bonding, whereas the amino acids and subgroups in the channel proteins are assembled with strong chemical bonds. Differences in charge densities and distributions may also account for the different electromechanical forces induced by the external electric field.

Reversibility of Na Channel Conformational Denaturation

It has been observed that electropores on the cell membrane induced by a high-intensity electric field can be divided into two categories in terms of reversibility: transient and stable. For the transient pores, there exist two groups with different recovery time scales—those sealing in microseconds to milliseconds and those taking minutes to close. Comparatively, our experimental results show that the reversibility of the high-intensity electrical field–induced reduction of the channel currents is far less than the reversibility of membrane electroporation.[9,25,26,28]

UNDERLYING MECHANISMS INVOLVED IN ELECTRICAL SHOCK-INDUCED DAMAGES IN THE MEMBRANE PROTEIN

Detailed mechanisms involved in membrane protein denaturation remain unknown. One of the possible mechanisms involved in electrical injury is Joule heating. The direct effects of Joule heating in damaging cells and tissues have long been recognized as denaturation of the cell membrane and membrane proteins. Another possibility is due to intensive electric field–induced proteins conformational damages. Among the supramembrane potential–induced electroconformational changes and the huge transmembrane current–mediated thermal effects, it is necessary to distinguish which is the dominant mechanism involved in membrane protein damage in electrical injury.

Joule Heating Effects Are Not the Main Factor in Denaturation of the Membrane Proteins

The characteristic of Joule heating–induced damages in cell membrane is that the degree of damage is proportional to electrical energy consumed in the cell membrane, which is generally directly related to the field-induced transmembrane currents.[29–31]

We have studied the relationship of the proteins' functional reduction and the Joule heating effects induced by the shock electric field.[28] We compared shock effects on the Na channels by −350-mV and +500-mV pulses with the same pulse duration. When compared with −350-mV shock, a +500-mV shock pulse drove transmembrane charges about 19 times and thereby generated 27 times Joule heating on the cell membrane, but only caused less than one-half of the reduction in the K^+ channel current.

Further studies show that the electrical shock–induced reductions in the K^+ channel functions are not directly related to either the globe transmembrane currents or the channel current.[27] Therefore, neither the cell membrane Joule heating nor the Joule heating on the K^+ channel proteins can be associated to the protein damages. In contrast, the protein damage level closely depends on the supramembrane potential, the potential magnitude, and polarity. In other words, the studies provide strong evidence that during an exposure to a high-intensity electrical field, the field-induced supramembrane potential (magnitude and polarity) plays a dominant role in damaging the K^+ channel proteins. The thermal effects caused by the field-induced huge transmembrane current and channel currents—accordingly the Joule heating effects—play a secondary, trivial role.

It is necessary to point out that the supramembrane potential we studied is up to 500 mV. We cannot rule out the possibility that when the intensity of the field further increases, the thermal effects may play a more and more important role in damaging the membrane proteins.

Supraphysiological Transmembrane Potential–Induced Changes in the Gating Systems of the K^+ Channel Proteins

As mentioned above, the amino acids or subgroups in channel protein functioning as voltage sensors may also be vulnerable to the high-intensity electrical field. A supramembrane potential may break down some chemical bonds or disable some movable particles in these voltage-sensor domains, resulting in a loss of their capability to open the channels.

Almers[32] and Hille[33] estimated the limiting number of gating charge particles in the ion channels. They employed a simple two-state model and studied the channel-open probability. They concluded that about six and five elementary movable charge particles in the Na^+ and K^+ channels, respectively, function as voltage-sensor. In order to better understand the mechanisms involved in the electrical shock–induced conformational damages in the channel proteins, we investigated the shock effects on the number of the gating charge particles in the K^+ channels.[25]

According to thermodynamics, the possibilities of channel opening can be described by the Boltzmann distribution of the movable charge particles in the gating system. There are two kinds of energy that dominate the distribution of the gating charge particles. One is the protein conformation energy difference between the closed state, E_c, and the open state, E_o. Another is electrical energy carried by the external stimulation pulses. The electrical energy is $-zeV$, where V is membrane potential of stimulation pulses, e is the elementary charge, and z is the limiting number of movable gating charge particles or dipole moments.[32,33] The total change in energy from the closed state to the open state responding to the stimulation pulse is E_o-E_c-zeV. We now assume a multiple-state model by which charged particles move from different closed states to the final closed state, where they are ready to move to the open state. This can be expressed by a simple first approximation,

$$z = z_0 + nV, \tag{1}$$

where z_0 is the free charged particles when the applied external membrane potential is zero and n is a coefficient constant. The number of free charge particles, z, is a linear function of the membrane potential, V. The probability of the open channels among the total number of channels, P, can be described by the Boltzmann equation,

$$P = \frac{1}{1 + \exp[(E_0 - E_c) - (z_0 + nV)/KT]}, \tag{2}$$

where K is the Boltzmann constant and T is absolute temperature.

Then, we fitted the data of the pre- and post-shock channel-open probabilities obtained from FIGURES 1 and 2 to Equation 2.[25] The two best-fit curves are shown in

FIGURE 5. The ratio of the number of gating charged particles, z, at a membrane potential of −50 mV is 4.97/6.53, where the nominator and denominator represent the post- and pre-shock K^+ channels, respectively. The ratio value is about 76%, indicating 24% of the channel gating charged particles was eliminated by the electric shock.

More interestingly, the fitted curves clearly show the trends of the relationship between the number of gating charge particles and the channel-open threshold. The smaller the number of gating charge particles, the more negative membrane potential is needed to obtain the same channel-open probability. The potential difference between the pre- and post-shock membrane potentials to open 5% (e^{-3}) of the K^+ channels is about 5 mV. These results are consistent with our experimental results shown in FIGURE 1, which clearly indicate that after electrical shock, not only the number of gating charged particles is reduced, but also the channel-open threshold shifts a few microvolts in the negative direction.

Suprahysiological Transmembrane Potential–Induced Changes in the Intramembrane Charge Movement Currents

The above results of the electrical shock–reduced number of gating particles in the channel proteins were made by analyzing the measured channel currents. To further confirm this conclusion, we directly monitored the changes in the intramembrane charge movement currents, which represent the movements or reorientation of the charged particles in the membrane proteins.[34]

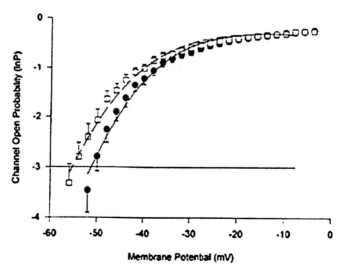

FIGURE 5. Channel-open probability as a function of membrane potential. The abscissa is the membrane potential in a linear scale; the ordinate is the channel-open probability expressed as a normalized channel conductance drawn in a logarithm scale. *Dark cycles* and *open squares* represent the pre- and post-shock channel-open probabilities, respectively. (From Chen.[24] Reproduced by permission.)

The upper panel of FIGURE 6 shows the measured charge movement currents responding to a series of stimulation pulses holding the membrane potential from −60 to 0 mV. The transient peak with an exponential decay is the fast component of the charge movement current, Q_β, while the slow component, Q_γ, is marked as a "hump" following the fast component. The vertical dotted line shows the position of the peak for each fast component, Q_β. The slow hump component, Q_γ, started to be noticeable at the second or third traces.

The lower panel of FIGURE 6 is the charge movement currents recorded after the fiber was shocked by a 4-ms, −350-mV pulse. The peak value of the slow component, Q_γ, is significantly reduced by the shock. The decrease in the peak of the hump component indicates that the moveable charged particles corresponding to Q_γ is reduced after the electrical shock. This can be shown as a downshift of the plateau of the post-shock Q-V curves in FIGURE 7. In contrast, the fast component, Q_β, in the post-shock current traces of FIGURE 7 shows very few changes.

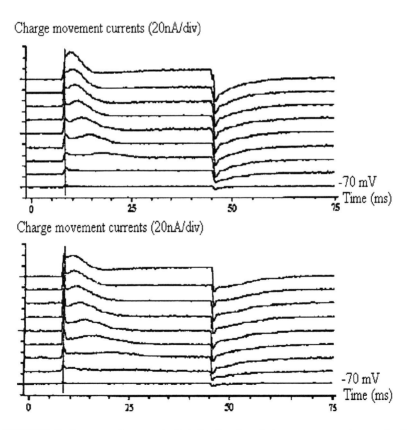

FIGURE 6. Supramembrane potential shock-reduced charge movement currents. The stimulation pulse sequence consists of nine 40-ms pulses holding the membrane potentials from −70 to 10 mV. The *upper* and *lower panels* represent the change movement currents recorded before and after shocked by a 4-ms, −350-mV pulse. (From Chen.[33])

When cells were shocked by a −500-mV pulse, both the fast and slow components decreased; therefore, the total charge movement or the movable charge particles were significantly reduced. This result clearly shows that the reduction in the charge movements depends on the magnitude of the supramembrane potential difference or the intramembrane field-strength. These studies lend further credence to our conclusion that supramembrane potential may affect the protein structures by altering the movability of the charge particles in the membrane proteins.

In addition to reduction in the magnitude of Q_γ, the kinetics of the hump component of the charge movement current was also changed due to the electric shock. The shape of the post-shock hump component became broader and the appearance of the hump became more delayed. From FIGURE 6, it is clearly seen that the time delay becomes longer in the post-shock charge movement currents than those in the pre-shock currents. These shock pulse-induced kinetics changes imply that after shock by a supramembrane potential-pulse, the moveable charge particles are more reluctant than those before the shock, if they are still moveable.

It is necessary to point out that the measured charge movement currents represent the movement of all of intramembrane charge particles in the membrane proteins, including the gating currents of the Na and K channels. The Na channel gating currents belong to the fast component, and the K channel gating currents to the slow components. The shock field–induced reduction in the intramembrane charge movement currents is consistent with our results that electric shock may cause denaturation in

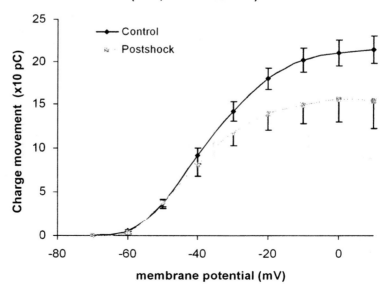

FIGURE 7. Steady-state I-V curves of the "on" charge movements: control (*circles*) and shocked by a 4-ms pulse of −350-mV (*squares*). (From Chen.[33] Reproduced by permission.)

the channel gating system. In addition, since the measured intramembrane charge movement currents included the contributions from other voltage-dependent membrane proteins, this result implies that those membrane proteins may also be vulnerable to the electric shock.

CLINICAL SIGNIFICANCE OF THE ELECTROCONFORMATIONAL DENATURATION IN MEMBRANE PROTEINS

Nerves and muscles use electrical potentials to convey information. A rapid signaling procedure, the action potential, conveys signals rapidly and efficiently over long distances. During the action potential, due to the opening and closing of ion channels, especially the Na and K channels, the membrane potential is quickly reversed and then returns to nearly the original resting potential, all within a few milliseconds. Features of the action potential, such as its magnitude, shape, frequency, latency time, and propagation speed, often determine the conveyed information. For example, shapes of the action potentials in nerves, skeletal muscle, and cardiac muscle differ significantly, and action potential frequency is often used to code signals. All of these features of the action potential depend on the functions of the Na and K channels. The functions of these Na and K channels are critical for many cells, especially for excitable cells.

The electrical shock–induced proteins' functional reductions and denaturation may cause membrane resting potential depolarization, and change the shape and duration of the action potentials. Our study explains the cellular mechanism of the *in vivo* study in rats in which electric shock reduced the magnitude of the action potential in muscles and sciatic nerves.[23] In neuromuscular junctions, the higher the magnitude and the frequency of the action potential that propagate into a nerve terminal, the higher the rate of transmitter secretion. In excitation contraction (E-C) coupling, the higher the frequency and the magnitude of the action potential is, the higher the tone and the stronger the fiber contraction. Any decrease in action potential magnitude and its frequency may result in dysfunction in those neuromuscular junctions and consequently the E-C coupling.

Common symptoms of an electrically shocked patient include shaking of the limbs, oversensitivity, and loss of loading capability. These neuromuscular symptoms may be partially explained by the functional reduction in the ion channels, especially the Na and K channels, the main determinant of an action potential.

The reason for there being no significant change in tissue appearance is because electroconformational damage in the membrane proteins mainly results from the high voltage of the shock field. There is very little, if any, current needed to alter the protein structure. This is significantly different from thermal damage and even membrane electroporation. Industrial workers electrocuted by a high-voltage field, and people shocked by lightning, are some typical examples.[35]

In summary, an intensive, brief electric shock–induced electroconformational damage or denaturation in membrane proteins, especially the voltage-dependent membrane proteins, strongly suggests a new underlying mechanism involved in electrical injury. For ion channels, these conformational damages or denaturation may affect the signal conveyed in excitable cells, resulting in functional reduction in those cells and tissues.

ACKNOWLEDGMENTS

This work is partially supported by research grants from the National Institutes of Health (2NIGM50785) and the National Science Foundation (PHY-0515787).

REFERENCES

1. PARSEGIAN, V.A. 1969. Energy of an ion crossing a low dielectrical membrane:solutions to four relevant electrostatic problems. Nature **221:** 844–846.
2. COLE, K.S. 1972. Membrane, Ions and Impulses. University of California Press. Berkeley. pp 569–590.
3. LEE, R.C. & M.S. KOLODNEY. 1987. Electrical injury mechanisms: electrical breakdown of cell membrane. Plast. Reconstr. Surg. **80:** 672–679.
4. LEE, R.C. & W. DOUGHERTY. 2003. Electrical injury: mechanisms, manifestations, and therapy. IEEE Trans Dielectrics and Electrical Insulation **10:** 810–819.
5. BIER, M., T.R. GOWRISHANKAR, W. CHEN & R.C. LEE. 2004. Electroporation of a lipid bilayer as a chemical reaction, Bioelectromagnetics **25:** 634–637.
6. BENZ, R. & U. ZIMMERMANN. 1980. Relaxation studies on cell membranes and lipid bilayers in the high electric field range. Bioelectrochem. Bioenerg. **7:** 723.
7. TSONG, T.Y. 1991. Electroporation of cell membrane. Biophys. J. **60:** 297–306.
8. JONES, J.L. & R.E. JONES. 1982. Effects of tetrodotoxin and verapamil on the prolonged depolarization produced by high-intensity electric field stimulation in cultured myocardial cells. Fed. Proc. **41:** 1383.
9. CHEN, W. & R.C. LEE. 1994. Altered ion channel conductance and ionic selectivity induced by large imposed membrane potential pulse. Biophys. J. **67:** 603–612.
10. WADA, A. 1976. The alpha-helix as an electric macro-dipole. Adv. Biophys. **9:** 1–63.
11. HOL, W.G.J., P.T. VAN DUNINEN & H.J.C. BERENDSEN. 1978. The alpha-helix dipole and the properties of proteins. Nature (Lond.) **273:** 443–446.
12. HOL, W.G.J. 1985. Effects of the alpha-helix dipole upon the functioning and structure of proteins and peptides. Adv. Biophys. **19:** 133–165.
13. ARMSTRONG, C.M. 1981. Sodium channels and gating currents. Physiol. Rev. **61:** 644–683.
14. DEWEER, P., D.C. GADSBY & R.F. RAKOWSKI. 1988. Annu. Rev. Physiol. **50:** 225–242.
15. LAUGER, P. 1991. Electrogenic Ion Pumps. Sinauer Associates, Inc. Sunderland, MA. pp. 168–225.
16. SCHNEIDER, M.F. & W.K. CHANDLER. 1973. Voltage dependent charge movement in skeletal muscle: a possible step in excitation-contraction coupling, Nature **242:** 244–246.
17. ADRIAN, R.H. & A.R. PERES. 1977. A gating signal for the potassium channel? Nature **267:** 800–804.
18. RIOS, E. & G. BRUM. 1987. Involvement of dihydropyridine receptors in excitation-contraction coupling in skeletal muscle. Nature **325:** 717–720.
19. TANABE, T., K.G. BEAM, J.A. POWELL & S. NUMA. 1990. Restoration of excitation-contraction coupling and slow calcium current in dysgenic muscle by dihydropyridine receptor complementary DNA. Nature **336:** 134–139.
20. TEISSIE, J. & T.Y. TSONG. 1980. J. Membr. Biol. **55:** 133–140.
21. TEISSIE, J., B. KNOX, T.Y. TSONG & J. WEHELE. 1981. Proc. Natl. Acad. Sci. USA **78:** 7473–7477.
22. KINOSITA, K. & T.Y. TSONG. 1977. Voltage induced pore formation and hemolysis of human erythrocyte membranes. Biochim. Biophys. Acta **471:** 227–242.
23. ABRAMOV, G.S., M. CAPELLI-SCHELLPFEFFER & R.C.LEE. 1996. Alteration in sensory nerve functions following electric shock. Burn **22:** 602–606.
24. CHEN, W. & R.C. LEE. 1994. An improved double Vaseline voltage clamp to study electroporated skeletal muscle fibers. Biophys. J. **66:** 700–709.
25. CHEN, W. 2004. Supra-physiological membrane potential induced conformational changes in K channel conducting system of skeletal muscle fibers. Bioelectrochemistry **62:** 47–56

26. CHEN, W., Z.S. ZHANG & R.C. LEE. 2005. Supramembrane potential-induced electroconformational changes in sodium channel proteins: a potential mechanism involved in electric injury. Burn. In press.
27. TUNG, L. & J.R. BORDERIES. 1992. Analysis of electric field stimulation of single cardiac muscle cells. Biophys. J. **63:** 371–386.
28. CHEN, W., Y. HAN, Y. CHEN & D.R. ASTUMIAN. 1998. Electric field-induced functional reductions in the K^+ channels mainly resulted from supramembrane potential-mediated electroconformational changes. Biophys. J. **75:** 196–206.
29. ROUGE, R.G. & A.R. DIMICK. 1978. The treatment of electrical injury compared to burn injury: a review of pathophysiology and comparison of patient management protocols. J.Trauma **18:** 43.
30. BINGHAM, H. 1986. Electrical burns. Clin. Plast. Surg. **13:** 75.
31. LEE, R.C., D.G. GAYLOR, D.L. BHATT & D.A. ISRAEL. 1988. Role of cell membrane rupture in the pathogenesis of electrical trauma. J. Surg. Res. **44:** 709.
32. ALMERS, W. 1978. Gating currents and charge movements in excitable membranes. Rev. Physiol. Biochem. Pharmacol. **82:** 96–190.
33. HILLE, H. 2001. Ionic Channels of Excitable Membrane. Sinauer Associates Inc. Sunderland, MA.
34. CHEN, W. 2004. Evidence of electroconformational changes in membrane proteins: field-induced reductions in intramembrane nonlinear charge movement currents. Bioelectrochemistry **63:** 333–335.
35. BIER, M., W. CHEN, E. BODNAR & R.C. LEE. 2005. Biophysical injury mechanisms associated with lightning injury. Neurorhabilitation **17:** 1–10.

Heat Injury to Cells in Perfused Systems

DENNIS P. ORGILL, STACY A. PORTER, AND HELENA O. TAYLOR

Division of Plastic Surgery, Brigham and Women's Hospital, Boston, Massachusetts, USA

> ABSTRACT: Tissue injury in response to excessive heat results in a clinical burn. Burns cause a range of physiologic derangements, including denaturation of macromolecular structures, leakage of cell membranes, activation of cytokines, and cessation of blood flow, all leading to tissue death. The purpose of this paper is to examine the mechanisms and consequences of burn injury and to discuss potential therapies based on these mechanisms. Knowledge of the thermal properties of tissues can predict the time–temperature relationship necessary to cause a specified thermal insult. Changes in cell membrane biochemistry and the stabilization of proteins through the heat-shock response can enable biomacromolecules to withstand supraphysiological temperatures. Mechanisms of cellular repair allow recovery of cellular function after thermal insult. An understanding of the response of proteins, cellular organelles, and cells to heat provides the foundation for understanding the pathophysiology and treatment of burn injury. The physics, biochemistry, and cellular biology behind the host response to thermal injury in perfused systems are reviewed.
>
> KEYWORDS: burns; thermal injury; protein denaturation, cellular injury; heat injury; membrane disruption

INTRODUCTION

Tissue injury to humans from excessive heat results in a clinical burn (FIG. 1). Burns cause a range of physiologic derangements including denaturation of macromolecular structures, leakage of cell membranes, activation of cytokines, and cessation of blood flow, all leading to tissue death. The purpose of this chapter is to examine the mechanisms and consequences of burn injury at the cellular level to help predict what happens in a perfused system following thermal injury. In order to understand burns in three-dimensional structures, a brief review of adaptations of other organisms to heat stress is warranted.

Life on earth depends critically on the thermodynamic properties of water. Most living organisms have high water content and depend on the aqueous phase of H_2O. Mammalian systems operate within an extremely narrow temperature range relative to the full range of temperatures recorded on earth. Despite the amazing operational capabilities of humans, core temperatures of 310°K ± 2% must be maintained for survival. This restricted temperature range is likely related to the narrow temperature

---

Address for correspondence: Dennis P. Orgill, M.D., Ph.D., Division of Plastic Surgery, Brigham and Women's Hospital, 75 Francis Street, Boston, MA 02115. Fax: 617-732-6387. dorgill@partners.org

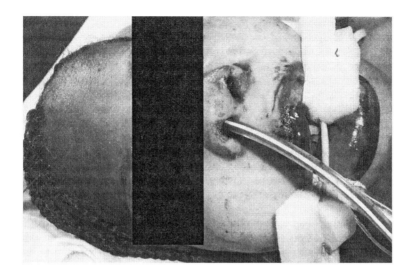

FIGURE 1. Deep second-degree flash burn to face after propane tank explosion. Deep burned areas have lost pigmentation as the epidermis is removed when the burn is cleaned. The white color of the skin in the central portion of the lip is indicative of poor tissue perfusion, while areas of red skin are indicative of increased perfusion.

profile for physiologic function of multiple biomacromolecules and supra-assemblies of these molecules. Exposure to supraphysiological temperatures leads to alteration in conformation or denaturation of these molecules, which can result in cell death and tissue necrosis.

Intriguingly, other organisms have adapted to survive at environmental extremes of heat, pressure, and acidity. There are four broad categories of microorganisms, based on the temperature range required for optimal growth: (1) hyperthermophilic (> 80°C); (2) thermophilic (35 to 55°C); (3) mesophilic (10 to 40°C); and (4) psychrophilic (−5 to 15°C).[1] Archaea has been observed at temperatures up to 113°C and some species can withstand 121°C for up to one hour. The successful adaptation of thermophiles to high temperatures suggests that there are mechanisms that stabilize macromolecular structures in the face of higher molecular energies. As thermophiles have evolved in several parallel systems, there are multiple mechanisms to confer thermal stability to organisms. Despite decades of research, however, the precise nature of these mechanisms remains enigmatic.

Enzymes and proteins in hyperthermophiles appear to function optimally at high temperatures as a result of critical amino acid substitutions.[2] They fold differently, allowing for additional stability in the face of the denaturing effects of heat.[3] Denaturation kinetics, which are used to compare hyperthermophilic proteins with their thermophilic and mesophilic counterparts, show that the *un*folding rate of the hyperthermophilic protein with its mesophilic counterpart differs by 2–3 orders of magnitude, while *re*folding kinetics between the two differ only slightly.[4] It has been suggested that protein stability is increased through the formation of ion pairs, increased hydrogen bonding, steric hinderance of chain flexibility, the presence of a number of salt bridges and by the densely packed hydrophobic interior nature of the proteins.[2] Hyperthermophiles exhibit significant metabolic flexibility, using sulfur, iron, and nitrogen compounds, CO_2, and organic material as energy sources.[1]

Thermophilic organisms have membranes that are rich in saturated fatty acids, which form strong hydrophobic bonds, thereby conferring stability and functionality at high temperatures.[3] tRNAs of hyperthermophiles derive most of their thermostability from numerous modified bases, which restrict bending at critical points.[5] Since DNA incorporates only four deoxyribonucleotides and the G + C content (which is more stable than A-T basepairs) of hyperthermophile DNA is not unusually high among prokaryotes, it appears that hyperthermophiles achieve DNA stability through mechanisms other than increased hydrogen bonding between strands. Intracellular salt may play a role in DNA stabilization by retarding heat-induced strand separation and depurination.[6] Small, basic proteins that bind duplex DNA and compact or bend DNA upon binding may also play a role in hyperthermophile DNA stabilization. Thermophiles have adapted to tightly coil their DNA using a gyrase enzyme. In addition, hyperthermophiles contain structurally novel polyvalent cations, which are effective thermal stabilizers. Hyperthermophiles may also stabilize DNA by preventing strand separation through positive supercoiling.[2] Extrinsic factors such as stabilizing solutes (or thermoprotectants), biofilm formation, and hydrostatic pressure may also enhance the stability of hyperthermophiles at super-optimal temperatures.[7]

In addition, hyperthermophiles make use of the heat-shock response (i.e., the synthesis of enzymes that prevent protein aggregation, reassemble damaged proteins, and degrade proteins beyond repair), which is common to all organisms. The TF55 chaperone protein (the major protein produced during heat shock) is abundant in hyperthermophiles under normal conditions, suggesting that it protects proteins from heat even when the cell is not under duress.[7]

Knowledge of the thermal properties of tissues can predict the time–temperature relationship necessary to cause a specified thermal insult. Changes in cell membrane biochemistry and the stabilization of proteins through the heat shock response can enable biomacromolecules to withstand supraphysiological temperatures. Mechanisms of cellular repair allow recovery of cellular function after thermal insult. An understanding of the response of proteins, cellular organelles, and cells to heat provides the foundation for understanding the pathophysiology and treatment of burn injury. This paper reviews the physics, biochemistry, and cellular biology behind the host response to thermal injury.

CLINICAL MANIFESTATION OF THERMAL INJURIES: CELLULAR AND MACROMOLECULAR IMPLICATIONS OF TISSUE EXPOSURE TO SUPRAPHYSIOLOGICAL TEMPERATURES

The spatial variation in temperature history that results from most burns provides important insights into the tissue response. Upon visual examination, three different zones of a burn can be appreciated immediately after injury (FIG. 2).[8] In the center is the zone of coagulation, where such extensive protein denaturation has occurred that the tissue's optical properties are altered. Since collagen is one of the more heat-stable macromolecules, when it is thermally denatured, one can expect that extensive

FIGURE 2. Deep second-degree flash burn to face after propane tank explosion. Deep burned areas have lost pigmentation as the epidermis is removed when the burn is cleaned. The white color of the skin in the central portion of the lip is indicative of poor tissue perfusion, while areas of red skin are indicative of increased perfusion.

cellular necrosis will occur. Accompanying the immediate and irreversible thermal injury in the zone of coagulation is a delayed and potentially reversible injury in the zone of stasis, which is characterized by established edema surrounded by an area of inflammation and active edema formation due to vasodilation and increased microvascular permeability.[9] Excessive local edema accumulation is followed by a reduction in perfusion, leading to more local tissue ischemia. The injury zone farthest away from the point of maximum heat exposure is called the zone of hyperemia. The nature of cellular injury in this zone is not fully described, but tissue in this area often survives.

Twelve to 24 hours post burn, local microcirculation is compromised to the worst extent.[10] Depth of superficial burns appears to remain stable with time, whereas that of deep dermal burns increases with time.[11] Depending on the treatment of the burn, thermally damaged cells can be repaired or can progress to cellular death. Decreased perfusion to the area, abnormal inflammatory or immunologic response, and the release of intracellular enzymes, hypoxia, and metabolic acidosis have all been proposed as factors contributing to burn deepening.[12] Because the outcome of burns is so dependent on the depth of injury and because burns continue to demarcate for hours to days after thermal assault, improvements in treatments that can reduce the depth of injury, if only by a fraction of a millimeter, are likely to have tremendous effects on the quality of life for burn victims. To accomplish this, it is important to examine tissue responses at the molecular level to develop strategies to reverse heat-generated alterations.

TEMPERATURE DISTRIBUTION WITHIN TISSUES

Moritz and Henriques[13] popularized the mathematical concept of heat transfer in a burn in 1946. This required an understanding of the thermal thresholds of living tissue, which depend on factors such as blood flow, tissue hydration, and the immune response to thermal injury. These factors are difficult to quantify and as such, prior mathematical models focused on the actual heat-transfer process that results in a time distribution of temperature related to the depth of skin. Moritz and Henriques utilized the Fourier rate equation to predict the time–temperature relationships in tissues exposed to a constant external heating source. In the two-dimensional case, the heat transferred equals the thermal conductivity (k) times the area (A) times the change in temperature with respect to the x-axis ($\Delta T/\Delta x$) (FIG. 3).

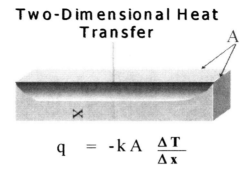

$$q = -kA \frac{\Delta T}{\Delta x}$$

FIGURE 3. Two-dimensional heat-transfer model considering only heat conduction.

After doing a heat balance equation on an infinitesimal block the Fourier heat transfer equation can be derived. This equation states that for any given coordinate, the change in temperature with respect to time is proportional to the second derivative of the temperature with respect to distance:

$$\rho c \frac{\partial T}{\partial t} = k \frac{\partial T}{\partial x^2} \tag{1}$$

There are several instances in which this differential equation can be solved in a closed form solution. In an infinite body at a steady state, the change in temperature with respect to time is zero and an ordinary second-order differential equation results:

$$k \frac{\partial^2 T}{\partial x^2} = 0 \tag{2}$$

In a model where a constant temperature is applied to the skin surface, a boundary condition can be derived that the heat transferred is proportional to the difference in temperature between the skin and the heat source:

$$q = H(T_s - T) \tag{3}$$

where q is the caloric uptake per min–cm^2, T_s is the temperature of the heat source, and H is the heat transfer coefficient.

Moritz and Henriques quantified an inverse relationship between thermal intensity and the time required to produce a tissue burn by measuring the time necessary for water between 44 and 100°C to cause blistering of the epidermis in porcine skin. Their experimental results yielded temperature range estimates for various degrees of damage and certain trends above and below the threshold temperatures. Their data showed an inverse relationship between temperature and the time required to produce a specific degree of thermal damage. As temperature increases, heat duration logarithmically decreases to produce equivalent burns. Since this relationship closely resembles the activation energy required for a chemical reaction, Moritz and Henriques used an Arrhenius integral to quantify the damage of the various depths of burn. A, E and R are the frequency factor, the activation energy and the gas constant, respectively. This function describes the rate of tissue damage and can be correlated with histologic examination. By integrating the damage rate over the burn, three injury thresholds $\Omega = 0.53$, 1 and 10^4 (first-, second-, and third-degree burns, respectively) were quantified.

$$\text{Arrhenius integral } \Omega = A \int_0^t e^{\frac{-\Delta E}{R(T_c + 273)}} dt \tag{4}$$

First-degree burn 0.53
Second-degree burn 1.0
Third-degree burn 10,000

The non-dimensional number (Ω) is calculated from the Arrhenius integral and has been correlated with the time–temperature relationship required to cause a particular degree of burn severity.

The analytical model used by Henriques and Moritz is limited in terms of boundary conditions and overall complexity. It does not allow for metabolic activity or blood flow nor does it allow for a variable input of heat into the system. Finite element and finite difference methods have enabled computers to solve complex heat-transfer problems by dividing a heat-transfer system into small elements and writing heat-transfer equations for each element. As these elements are linked together, the entire system of equations can be solved, yielding time–temperature distributions in both time and space. The *bio-heat-transfer equation* takes into account the various heat-transfer elements in biological systems. As expressed below, the change in temperature with respect to time is equal to the temperature change due to conduction plus the temperature change due to blood flow. In this particular expression, the heat generated by cellular metabolism is omitted since this is generally negligible for thermal injuries that occur over a short period of time, but this term is easily added. The equation reads:

$$\rho c \frac{\partial T}{\partial t} = \nabla k \nabla T + \omega_b \rho_b c_b (T - T_b), \tag{5}$$

where ρ is tissue density, c is heat capacity, k is conductivity, and ω_b is the normalized blood perfusion of tissue.

In 1983, Diller and Hayes[14] proposed a finite element model of burn injury using the above bio-heat-transfer equation. Their model consisted of a cylindrical disk, 1.0 cm in diameter, confined to a natural (air) convection boundary condition for a determined temperature and time. Cylindrical coordinates (r and z) instead of the Cartesian coordinates used by Henriques and Moritz were employed to account for the cylindrical symmetry of the model. The skin is divided into three distinct layers: the epidermis, dermis, and subcutaneous fat. Each layer is considered individually with its own heat capacity, thermal conductivity, and all other pertinent properties. With the appropriate initial and boundary conditions, a finite element solution can be determined for the bio-heat transfer equation.

Using their two-dimensional finite element program, Diller and Hayes were able to predict the extent and severity of burn injury within the specified parameters. One of the greatest advantages of the finite element method is that there is no loss of accuracy near the boundaries while convective boundaries, temperature, and heat flux can all be varied. Curved geometry and flux patterns can also be accurately resolved. The finite element method is flexible with respect to a grid mesh of geometry. In areas where elements are of high concentration, such as the layer interfaces, nonlinear networks can be produced.

CELLULAR ALTERATIONS IN RESPONSE TO HEAT

When tissue temperature is raised well above physiological values, alteration in the structure of biological proteins and organelles follows due to high-energy collisions with solvent water. Cytoskeletal cell components are disrupted and changes in membrane permeability lead to an increase in intracellular Na^+, H^+, and Ca^{2+}. DNA synthesis and transcription are inhibited, as is RNA processing and translation. Progression through the cell cycle halts as proteins are denatured and misaggregated and finally degraded through both proteasomal and lysosomal pathways.[15] Each of these processes depends on a multitude of factors, including the strength of chemical bonds, the concentration of macromolecules in the cytoplasm, and interactions with molecular chaperones and immune components, which provide cellular components with the ability to resist some thermal insults.[16]

Because proteins and other organelles have unique functional structures, some are more susceptible to denaturation at supraphysiological temperatures than others. Since the time–temperature relationship for burn injury is well-described by an Arrhenius integral, it is either the case that tissue death following a burn is due to a multitude of subordinate processes cumulatively behaving like a single Arrhenius process (as described by the central limit theorem in statistics) or to the action of a few principal molecular mechanisms critical for cell survival. Despa *et al.*[17] suggest that the latter hypothesis is correct, using calorimetric measurements of the level of denaturation of 12 proteins, RNA, DNA, and cell membrane components when exposed to excessive heat, to show that the lipid bilayer and membrane-bound ATPases are the least thermally stable cell components and are most responsible for heat-induced tissue necrosis. Their findings support the proposition that cell membrane permeability is the most important pathophysiologic event leading to tissue death[18] and raise the possibility that therapeutic interventions aimed at stabilizing the cell membrane may improve tissue survival following thermal injury.

The cell membrane has long been of interest to researchers owing to its inherent capacity for repair, an adaptive mechanism likely due to the frequent exposure of plasma membrane to mechanical disruption.[19] Because the bilayer lipid component of the cell membrane is held together only by forces of hydration, the cell membrane is particularly vulnerable to thermal trauma. Even at temperatures only 6°C above normal (i.e., 43°C), the kinetic energy of the molecules in the cell membrane can exceed the hydration energy barrier, which holds phospholipids in the membrane as a supramolecular assembly.[20] In effect, the warmed membrane goes into solution in the surrounding water, rendering the membrane freely permeable to small ions.

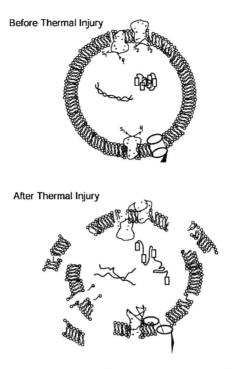

FIGURE 4. Disruption of plasma membrane and denaturation of cellular proteins as a result of heating causes leakage of electrolytes and reduction of the electrochemical gradient.

Plasma membrane stress failure can lead to cell death, as the membrane's primary role as a barrier is compromised. Many cells, however, exhibit the capacity to rapidly repair cell membrane rupture. Until recently, this repair was thought to be a simple, thermodynamically favored process that removed hydrophobic domains from the aqueous environment. It now appears that membrane re-sealing is much more complex and dynamic, particularly for large disruptions.[19] An influx of Ca^{2+} into the cytoplasm initiates the rapid fusion of lysosomal vesicles with one another and then with the surrounding cytoskeleton through exocytosis to form a "patch" over the disruption site.[21] This restores Ca^{2+} homeostasis and prevents immediate cell death. Repair of deeper damage to the cortical cytoskeleton is then initiated by the assembly of a ring of concentrated F-actin and myosin 2 around the wound, which constricts to pull the wound cytoplasm closed.[22] Finally, "fingers" of F-actin form along the sides of the wound border. Similar actin structures have been observed during dorsal closure in *Drosophila*, suggesting parallels between the repair of tissue and of *individual* cells.

Cell membrane rupture and the denaturation of proteins and other cellular components in response to thermal assault occur in a dynamic cellular environment. In response to stress, all cells increase production of a class of proteins dubbed heat-shock proteins (HSPs).[23] These proteins act largely as molecular chaperones, synthesizing enzymes that prevent protein aggregation, reassembling damaged proteins, and degrading proteins beyond repair.[24] Heat-shock proteins are among the most phylogenetically conserved proteins in existence; they are present, and can be induced, in all species.[25] HSPs are grouped into five protein families by approximate molecular mass, with the Hsp70 family the most prominent group in eukaryotes. Many HSPs are present under normal physiological conditions (making up 5–10% of total protein weight) and function as molecular chaperones, involved in the assembly and folding of oligomeric proteins and in the degradation of damaged proteins.[25,26] When induced by heat or other cellular insults (oxidative stress, viral infection, ultraviolet irradiation, chemical exposure, etc.) that cause protein unfolding, misfolding, or aggregation, HSPs can increase to 15% of total protein weight. Increased generation of HSPs, however, is transient even if heat exposure occurs over an extended period of time.[26] Heat-shock protein synthesis increases during moderate heat exposure, but above a distinct threshold temperature, inhibition of HSP synthesis occurs.[27]

It has been observed that short preincubation of cells at sublethal heat-shock temperatures greatly increases their survival under severe heat stress.[28] This phenomenon of acquired thermotolerance depends on the ClpB chaperone in bacteria and Hsp104 in yeast working in concert with the Hsp70 system to solubilize and refold aggregated proteins. Although the precise mechanism of action remains unclear, it appears that adaptive thermotolerance requires the reactivation of aggregated proteins and not simply their disaggregation and removal.

Clinical investigations of hyperthermia in cancer treatment have added to our knowledge of the cytotoxic effects of heat in the cellular environment. Application of temperatures between 40–43°C, usually in conjunction with another cancer treatment modality (most often radiotherapy or chemotherapy), has been shown to improve survival rates for several types of cancer, particularly those in the abdominal region.[27] The rationale for the use of hyperthermia comes from *in vitro* and early clinical studies, suggesting that tumor cells may be more sensitive to heat than normal cells, that heat may enhance the effects of chemotherapy or radiation and overcome acquired drug resistance, and that high temperatures can stimulate the immune system. Heat appears to be particularly effective against two types of tumor cells—those that are making DNA in preparation for division and those that are acidic and poorly oxygenated.[29]

Morphologic changes associated with hyperthermia include endothelial swelling, the movement of plasma fluid into the interstitium, microthrombosis due to activation of hemostasis, and changes of the viscosity of blood cell membranes. Increased fluidity of cell membranes is observed in thermosensitive, but not thermotolerant cells, suggesting that a component of thermotolerance is enhanced membrane stability.

"Moderate" hyperthermia involving temperatures below 42°C seems to increase blood flow to the tumor, while temperatures above 42°C have been shown to decrease tumor blood flow, impairing oxygen and nutrient supply and inducing acidosis. This corresponds with the observation that superficial second-degree burns have perfusion values greater than those of normal skin, while perfusion in deep second- and third-degree burns is compromised. The behavior of tumor vascularization under hyperthermic conditions seems to depend on the method of hyperthermia applied. Hyperthermia can induce either apoptosis (programmed cell death) or necrosis, depending on the temperature used and the susceptibility of different types of cells.

Cells have different heat sensitivities in different phases of the cell cycle.[27] RNA and protein synthesis recover rapidly after heat exposure ceases; DNA synthesis is inhibited for a longer period.

SYSTEMIC RESPONSE TO THERMAL INJURY

Burns are dynamic wounds that are in a state of flux for up to 72 hours after injury. Thermal insult activates an inflammatory cascade characterized by the infiltration of specialized cells into the injury site. Blood-borne immune cells migrate through the extracellular matrix (ECM), which is composed of macromolecules such as collagen, fibronectin, and glycosaminoglycans, to inflammatory sites.[30] The ECM serves as both a reservoir for inflammatory mediators (cytokines, heat-shock proteins, and chemokines) and as a structural scaffold, providing tissue integrity and cell adhesion. Inflammation dynamically modifies the composition of the ECM, as cytokines and other growth factors signal the need for an influx of connective tissue cells and a new blood supply. Fibroblasts provide the collagen and ECM disposition needed for tissue repair. If *too much* collagen is deposited in the wound site, fibrosis results and function may be impaired. If *too little* collagen is deposited, wound healing is delayed.

Thermal injury activates macrophages, which then affect T or B lymphocytes either directly by intercellular contact or by producing cytokines, chemokines, and other bioactive substances such as interleukin-1 (IL-1), tumor necrosis factor-α (TNF-α), MCP-1, nitric oxide, and prostaglandin E_2 (PGE_2). The release of pro-inflammatory mediators results in macrophage hyperstimulation, which, in turn, produces an inhibitory cascade after several days. In a mouse burn model, the suppression of T cell activity occurred 4 days after injury, reaching maximum suppression at 7 days, and still persisting at 14 days.[31] The mechanisms behind T-cell suppression specifically and immunosuppression following burns more generally, remain speculative.[32] There is, however, a clear correlation between the area of thermal injury and the degree of T cell dysfunction.[33]

It is generally appreciated that macrophage hyperactivity and functional lymphocyte deficiency characterize thermal injury.[32] Immunosuppression in burn patients has been attributed to a variety of factors, including loss of barrier functions, tissue ischemia and destruction, foreign body introduction, neutrophil dysfunction, abnormality in opsonic activity, depletion of the complement cascade, helper-cell dysfunction, macrophage dysfunction, production of soluble immunosuppressive substances, stress-associated hormones, cytokine disarray, and an increase in the PGE-2 level.[34]

Increased expression of inducible nitric oxide synthase (iNOS) is associated with cytotoxic activity via inhibition of ribonucleotide reductase activity, mitochondria respiration, and induction of apoptosis.[35] Schwacha and Chaundry[31] propose a model of immunosuppression whereby thermal injury (via multiple factors including tissue injury/ishemia, bacterial translocation, hypermetabolic response, etc.) leads to the activation of γ/δ T cells in the epithelial layers (skin, gut), which then "prime" macrophages for a hyperactive response (possibly by interferon-γ) to a secondary stimulus such as sepsis. The increased production of immunosuppressive mediators (nitric oxide, PGE_2, and IL-10) act to suppress α/β T cell function.

Heat-shock proteins (HSPs) have been shown to be released into peripheral circulation by both necrotic and intact cells in culture.[36] The physiological role of these proteins in circulation is unknown, although it is now widely accepted that HSPs have an intercellular signaling role as well as a chaperone function. HSPs produce many immunologic effects, including the induction of pro-inflammatory cytokine secretion and adhesion molecule expression. HSPs have also been shown to induce macrophage secretion of interleukin 6 (IL-6), tumor necrosis factor-α (TNF-α), and nitric oxide. Additionally, HSPs have been shown to modify allograft rejection in animals and have been utilized in organ transplantation, where a prior heat treatment decreases the rate of rejection by the host organism.[25,26] Immunizing mice with self-Hsp60 or Hsp60 peptides delays allograft rejection, suggesting that rather than being pro-inflammatory, self-Hsp T cell reactivity could be part of the normal immunoregulatory T cell response.[37]

DISCUSSION

Advances in burn care include interventions such as fluid resuscitation and plasma exchange, early excision, and the administration of anti-inflammatory agents such as NSAIDS and antibiotics when infection or endotoxin are present.[38] Results from this review suggest two time points for intervention in the future to reduce the clinical effects of thermal injury: The first involves strategies to stabilize macromolecules and possibly re-nature biomolecules after thermal injury. For example, it has been observed that fibroblasts reseal a second membrane disruption more rapidly than one made 10 minutes earlier, likely due to already enhanced vesicle production.[39] It appears that membrane resealing depends on a decrease in surface tension, which, under normal conditions, is provided by Ca^{2+}-dependent exocytosis of the new membrane near the disruption site. Poloxamer-188, a non-ionic block copolymer surfactant of polyoxyethylene and polyoxypropylene, has been shown to seal the plasma membrane of both electropermeabilized skeletal muscle *in vivo*[18] and heat-shocked fibroblasts *in vitro*.[40] P-188 intravenously injected after thermal insult reduced the area of the zone of coagulation at 24 hours, while P-188 administered prior to burn injury prevented the formation of the zone of stasis within 2 hours.[12] Although the mechanisms remain unclear, it appears that P-188 prevents the reduction in blood flow that characterizes the zone of stasis by preserving the structure of endothelial cells and preventing endothelial swelling, thereby reducing capillary obstruction.[41] P-188 has also been shown to inhibit leukocyte chemotaxis, adhesion, and migration, which may help prevent the formation of the zone of stasis by limiting neutrophil recruitment.[42]

Other macromolecules also show potential to mediate the cytokine cascade initiated by thermal injury. Galactin-1, a β-galactoside-binding protein, has been shown to inhibit T cell adhesion and pro-inflammatory cytokine secretion in the ECM.[30] In general, preventing the activation of macrophages and neutrophils in burn patients is preferable to regulating their runaway activity.[43]

The second potential time point on which to focus efforts is the development of novel strategies to avoid the deepening of burns that occurs for up to 72 hours following the initial thermal injury. Clinicians who care for burn patients often underestimate the eventual depth of a burn based on its initial presentation. In our experimental model with porcine burns, we noted that burn depth more than doubled from one hour to 24 hours in the absence of any systemic treatment. Much of this is likely due to the zone of stasis,[17] which never regains blood flow to the area. After a burn there is a loss of stability of capillary integrity with leaking of fluid into the extracellular space. Without critical blood flow to the skin, particularly in a time of injury, adequate nutrients and molecules cannot be supplied to the cells in need of repair. Strategies that will allow and maintain blood flow to this area will be critical in reducing the deepening of clinical burns. Currently, vacuum-assisted closure (VAC) therapy appears to be such a technology, but further study is needed. The VAC device utilizes on open-pore polyurethane foam covered by an occlusive dressing and connected to suction. It likely reduces swelling in the burn victim that may result in enhanced blood flow to skin.[44]

Other pharmacologic strategies that can rapidly restore capillary integrity, including stabilization of the immune and complement systems, may also be effective in reducing burn deepening.

REFERENCES

1. STETTER, K.O. 1995. Microbial life in hyperthermal environments. Am. Soc. Microbiol News **61:** 285–290.
2. GROGAN, D.W. 1998. Hyperthermophiles and the problem of DNA instability. Mol. MicroBiol. **28:** 1043–1049.
3. Russell, A.D. 2003. Lethal effects of heat on bacterial physiology and function. Sci. Progr. **86:** 115–137.
4. JAENICKE, R. 2000. Do ultrastable proteins from hyperthermophiles have high or low conformational stability? Proc. Natl. Acad. Sci. USA **97:** 2962–2964.
5. KOWALAK, J.A., J.J. DALLUGE, J.A. MCCLOSKEY & K.O. STETTER. 1994. The role of posttranscriptional modification in stabilization of transfer RNA from hyperthermophiles. Biochemistry **33:** 7869–7786.
6. ADAMS, M.W.W. 1993. Enzymes and proteins from organisms that grow near and above 100°C. Annu. Rev. Microbiol. **47:** 627–658.
7. HOLDEN, J.F., M.W.W. ADAMS & J.A. BAROSS. 1999. Heat shock response in microhyperthermophilic microorganisms. Proceedings of the International Symposium on Microbial Ecology.
8. JACKSON, D.M. 1953. The diagnosis of the depth of burning. Br. J. Surg. **40:** 588–596.
9. PAPE, S.A., C.A. SKOURAS & P.O. BYRNE. 2001. An audit of the use of laser Doppler imaging (LDI) in the assessment of burns of intermediate depth. Burns **27:** 233–239.
10. NANNEY, L.B., B.A. WENCZAK & J.B. LYNCH. 1996. Progressive burn injury documented with vimentin immunostaining. J. Burn Care Rehabil. **17:** 191–198.
11. WATTS, A.M.I., M/P.H. TYLER, M.E. PERRY, *et al.* 2001. Burn depth and its histological measurement. Burns **27:** 154–160.
12. BASKARAN, H., M. TONER & M.L. YARMUSH. 2001. Poloxamer-188 improves capillary blood flow and tissue viability in a cutaneous burn wound. J. Surg. Res. **101:** 56–61.
13. MORITZ, A.R. & F.C. HENRIQUES. 1947. Studies in thermal injury: the relative importance of time and surface temperature in the causation of cutaneous burns. Am. J. Pathol. **23:** 695–720.
14. HAYES, L.J. & K.R. DILLER. 1983. Implementation of phase change in numerical models of heat transfer. J. Energy Resour. Technol. **105:** 431–435.
15. SONNA, L.A., J. FUJITA, S.L. GAFFIN & C.M. LILLY. 2002. Effects of heat and cold stress on mammalian gene expression. J. Appl. Physiol. **92:** 1725–1742.
16. WELCH, W.J. 1992. Mammalian stress response. Physiol. Rev. **72:** 1063–1081.
17. DESPA, F., D.P. ORGILL, J. NEUWALDER & R.C. LEE. 2005. The relative thermal stability of tissue macromolecules and cellular structure in burn injury [review]. Burns **31:** 568–577. E-pub Apr. 7, 2005.

18. LEE, R.C., L.P. RIVER, F. PAN, et al. 1992. Surfactant-induced sealing of electropermeabilized skeletal muscle. Proc. Natl. Acad. Sci. USA **89:** 4524–4528.
19. MCNEIL. P.L. & R.A. STEINHARDT. 1997. Loss, restoration, and maintenance of plasma membrane integrity. J. Cell Biol. **137:** 1–4.
20. GERSHFELD, N.L. & M. MURAYAMA. 1968. Thermal instability of red blood cell membrane bilayers: temperature dependence of hemolysis. J. Memb. Biol. **101:** 62–72.
21. MCNEIL, P.L. 2002. Repairing a torn cell surface: make way, lysosomes to the rescue. J. Cell Sci. **115:** 873–879.
22. WOOLLEY, K. & P. MARTIN. 2000. Conserved mechanisms of repair: from damaged single cells to wounds in multicellular tissues. BioEssays **22:** 911–919.
23. GETHING, M.J. & J. SAMBROOK. 1992. Protein folding in the cell. Nature 355: 33–45.
24. WELCH, W.J. 1992. Mammalian stress response. Physiol. Rev. **72:** 1063–1081.
25. CSERMELY, P., C. SOTI, E. KALMAR, et al. 2003. Molecular chaperones, evolution and medicine. J. Mol. Struct. **666:** 373–380.
26. POCKLEY, A.G. 2001. Heat shock proteins in health and disease: therapeutic targets or therapeutic agents? Exp. Rev. Mol. Med. Sept. 21: 1–21.
27. HILDENBRANDT, B., P. WUST, O. AHLERS, et al. 2002. The cellular and molecular basis of hyperthermia. Crit. Rev. Oncol. Hematol. **43:** 33–56.
28. WEIBEZAHN, J., P. TESSARZ, C. SCHLIEKER, et al. 2004. Thermotolerance requires refolding of aggregated proteins by substrate translocation through the central pore of ClpB. Cell **119:** 653–665.
29. VAN DER ZEE, J. 2002. Heating the patient: a promising approach? Ann. Oncol. **13:** 1173–1184.
30. RABINOVICH, G.A., A. ARIEL, R. HERSHKOVIZ, et al. 1999. Specific inhibition of T-cell adhesion to extracellular matrix and proinflammatory cytokine secretion by human recombinant galectin-1. Immunology **97:** 100–106.
31. SCHWACHA, M.G. & I.H. CHAUDRY. 2002. The cellular basis of post-burn immunosuppression: macrophages and mediators. Int. J. Mol. Med. **10:** 239–243.
32. OPAL, S.M. 2002. Insights into the immune dysfunction associated with thermal injury. Crit. Care Med. **30:** 1651–1652.
33. SPARKES, B.G. 1997. Immunological responses to thermal injury. Burns **23:** 106–113.
34. MACK, V.E., M.D. MCCARTER, H.A. NAAMA, et al. 1996. Dominance of T-helper 2-type cytokines after severe injury. Arch. Surg. **131:** 1303–1309.
35. MASSON, I., J. MATHIEU, X.B. NOLLAND, et al. 1998. Role of nitric oxide in depressed lymphoproliferative responses and altered cytokine production following thermal injury in rats. Cell Immunol. **186:** 121–132.
36. HIGHTOWER, L.E. & P.T. GUIDON. 1989. Selective release from cultured mammalian cells of heat-shock (stress) proteins that resemble glia-axon transfer proteins. J. Cell Physiol. **138:** 257–266.
37. BIRK, O.S., S.L. GUR, D. ELIAS, et al. 1999. The 60-kDa heat shock protein modulates allograft rejection. Proc. Natl. Acad. Sci. USA **96:** 5159–5163.
38. MUNSTER, A.M. 1994. Alteration of the immune system in burns and implications for therapy. Eur. J. Pediatr. Surg. **4:** 231.
39. TOGO, T., J.M. ALDERTON, G.Q. BI, et al. 1999. The mechanism of facilitated cell membrane resealing. J. Cell Sci. **112:** 719–731.
40. MERCHANT, F.A., W.H. HOLMES, M. CAPELLI-SCHELLPFEFFER, et al. 1998. Poloxamer 188 enhances functional recovery of lethally heat-shocked fibroblasts. J. Surg. Res. **74:** 131–140.
41. FORMAN, M.B., D.W. PUETT, S.E. BINGHAM, et al. 1987. Preservation of endothelial cell structure and function by intracoronary perfluorochemical in a canine preparation of reperfusion. Circulation **76:** 469.
42. LANE, T.A. & G.E. LAMKIN. 1984. Paralysis of phagocyte migration due to an artificial blood substitute. Blood **64:** 400–405.
43. SPARKES, B.G. 1997. Immunological responses to thermal injury. Burns **23:** 106–113.
44. MOLNAR, J.A. 2004. Applications of negative pressure wound therapy to thermal injury. Ostomy Wound Manag. **50**(4A Suppl.): 17–19.

Cryo-Injury and Biopreservation

ALEX FOWLER[a,b] AND MEHMET TONER[b]

[a]*Department of Mechanical Engineering, University of Massachusetts, Dartmouth, Dartmouth, Massachusetts, USA*

[b]*Center for Engineering in Medicine, and Surgical Services, Massachusetts General Hospital, Harvard Medical School, Boston, Massachusetts, USA*

> ABSTRACT: Mammalian cells appear to be naturally tolerant to cold temperatures, but the formation of ice when cells are cooled leads to a variety of damaging effects. The study of cryo-injury, therefore, becomes the study of when and how ice is formed both inside and outside the cell during cooling. Protectant chemicals are used to control or prevent ice formation in many preservation protocols, but these chemical themselves tend to be damaging. Cooling and warming rates also strongly affect the amount and location of ice that is formed. Through careful modification of these parameters successful cold preservation techniques for many cell types have been developed, but there are many more cell types that have defied preservation techniques, and the extension of cell-based techniques to tissues and whole organs has been very limited. There are many aspects to the damaging effects of ice in cells that are still poorly understood. In this brief article we review our current understanding of cellular injury and highlight the aspects of cellular injury during cryopreservation that are still poorly understood.
>
> KEYWORDS: cryopreservation; freezing; vitrification; cell injury

INTRODUCTION

Cellular injury that results from freezing and thawing has been the object of scientific study for more than 60 years.[1] The concept of preserving cells and tissues for long periods of time by freezing them has led to a continual interest in how and why cells are damaged when they are exposed to cold temperatures. Successful preservation of human red blood cells by freezing in the early 1950s made successful organ preservation and possibly even suspended animation appear to be within reach. Researchers increased their efforts to understand the sources of damage to cells that experienced freezing and ways to prevent that damage. It turns out that preventing cellular damage during freezing is far more difficult than most people imagined in 1950. In this brief article we will review our current understanding of cellular injury and highlight the aspects of cellular injury during cryopreservation that are still poorly understood.

Address for correspondence: Mehmet Toner, Ph.D., Massachusetts General Hospital, Center for Engineering in Medicine, 114 16th Street, Room 1401, Charlestown, MA 02129-4404. Voice: 617-724-3044; fax: 617-573-8471.
 mtoner@sbi.org
 mtoner@partners.org

The motivation for cellular preservation by freezing is fairly clear. Many people understand the benefits of being able to freeze blood, and can understand the benefits of being able to store a kidney, heart, or liver for long periods of time. The goal of freezing whole organs for future use remains one of the main goals of the field of cryopreservation, but the many difficulties researchers have encountered in trying to reach that goal have made whole-organ preservation seem like an ever-receding horizon. Techniques and lessons learned from successful preservation of suspended and cultured cells have not translated well to complex tissues. Research into whole-organ preservation continues, and may someday be successful, but no one expects to see a whole liver cryopreserved anytime soon.

The field of cryopreservation has gained renewed momentum and garnered new interest recently, however, because of the emerging interest in cell-based therapies. Cryopreservation has achieved reasonable success in establishing protocols for the frozen preservation of cells in culture; and many new therapies are emerging that make use of living cells as the basis for treatment.[2,3] Tissue engineering, gene therapy, and cellular implantation all rely on the ability to store and transport cells and tissues in order to be clinically successful. While whole-organ preservation remains an important goal for cryopreservation, the ability to preserve and maintain a wide variety of cell types in suspension and culture has grown increasingly important on its own.

It appears to be the case that cells are not damaged by exposure to cold temperatures. Extensive evidence indicates that ice-formation inside or outside of the cell leads to the damage that cells experience when exposed to cold, not the cold temperature itself. This can be contrasted directly with hyperthermic exposure, where the proteins within the cells begin to denature as soon as the cell temperature is raised a few degrees above the physiological norm. Patients undergoing major surgery are routinely cooled 15 to 20 degrees below normal in order to slow physiological degradation during the surgery; and this does not appear to harm any of the cells in the human body. Whole organs are also cooled to hypothermic temperatures in order to store them for a short period of time prior to transplantation. Since it is well established that cooling slows down metabolic and degradation processes in cells and tissues, the fact that cells are not injured "immediately" by cooling suggests that if one cooled them to low enough temperatures one could virtually stop the degradative processes and preserve the cells for long periods of time. This is the concept behind cryopreservation efforts in both medicine and food science.

When cells are cooled below the freezing point of water, however, ice can form. The formation of ice can lead to lethal damage to cells and tissues. It has been observed in countless studies that many cell types can survive freezing, but the number of cells that survive depends critically on the rate at which the cells are cooled and the rate at which they are warmed. This is because the cooling and warming rates critically affect whether ice is formed inside the cell, outside the cell, or is never formed at all. If ice is not formed, then the cells and tissues generally remain undamaged.

VITRIFICATION

The formation of ice in pure water becomes energetically favorable when water at atmospheric pressure is cooled to below 0°C. In order for ice formation to actually occur, however, a group of water molecules has to arrange itself into a stable crystalline nucleus.[4] Once a stable ice nucleus is formed, additional molecules of water will attach and the crystal will grow as long as the temperature remains below the equilibrium melting point. The formation of the initial crystal nucleus is stochastic,[5] resulting from random motion of the liquid molecules. Nucleation becomes increasingly likely as temperature decreases because the number of molecules that have to coordinate to form a stable crystal nucleus decreases with decreasing temperature. Usually the formation of a crystal nucleus is abetted by the presence of a non-water molecule that imposes a certain amount of local order on the surrounding water molecules, thereby increasing the probability that they will form an ordered crystal.[6] These types of objects are heterogenous nucleators, and they usually cause ice to form at temperatures between −10 and −18°C. Even if no heterogeneous nucleators are present, however, a nucleating crystal will form by random action of the water molecules at about −40°C in most circumstances. If the water is able to reach a temperature of about −138°C without crystal formation, then a glass-phase transition occurs and the water will remain in an amorphous solid state for as long as it is held below the glass-phase transition temperature.[7] The ability to reach −138°C without crystal ice formation should provide the perfect cryopreservation method. There are two known methods for reaching cold temperatures without ice formation: one can freeze really fast or one can add solutes and molecular ice inhibitors so that crystals will not form.

The creation of a non-crystalline glass-phase solid from a liquid is called "vitrification."[8] Glass formation is a second-order phase transition in which the specific heat and the viscosity of the substance change significantly. The viscosity of glass is so great that inter-molecular relaxation and diffusion will not occur in ordinary laboratory time scales. In a glass, crystals do not form, even though they are energetically favorable, because the molecules are no longer free to arrange themselves into a crystal structure. In addition, chemical reactions, such as those necessary for cellular degradation, become virtually impossible due to molecular immobility.

For pure water, glass forms at −138 °C. Glass-phase water cannot ordinarily be formed because ice crystals form at temperatures much higher than that, but vitrification of pure water can be achieved by very rapid cooling (10^6 °C/sec) so that ice crystals do not have time to form. Cooling rates this high are achievable for regions up to about 10 microns in depth if the water surface is instantaneously exposed to liquid nitrogen temperatures. The most successful technique for the vitrification of pure water has been supersonic deposition of droplets onto cryogenically cooled surfaces.[9] Cooling rates this high have only rarely been achieved using cells[10,11]; and there does not seem to be any way to achieve cooling rates this high for tissues or whole organs.

Vitrification can be achieved at more modest cooling rates by adding cryoprotectant chemicals that inhibit the formation or growth of ice crystals or increase the glass-phase temperature of the solution. FIGURE 1 illustrates a number of techniques that have been used in conjunction with vitrification-promoting chemicals to achieve vitrification of cells.[12–14] The overall approach is to minimize thermal mass of the holder in order to achieve high cooling rates and thus lower the likelihood of deleterious ice formation. One can see that these techniques are still 100 times too slow to achieve vitrification of pure water, but they can be used to vitrify cryoprotectant solutions. If the concentration of cryoprotectants is high enough, then the system will

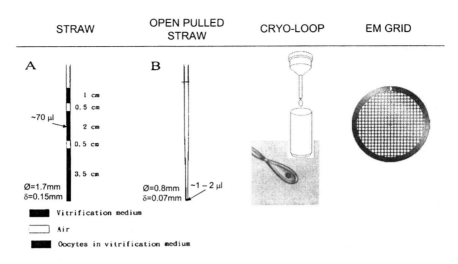

FIGURE 1. Schema of various vitrification techniques. Straw technique is the most commonly used approach for oocyte and embryo vitrification using cryoprotectant cocktails of 5 to 7M in concentration. Typically such high cryoprotectant concentrations are very toxic and difficult to load into and remove from cells. Open pulled straw, cryo-loop, and EM (electron microscopy) grid techniques have been developed to achieve rapid cooling to increase the probability of vitrification at a lower cryoprotectant concentration.

form a glass regardless of cooling rate.[8] Solutions containing 41–50% of propanediol are reported to have heterogenous nucleation temperatures that are below their glass-phase transition temperatures.[15] This means that ice crystals will not form at temperatures above the glass-transition temperature. A non-crystalline glass phase, therefore, can be achieved at any cooling rate, which makes it possible to vitrify large samples. Luyet suggested in 1937 that vitrification might provide an excellent method for cryopreservation,[8] and recent articles have touted its potential for preserving whole organs.[16] If a solution can be put into the glass phase without any possible chance of ice formation, then the ice will not kill the cells. Whole organs or even whole bodies should be possible to preserve.

The problem is the incredibly high concentration of cryoprotectants necessary to achieve vitrification at ordinary cooling rates.[15,17,18] Cryprotectants are generally toxic to cells, and at the concentrations typically used for vitrification, almost anything except water is toxic to cells. Even the simple osmotic stresses associated with addition and removal of such massive amounts of additives can be lethal.[15,17,19] The challenge of finding vitrification solutions that do not kill the cells they are intended to preserve, and the challenge of introducing those solutions into every part of a complex tissue without causing severe damage, remain formidable. Vitrification is currently a very promising cryopreservation technique for the possible preservation of whole organs and complex tissues, but its success at this point is limited to a few simple tissues[16,20] and it is not yet clear whether it will become the answer to the ultimate challenges of biopreservation.

SLOW-FREEZING INJURY

The curve of survival versus cooling rate for many cells takes on the shape of an inverted U, in which a maximum survival is achieved at an intermediate cooling rate.[21–23] This optimal cooling rate depends strongly on cell type and also on the warming rate used to thaw the cells. Studies have revealed that one of the parameters that most strongly effects the optimal cooling rate is the membrane permeability of the cell (often symbolized as L_{pg}), which is a measure of how quickly water can move in and out of the cell through the cellular membrane. Cells with highly permeable membranes, such as erythrocytes, have very high optimal cooling rates, and cells with relatively low permeabilities, such as oocytes, tend to have low optimal cooling rates.[24]

The inverted U phenomenon has been shown to be a direct function of exactly when and where ice is formed within the system. There do not appear to be any "strong" heterogeneous nucleators present in the cellular interior. This means that if a cell is in isolation, ice will not form inside the cell until the cell reaches a temperature of about −40°C. Ice will usually form outside the cell, therefore, where heterogeneous nucleators are almost always present, before ice forms inside the cell.[25] In most practical approaches, ice is seeded in a controlled manner in the extracellular solution such as by touching with a spatula cooled to cryogenic temperatures. One of the unusual properties of water is its very strong solute rejection.[4] When ice forms, very little solute is incorporated into the ice. Since cells are always frozen in solutions, rather than in pure water, solute rejection occurs as soon as ice is formed and the solute concentration in the remaining water increases. The increased solute

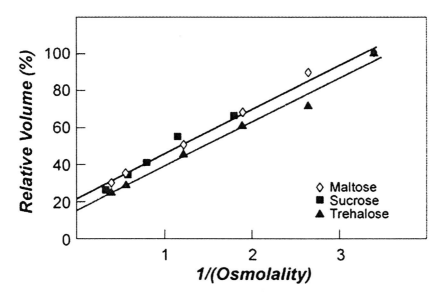

FIGURE 2. Osmometric behavior of human oocytes as a function of increasing extracellular osmolality. (From McWilliams et al.[72] Reproduced by permission.)

concentration has two effects: it lowers the freezing point of the remaining solution and it creates an osmotic pressure gradient across the cellular membrane. FIGURE 2 illustrates the relationship between cellular volume and external osmolality for human oocytes. One can see that increasing the osmolality of the external solution by the addition of sugars (which do not enter the cell) can cause the cells to shrink to about 20% of their initial volume.

During freezing the formation of extracellular ice causes an increase in extracellular osmolality. As soon as ice forms outside of a cell in solution, the cell begins to dehydrate. If the cell membrane is highly permeable to water, then the intracellular solution will maintain osmotic equilibrium with the extracellular solution as the temperature continues to drop and progressively more ice is formed. The intracellular and extracellular solution concentrations will continue to increase as temperature drops until a eutectic point is reached and the remaining solution solidifies, or more commonly in cryobiological systems, a glass transition occurs and the remaining unfrozen solution is vitrified. At this point the system can be cooled to liquid nitrogen temperatures without significant changes to the chemical structure inside or outside the cell.

During very slow cooling, the process of gradually dehydrating the cell as ice is slowly formed outside the cell leads to extensive cellular damage. A complete review and analysis of the literature relating to this type of damage is given in a recent review by Mazur.[24]

It has been clearly established that one source of cellular injury in the slow-freezing process is the lethal concentration of electrolytes to which the cell is exposed.[26] Since electrolytes are almost always present both inside and outside the cell, their concentrations will inevitably go up as ice is formed extracellularly. High concentrations of electrolytes, such as those seen by cells during freezing, have been shown to be lethal. The damage seen during slow freezing has been well correlated with the damage seen when cells are exposed to equivalent electrolyte concentrations without freezing.[24,27] This mechanism of damage, usually called solute damage or solute effect, is the one most cryobiologists consider when thinking about the cause of slow-freezing death to cells.

There is evidence, however, that slow-freezing damage is not solely a function of increased electrolyte concentration. Mazur and Cole have shown that the survival of slowly frozen red blood cells depends on electrolyte concentration and the unfrozen fraction of water.[27] The unfrozen fraction is the fraction of water in the system that does not form ice, but instead is incorporated into a eutectic solid or into a glass. By adding glycerol to the cellular solution they controlled the electrolyte concentration independently of the unfrozen fraction. They found that blood cell survival depends strongly on electrolyte concentration as long as there is a sufficiently large fraction of unfrozen water. If the unfrozen fraction is less than 20%, however, the survival depends only on the unfrozen fraction, and survival decreases dramatically as the unfrozen fraction decreases. They believe this damage is caused by the aggregation of the cells into small channels as they are excluded from the ice, and they are forced into some type of adverse cell–cell interaction. Further evidence for adverse cell interactions as a source of cellular injury is provided by studies involving the dense packing of cells during freezing.[28]

Other researchers have suggested that the total volume change the cells experience as they undergo shrinkage during slow freezing may cause membrane damage, and still others have proposed that there may be a critical minimum volume for cells, and if cells are shrunken to sizes smaller than that critical level during slow freezing they may suffer irreparable damage.[24] None of the evidence for any one of these hypotheses is conclusive; but it is clear and well established that if cells are frozen slowly without cryoprotectant, they will shrink as ice forms, and they will almost always die.

Excessive cellular shrinkage can be reduced by the addition of penetrating cryoprotectants. Cryoprotectants are chemicals that are added to the cellular solution to try and protect the cell during freezing or vitrification. There are hundreds of cryoprotectants of different types. Some of the most commonly used are dimethyl sulfoxide (DMSO), glycerol, sucrose, and trehalose. There are two fundamental types of cryoprotectant: penetrating cryoprotectants and non-penetrating cryoprotectants. Penetrating cryoprotectants pass through the cellular membrane unaided, and non-penetrating ones do not. When considering slow freezing damage, which seems to be caused by cellular shrinkage and increased electrolyte concentration, the two types of cryoprotectant act very differently. The addition of non-penetrating cryoprotectants increases the osmolality of the extracellular solution and causes the cells to shrink before freezing as shown in FIGURE 2. Freezing may be harder to initiate due to the presence of the cryoprotectant (at the very least a higher osmolality will lead to a lower equilibrium freezing temperature), but once freezing is initiated solute rejection will commence and the cell will begin to shrink as the osmolality of the unfrozen solution continues to increase. The liquidus curve of the extracellular solution will depend on its components, so the osmolality of the system as a function of temperature may be different depending on the cryoprotectant used, and the final state of the extracellular unfrozen fraction may be altered. The extracellular solution may form a glass rather than a water and salt eutectic. The composition of the intracellular solution, however, will not be altered by non-penetrating cryoprotectants, so the electrolyte concentration of the intracellular solution and the cell size will both be functions of the extracellular osmolality exactly as if the cryoprotectant were not added. If water is removed from the cell until the intracellular solution forms a glass or eutectic solid, then the end state inside the cell will be the same whether non-penetrating cryoprotectant is added or not. In most cases, unless the cryprotectant dramatically alters the liquidus curve of the extracellular solution, the addition of non-penetrating cryoprotectants will not prevent injury to slowly frozen cells.

Penetrating cryoprotectants, on the other hand, pass through the cellular membrane. Adding these cryoprotectants, of which DMSO and glycerol are the most common, causes the concentration of electrolytes to decrease for a given temperature during freezing, since for any given osmolality a certain fraction of the solutes inside and outside the cell are cryoprotectants. They also increase the final size of the cell when dehydration due to freezing is complete, because in addition to the usual intracellular components, there is an additional non-water volume component to the intracellular volume. The unfrozen fraction of the solution is also increased, as is the case for non-penetrating cryoprotectants. The effects of DMSO addition on the amount of ice formed and on the electrolyte concentration are illustrated in FIGURE 3. One can see that the addition of DMSO substantially decreases the amount of ice that is formed and the electrolyte concentration as the sample is cooled. Since these are the leading causes of cellular damage during slow cooling it is clear that the addition of DMSO can mitigate this damage; but DMSO is toxic to cells.[19] The amount

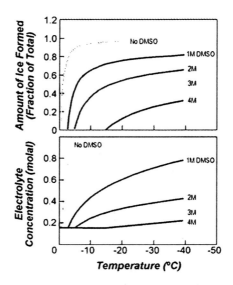

FIGURE 3. Amount of ice and electrolyte in frozen DMSO solutions.

of DMSO that cells can tolerate varies by cell type. Often the amount of DMSO or other penetrating cryoprotectants required to prevent electrolyte damage during slow cooling is lethal to cells.

The end result of a slow-freezing process is one in which the cells are shrunken and trapped in the unfrozen fraction of the extracellular media. With the addition of penetrating cryoprotectants, the interior of cells are usually in a glass-phase solid or in a partially crystallized glassy matrix. During warming the cells are rehydrated as the ice melts. In general the cell survival for slowly frozen cells does not depend strongly on the rate at which they are warmed.[23] Unfortunately most cells die when they are frozen too slowly.

FAST-FREEZING INJURY

In order to avoid solute effects and excessive cellular shrinkage one can freeze the cells more quickly. Since the cellular membrane resists water transport, water can be retained inside cells during cooling if one cools the cells at a high rate. FIGURE 4 illustrates the two possible outcomes for frozen cells: if the cooling is slow, then the cells dehydrate and ice is formed only outside the cells, but if cooling is fast, intracellular water will be retained and freezing will occur both inside and outside the cells. Many studies have shown that the presence of ice outside the cell promotes ice formation inside the cell.[29–31] FIGURE 5 illustrates the two mechanisms that can lead to ice inside a suspended cell. The precise mechanism by which ice outside the cell promotes ice formation inside the cell is unknown: three widely held theories are (1) that ice manages to propagate through pores in the cell membrane; (2) that ice on one

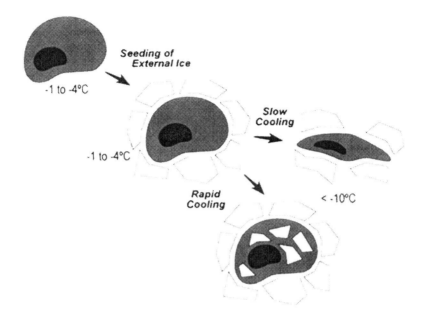

FIGURE 4. Schema of the key events during freezing of cells. Typically, the extracellular ice is seeded between −1 and −4°C, and then the cooling process starts.

side of the membrane causes conformational changes in the membrane or on the water just inside the membrane that make ice formation more likely; or (3) that the cell membrane is damaged by the formation of extracellular ice and ceases to be a barrier to ice propagation into the cell.[25,30,31] Whatever the mechanism, it is clear that ice forms at much higher temperatures inside the cell if there is ice already present outside the cell.

Whether or not ice is formed inside the cell is more complicated when dealing with tissues rather than with suspended cells. Studies of intracellular ice formation (IIF) for confluent mammalian cells in culture have shown that ice nucleated in one cell can propagate into adjacent cells.[32,33] Recent studies by Irimia and Karlsson have enabled quantification of the rate of intracellular ice propagation from one cell to another.[34] Using microfabrication they were able to precisely control the degree of cell–cell interaction experienced between cells and to study its effect on ice formation. They found that cell–cell interactions did not affect the rate at which ice nucleation occurred within an independent cell, but that intracellular ice did propagate from one cell to another. The rates of cell-to-cell ice propagation were greater for cells that had been in contact with each other for longer periods of time. These studies suggest that intracellular ice is much more likely to form in complex tissues than in suspended cells, because once ice forms in any cell of the tissue that ice can propagate into adjacent cells. This means that freezing protocols that seek to prevent intracellular ice formation may be very difficult to implement for large tissues.

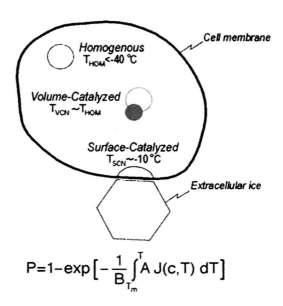

FIGURE 5. Schematic of multi-modal mechanisms of ice nucleation in cells.

A hydrated cell, therefore, will experience ice formation inside the cell during freezing. Most cryobiologists use as a working hypothesis that intracellular ice kills cells. In almost all studies it has been found that if cells are frozen fast enough to form intracellular ice, the cells die. This leads to the famous inverted "U" curve for cell survival versus cooling rate which is illustrated in FIGURE 6. The figure illustrates Mazur's two-factor hypothesis[35]: freezing too slowly kills cells owing to one effect and freezing too quickly kills cells owing to another, but cooling just right may get them to live.

FIGURE 7 illustrates the cooling rates for a wide variety of cells that lead to 50% cell survival and 50% of the cells having intracellular ice. One can see that the cooling rates necessary to get 50% survival vary by three orders of magnitude, and that there is a very strong correlation between the 50% IIF points and 50% survival points. Data such as these clearly suggest that IIF is lethal.

Most successful fast-freezing protocols require that the cells be dehydrated by some means prior to initiating fast freezing so that intracellular ice can be avoided.[36] Two-step methods rely on the seeding of extracellular ice. In this step an ice crystal is formed at a particular temperature by adding an ice crystal, an ice nucleator, or simply touching the sample with something small and very cold. This causes ice to form at a precise temperature rather than at a randomly determined temperature and allows the extracellular ice to cause cellular dehydration so that intracellular ice can be avoided. Optimal freezing protocols cool the cells quickly enough so that some water is retained inside the cell in order to avoid slow-freezing damage, but slowly enough for some cellular dehydration to occur so that intracellular ice will not form.

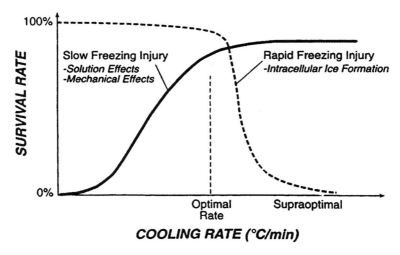

FIGURE 6. Inverted-U hypothesis of cell injury during freezing of cells.

For highly permeable cells such as erythrocytes, the optimal cooling rate is more than 1000°C/min, but for relatively impermeable stem cells the optimum is about 1°C/min.

Determination of optimal cooling rates has been a highly fruitful area of research both empirically and theoretically.[37] The original concept for defining an optimum was proposed by Mazur[36] as the fastest cooling rate possible that would not promote intracellular ice formation. He proposed that intracellular ice would not form if the cell had less than 10% of its initial water when it entered the ice nucleation zone of temperature. Bischof and co-workers refined these values to 5% of the initial water content when the cell reached −30°C.[38] On the basis of this definition and water transport equations originally developed by Mazur,[36] Thirumala and Devireddy developed a theory for determining optimal cooling rates for cell survival.[38] They found they could predict optimal cooling rates that agreed very well with experimental values for a wide variety of cells solely based on knowledge of the cell membrane permeability and the cell's initial surface area to volume ratio. This provides strong evidence that the main sources of cellular injury during freezing are a result of the kinetics of membrane water transport. If too little water is removed, intracellular ice forms, and if too much is removed, the cells die from dehydration effects.

The role of cryoprotectants in preventing injury during fast cooling is very different from their role in slow freezing. Non-penetrating cryoprotectants can be used to partially dehydrate the cell prior to freezing, thereby reducing the chances that ice will nucleate in the cellular interior. Cryprotectants can raise the glass-phase transition temperature so that the cells have a greater chance to reach glass phase without nucleating intracellular ice.[39] Cryoprotectants also change fluid properties and can thereby directly reduce the chances for ice nucleation.[40,41,36]

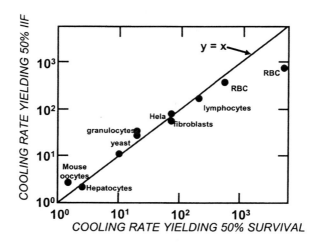

FIGURE 7. Relationship between the cooling rate yielding 50% survival and the cooling rate yielding 50% intracellular ice formation (IIF).

While it is extremely clear from countless studies that the formation of intracellular ice usually leads to cell death, it is not at all clear that the formation of intracellular ice causes cell death.[42,43,44] When one observes cells that have been frozen very rapidly so that intracellular ice has formed, one sees cells that look perfectly intact, although they are encased in ice. During warming, many cells with intracellular ice suddenly lyse. Many studies indicate that the survival of rapidly frozen cells, cells that have been cooled at rates faster than their optimum, depends strongly on the rate of warming.[22,45-48] The faster the cells are warmed, the more of them survive. A lot of evidence seems to suggest that the lethal damage caused by intracellular ice is not caused when the ice is formed, but by some process that occurs if intracellular ice is present as the cell is warmed.[10,49,50] An elegant set of studies by Rall et al. showed that damage caused by IIF in rapidly frozen and slowly warmed mouse embryos was not inflicted until the mouse embryos were warmed to a temperature of about −65°C.[51,52] This was after the point at which ice recrystallized from vitrified regions of the sample at about −85°C. No mechanism has been precisely identified for this damage during rewarming, but cells in which IIF could be clearly seen at −85°C could be successfully recovered if they were warmed rapidly through the critical −65°C range. This proves that the formation of IIF during warming was not the lethal event, and that IIF is not necessarily lethal.

The mechanism by which intracellular ice damages cells is not clear. A complete review of the literature relating to the damage caused by intracellular ice is provided by Muldrew et al.[53] If the damage actually occurs during warming and not during freezing, then the most likely mechanism involves recrystallization, as proposed by Mazur.[24] When ice is formed intracellularly during rapid freezing, many nucleation sites are formed, as there is not much time during freezing for crystal growth. The frozen cell is filled with many tiny crystals. During warming the small crystals reorganize into larger crystals, which may damage organelles or other cellular structures. How quickly this happens is not known; but if lethal damage to cells with intracellular ice occurs during warming it is probably the formation of large crystals inside the cell that leads to the damage. Indirect evidence in support of the theory that recrystallization may play a major role in cellular injury is provided by naturally freeze-tolerant organisms, some of which generate recrystallization inhibitors in response to freezing stress.[54,55]

Some of the newest and most interesting findings involving intracellular ice come from a series of papers by Acker *et al.*,[32,56-58] who show that intracellular ice propagates from cell to cell in confluent layers of V79W cells, and that although cells that nucleate intracellular ice die, cells into which ice propagates from an adjacent cell do not die. A mechanism for why this would be true has not been determined. One speculation is that nucleated ice may form in all regions of the intracellular space, whereas propagated ice may form in specific non-lethal regions. Another possibility is that some membrane-damaging event leads to the initiating ice nucleating events. Whatever the mechanism, these papers indicate once again that not all intracellular ice is lethal.

SUMMARY OF CELLULAR INJURY DURING CRYOPRESERVATION

If ice can be avoided it appears that cellular injury can be avoided. Unfortunately the solutes and chemicals that prevent the formation of ice are typically toxic at high concentrations. Ice can also be avoided by very rapid cooling, but the cooling rates required to vitrify pure water or unadulterated cytoplasm seem impossible to achieve for samples that are larger than a single cell. The study of cellular injury during vitrification, therefore, is the study of limiting toxicity and osmotic stress caused by cryoprotectants in order to enable vitrification at achievable cooling rates. A nontoxic penetrating glass-former would be the ideal solution.

The type of cellular injury induced during freezing is a function of water transport. Slow-freezing injury definitely results when electrolyte concentrations reach damaging levels as the formation of extracellular ice dehydrates the cell. There are also damaging effects that seem to result from the constriction of cells into the small unfrozen fraction that can exist during some freezing protocols. Penetrating cryoprotectants can reduce both of these damaging effects, but as in the case of vitrification, the cryoprotectants are toxic. Slow freezing and vitrification are almost identical from a physiochemical perspective. In both cases the cells do not experience freezing and they end up in a vitrified state. The difference is that slow freezing uses the solute-exclusion property of ice to drive up solute concentrations inside and outside the cell. Vitrification protocols add more cryoprotectant, both penetrating and non-penetrating, to control the dehydration and solute concentration increase.

Cellular injury that results from intracellular ice formation is not well understood. It is clear that the formation of intracellular ice usually leads to cell death; but that damage can be mitigated by warming very rapidly. In addition there is a lot of evidence that the lethal damage occurs owing to recrystallization during warming, not during ice formation, which would explain why very rapid warming might mitigate the damage. No understanding yet exists for why ice that propagates from adjacent cells in cell monolayers is not damaging and no one knows what it is about recrystallization that causes the damage. A possible future for cryopreservation research may lie in the development of cryopreservation protocols that manage intracellular ice so that it is non-lethal, rather than attempt its prevention. If we could understand more fully why some intracellular ice is damaging and other ice is not, this may become possible.

EMERGING TECHNIQUES IN BIOPRESERVATION

Efforts continue to avoid intracellular ice without damaging the cell. One type of glass-former that is relatively non-toxic is sugar. Many sugars are excellent glass-formers, and cells tolerate them to a high degree. Unfortunately most sugars are non-penetrating, which means they cannot promote glass formation inside the cell, nor prevent the concentration build-up of electrolytes within the cell during freezing. Techniques to get non-permeating sugars inside cells, however, have been developed recently,[59,60,61] and they have been very successful in promoting cell survival after freezing even at concentrations low enough that the cells can be cultured without removing the cryoprotectant.[62,63] A non-metabolizable glucose has also been developed that is actively transported inside the cell by the glut transporters. This sugar (3OMG) has also provided substantial survival benefit during freezing at relatively low concentrations.[64]

Freezing under high pressure has long been recognized as a possible method for decreasing ice nucleation.[65] A recent study by Rubinsky and co-workers has revisited the idea of using isochoric (constant volume) freezing to increase pressure in the system as ice forms.[66] He has shown that by means of this method the solute concentrations as a function of temperature can be decreased by almost an order of magnitude. This would allow for the use of significantly less cryprotectant in slow-freezing protocols.

Finally, the recognition that slow-freezing protocols are actually a method for dehydrating cells, in combination with the observation that many natural organisms survive dehydration, has led to a flurry of research into dried preservation of cells and tissues.[61,67–71] In these studies glass-forming sugars are introduced into the cells by means of a variety of techniques before the cells are air- or freeze-dried.[67] Research into the mechanisms of cellular injury for cells during drying and storage has just begun.

ACKNOWLEDGMENTS

This work was partially supported by the National Institutes of Health grants.

REFERENCES

1. MERYMAN, H. 2004. Foreword. *In* Life in the Frozen State. B.J. Fuller, N. Lane & E.E. Benson, Eds. CRC Press. Boca Raton, FL.
2. LANGER, R. & J.P. VACANTI. 1996. Tissue engineering: the development of functional substitutes for damaged tissue. Science **260:** 920–926.

3. KARLSSON, J.O.M. & M. TONER. 2000. Cryobiology: foundations and applications in tissue engineering. *In* Principles of Tissue Engineering, 2nd ed. R. Lanza, R. Langer & J.P. Vacanti, Eds.: 293–307. Academic Press. New York.
4. HOBBS, P.V. 1974. Ice Physics. Oxford. Clarendon Press, UK.
5. TURNBULL, D. 1956. Phase changes. *In* Solid State Physics, Vol. 3. F. Seitz & D. Turnbull, Eds.: 225–306. Academic Press. New York.
6. TURNBULL, D. 1962. On the relation between crystallization rate and liquid structure. J. Phys. Chem. **66:** 609–613.
7. CHEN, T. 2000. Literature review: supplemented phase diagram of trehalose-water binary mixture. Cryobiology **40:** 277–282.
8. LUYET, B.J. 1937. The vitrification of organic colloids and of protoplasm. Biodynamica **29:** 1–14.
9. JOHARI, G.P. *et al.* 1996. Two calorimetrically distinct states of liquid water below 150 Kelvin. Science **273:** 90–91.
10. FOWLER, A.J. & M. TONER. 1998. Prevention of hemolysis in rapidly frozen erythrocytes by using a laser pulse. Ann. N.Y. Acad. Sci. **858:** 245–252.
11. GOETZ, A. & S.S. GOETZ. 1938. Vitrification and crystallization of organic cells at low temperatures. J. Appl. Phys. **9:** 718–729.
12. KULSHOVA, L. *et al.* 1999. Birth following vitrification of a small number of human oocytes. Human Reprod. **14:** 3077–3079.
13. VAJTA, G. *et al.* 1997. Vitrification of porcine embryos using the open pulled straw (OPS) method. Acta Vet. Scand. **38:** 49–52.
14. MARTINO *et al.* 1996. Development into blastocysts of bovine oocytes cryopreserved by ultra-rapid cooling. Biol. Reprod. **54:** 1059–1069.
15. TAYLOR, M.J. *et al.* 2004. Vitrification in tissue preservation: new developments. *In* Life in the Frozen State. B.J. Fuller, N. Lane & E.E. Benson, Eds.: 603–641. CRC Press. Boca Raton, FL.
16. KAISER, J. 2002. New prospects for putting organs on ice. Science **295:** 1015.
17. FAHY, G.M. & E.A. SUJA. 1997. Cryopreservation of the mammalian kidney. Cryobiology **35:** 114–131.
18. MEHL, P.M. 1993. Nucleation and crystal growth in a vitrification solution tested for organ cryopreservation by vitrification. Cryobiology **30:** 509–518.
19. SONG, Y.C. *et al.* 1995. Cryopreservation of the common carotid artery of the rabbit: optimization of dimethyl sulfoxide concentration and cooling rate. Cryobiology **32:** 405–421.
20. SONG, Y.C. *et al.* 2000. Vitreous cryopreservation maintains the function of vascular grafts. Nature Biotechnol. **18:** 296–299.
21. BANK, H.L. *et al.* 1979. Cryogenic preservation of isolated rat islets of Langerhans: effect of cooling and warming rates. Diametologia 16: 195–199.
22. LEIBO, S.B. *et al.* 1970. Effects of freezing on marrow stem cell suspensions: interactions of cooling and warming rates in the presence of PVP, sucrose, or glycerol. Cryobiology **6:** 315–332.
23. LEIBO, S.P. & P. MAZUR. 1971. The role of cooling rates in low-temperature preservation. Cryobiology **8:** 447–452.
24. MAZUR, P. 2004. Principles of cryobiology. *In* Life in the Frozen State. B.J. Fuller, N. Lane & E.E. Benson, Eds.: 3–65. CRC Press. Boca Raton, FL.
25. TONER, M. & E.G. CRAVALHO. 1990. Thermodynamics and kinetics of intracellular ice formation during freezing of biological cells. J. Appl. Phys. **67:** 1582–1593.
26. LOVELOCK, J.E. 1953. The mechanism of the protective action of glycerol against haemolysis by freezing and thawing. Biochim. Biophys. Acta **11:** 28–36.
27. MAZUR, P. & K.W. COLE. 1989. Roles of unfrozen fraction, salt concentration, and changes in cell volume in the survival of frozen human erythrocytes. Cryobiology **26:** 1–29.
28. PEGG, D.E. *et al.* 1984. The effect of cooling and warming rate on the packing effect in human erythrocytes frozen and thawed in the presence of 2 M glycerol. Cryobiology **21:** 491–502.
29. KARLSSON, J.O.M. *et al.* 1993. Nucleation and growth of ice crystals inside cultured hepatocytes during freezing in the presence of dimethyl sulfoxide. Biophys. J. **65:** 2524–2536.

30. MAZUR, P. 1965. The role of cell membranes in the freezing of yeast and other single cells. Ann. N.Y. Acad. Sci. **125:** 658–676.
31. MAZUR, P. 1984. Freezing of living cells: mechanisms and implications. Am. J. Physiol. **247:** 125–142.
32. ACKER, J.P. *et al.* 1999. Intracellular ice formation is affected by cell interactions. Cryobiology **38:** 363–371.
33. BERGER, W.K. & B. UHRIK. 1996. Freeze-induced shrinkage of individual cells and cell-to-cell propagation of intracellular ice in cell chains from salivary glands. Experientia **52:** 843–850.
34. IRIMIA, D. & J.O.M. KARLSSON. 2002. Kinetics and mechanism of intercellular ice propagation in a micropatterned tissue construct. Biophys. . **82:** 1858–1868.
35. MAZUR, P. 1970. Cryobiology: the freezing of biological systems. Science **168:** 939–949.
36. MAZUR, P. 1990. Equlibrium, quasi-equilibrium, and non-equilibrium freezing of mammalian embryos. Cell Biphys. **17:** 53–92.
37. KARLSSON, J.O.M. *et al.* 1996. Fertilization and development of mouse oocytes cryopreserved using a theoretically optimized protocol. Human Reprod. **11:** 1296–1305.
38. THIRUMALA, S. & R.V. DEVIREDDY. 2005. A simplified procedure to determine the optimal rate of freezing biological systems. J. Biomech. Eng. **127:** 295–300.
39. LUYET, B. & D. RASMUSSEN. 1968. Study by differential thermal analysis of the temperatures of instability of rapidly cooled solutions of glycerol, ethylene glycol, sucrose and glucose. Biodynamica **10:** 167–191.
40. MYERS, S.P. *et al.* 1989. Characterization of intracellular ice formation in *Drosophilia melanogaster* embryos. Cryobiology **26:** 472–484.
41. RALL, W.F. *et al.* 1983. Depression of the ice-nucleation temperature of rapidly cooled mouse embryos by glycerol and dimethyl sulfoxide. Biphys. J. **41:** 1–12.
42. ALBRECHT, R.M. *et al.* 1973. Survival of certain microorganisms subjected to rapid and very rapid freezing on membrane filters. Cryobiology **10:** 233–239.
43. SHIMADA, K. & E. ASAHINA. 1975. Visualization of intracellular ice crystals formed in very rapidly frozen cells at $-27\,°$C. Cryobiology **12:** 209–218.
44. DAVIS, D.J & R.E. LEE. 2001. Intracellular freezing, viability, and composition of fat body cells from freeze-intolerant larvae of *Sarcophagi crassipalpis*. Arch. Insect Biochem. Physiol. **48:** 199–205.
45. KOSHIMOTO, C. & P. MAZUR. 2002. Effects of cooling and warming rate to and from $-70°$C to $-196\,°$C on the motility of mouse spermatozoa. Biol. Reprod. **66:** 1477–1484.
46. KEARNEY, J.N. 1991. Cryopreservation of cultured skin cells. Burns **17:** 380–383.
47. AKHTAR, T. *et al.* 1979. The effect of cooling and warming rates on the survival of cryopreserved L-cells. Cryobiology **16:** 424–429.
48. HARRIS, L.W. & J. B. GRIFFITHS. 1977. Relative effects of cooling and warming rates on mammalian cells during the freeze-thaw cycle. Cryobiology **14:** 662–669.
49. FARRANT, J. 1977. Water transport and cell survival in cryobiological procedures. Phil. Trans. R. Soc. Lond. B. **278:** 191–205.
50. GOETZ, A. & S.S. GOETZ. 1938. Death by devitrification in yeast cells. Biodynamica **43:** 1–8.
51. Rall, W.F. *et al.* 1984. Analysis of slow-warming injury of mouse embryos by cryomicroscopical and physiochemical methods. Cryobiology **21:** 106–121.
52. RALL, W.F. *et al.* 1980. Innocuous biological freezing during warming. Nature **286:** 511–514.
53. MULDREW, K. *et al.* 2004. The water to ice transition: implications for living cells. *In* Life in the Frozen State. B.J. Fuller, N. Lane & E.E. Benson, Eds.: 67–108. CRC Press. Boca Raton, FL.
54. KNIGHT, C.A. *et al.* 1988. Solute effects on recrystallization: an assessment technique. Cryobiology **25:** 55–60.
55. RAMLOV, H. *et al.* 1996. Recrystallization in a freezing-tolerant Antarctic nematode, *Pnagrolaimus davidi*, and an alpine weta, *Hemideina maori* (orthoptera; stenopelmatidae). Cryobiology **33:** 607–613.
56. ACKER, J.P. & L.E. MCGANN. 2000. Cell-cell contact affects membrane integrity after intracellular freezing. Cryobiology **40:** 54–63.

57. ACKER, J.P. & L.E. MCGANN. 2003. Protective effect of intracellular ice during freezing. Cryobiology **46:** 197–202.
58. ACKER, J.P. *et al.* 2001. Intercellular ice propagation: experimental evidence for ice growth through membrane pores. Biophys. J. **81:** 1389–1397.
59. ACKER, J.P. *et al.* 2003. Measurement of trehalose loading of mammalian cells porated with a metal-actuated switchable pore. Biotechnol. Bioeng. **82:** 525–532.
60. EROGLU, A., *et al.* 2003. Quantitative microinjection of trehalose into mouse oocytes and zygotes, and its effect on development. Cryobiology **46:** 121–34.
61. CROWE, J.H. *et al.* 2003. Stabilization of membranes in human platelets freeze-dried with trehalose. Chem. Phys. Lipids **122:** 41–52.
62. EROGLU, A. *et al.* 2000. Intracellular trehalose improves the survival of cryopreserved mammalian cells. Nat. Biotechnol. **18:** 163–167.
63. EROGLU, A. *et al.* 2002. Beneficial effect of microinjected trehalose on the cryosurvival of human oocytes. Fertil. Steril. **77:** 152–158.
64. SUGIMACHI, K., K.L. ROACH, D.B. RHODS, R.G. TOMPKINS & M. TONER 2005. Nonmetabolizable glucose compounds impart cryotolerance to primary rat hepatocytes. Tissue Engineering. In press.
65. SARTORI, N. *et al.* 1993. Vitrification depth can be increased more than 10-fold by high-pressure freezing. J. Microsc. **172:** 55–61.
66. RUBINSKY, B. *et al.* 2005. The thermodynamic principles of isochoric cryopreservation. Cryobiology **50:** 121–138.
67. ACKER, J.P., T. CHEN, A. FOWLER & M. TONER. 2004. Engineering desiccation tolerance in mammalian cells: tools and techniques. *In* Life in the Frozen State. B. J. Fuller, N. Lane & E. Benson, Eds.: 563–580. CRC Press. New York.
68. CROWE, J.H., L.M. CROWE, F. TABLIN, *et al.* 2004. Stabilization of cells during freeze-drying: the trehalose myth. B. J. Fuller, N. Lane, & E. Benson, Eds.: 581–601. CRC Press. New York.
69. BHOWMICK S., L. ZHU, L. MCGINNIS, *et al.* 2003. Desiccation tolerance of spermatozoa dried at ambient temperature: production of fetal mice. Biol. Reprod. **68:** 1779–1786.
70. ACKER J.P., A. FOWLER, B. LAUMAN, *et al.* 2002. Survival of desiccated mammalian cells: beneficial effects of isotonic media. Cell Pres. Technol. **1:** 129–140.
71. CROWE, J.H. & F.A. HOEKSTRA. 1992. Anhydrobiosis. Annu. Rev. Physiol. **60:** 73–103.

Oxidative Reactive Species in Cell Injury Mechanisms in Diabetes Mellitus and Therapeutic Approaches

LEONID E. FRIDLYAND AND LOUIS H. PHILIPSON

Department of Medicine, University of Chicago, MC-1027, Chicago, Illinois 60637, USA

ABSTRACT: Mammalian cells are continuously subject to insult from reactive species. Most of the pathogenic mechanisms that have been considered to date reflect overproduction of reactive oxygen species (ROS) or a peculiar failure in intracellular defenses against ROS. We have attempted to consider briefly the most important mechanisms of ROS production, defense, and reactive species–induced cell damage and approaches to therapy, focusing on the example of diabetes mellitus. An improved understanding of these mechanisms should facilitate development of antioxidant intervention strategies leading to reduction in diseases associated with oxidative stress.

KEYWORDS: beta-cell; insulin resistance; mitochondria; diabetes mellitus

INTRODUCTION

Mammalian cells are continuously subject to insult from oxidative reactive species (ORS). The damage inflicted by ORS has been implicated in numerous pathogenetic processes, including inflammation, age-related degeneration, tumor formation, and diabetes mellitus.[1–3] The multiple mechanisms leading to ORS-induced cell damage underlie a complexity of disposition of reactive species at biological sites, leading to conceptual and practical difficulties in establishing appropriate interventions for prevention or treatment. An improved understanding of these mechanisms should facilitate development of antioxidant intervention strategies leading to reduction in diseases associated with oxidative stress, although progress thus far has been disappointing. The hope, of course, is that prevention strategies may delay the appearance of related diseases, as well as retard progression.

Most of the pathogenic mechanisms that have been considered to date reflect overproduction of reactive oxygen species (ROS) or a peculiar failure in intracellular defenses against ROS.[2–5] In this chapter, we attempt to consider briefly the most important mechanisms of reactive species–induced cell damage and approaches to therapy, focusing on the example of diabetes mellitus.

Address for correspondence: Louis H. Philipson, Department of Medicine, MC-1027, The University of Chicago, 5841 S. Maryland Ave., Chicago, IL 60637. Voice: 773-702-9180; fax: 773-702-9194.

l-philipson@uchicago.edu

REACTIVE SPECIES PRODUCTION, INTRACELLULAR DEFENSE, AND CELL DAMAGE

Some of the important mechanisms related to production of ROS and intracellular defense mechanisms are summarized in FIGURE 1 (for details, see reviews Refs. 5–7). Mitochondria generate cellular energy through the process of oxidative phosphorylation, which involves synthesis of reducing equivalents [NAD(P)H and FADH$_2$] that are reoxidized as electrons that pass through the electron transport chain (ETC). In this process, the ETC complexes in the inner mitochondrial membrane pump protons out of the mitochondrial inner membrane, creating the trans-

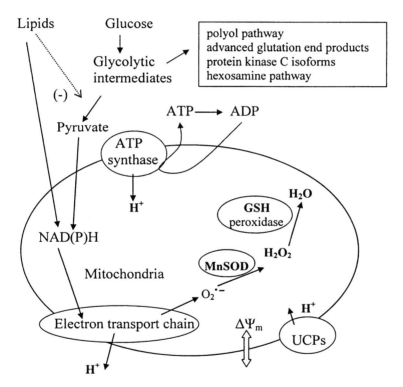

FIGURE 1. Pathways of reactive oxygen species production, clearance, and action. Glucose and fatty acid metabolism leads to an increase in NAD(P)H and FADH$_2$. The transfer of electrons from these reducing equivalents through the mitochondrial electron transport chain is coupled with the pumping of protons from the mitochondrial matrix to the intermembrane space and with hyperpolarization of the mitochondrial membrane. The resulting transmembrane electrochemical gradient drives the ATP synthesis at ATP-synthase. Part of the protons may leak through uncoupling proteins (UCPs) and other ways. The respiratory chain also produces superoxide anions ($O_2^{\bullet-}$), initiating ROS production. $O_2^{\bullet-}$ are enzymatically converted to hydrogen peroxide (H_2O_2) by a manganese-superoxide dismutase (MnSOD) within mitochondria. H_2O_2 is disposed of by the mitochondrial enzyme glutathione (GSH) peroxidase. Mitochondrial overproduction of superoxide activates four major pathways of hyperglycemic damage, which can begin from glycolytic intermediates.

membrane electrochemical gradient. This gradient itself is the driving force to make ATP from ADP and P_i, driven by proton movement back through the ATP synthase complex that essentially functions to release formed ATP. Under normal conditions, the proton gradient is also diminished by the H^+ backflow to the matrix. The backflow occurs either via non-protein membrane pores, protein/lipid interfaces (H^+ leak), or by uncoupling proteins (UCPs).

However, mitochondria can also play a major role in non-enzymatic ROS and reactive nitrogen species production because these reactive species are an unavoidable evil generated by oxidative phosphorylation.[1,5–7] Superoxide anions ($O_2^{\bullet}$) are an inevitable byproduct of single electron reduction of ubiquinone on the way to reduction of molecular oxygen primarily in complex III and possibly also in complex I of the mitochondrial ETC. Furthermore, these anions are the major contributors to other reactive species inside mitochondrion.[5–7] $O_2^{\bullet}$ may react with other radicals producing oxidants; for example, the reaction of $O_2^{\bullet}$ with nitric oxide produces peroxynitrite. Superoxide anions, peroxynitrite, and other reactive species are very powerful oxidants.[1,5,7]

$O_2^{\bullet}$ production depends upon the reduction state of the ETC complexes because the ETC carriers in a more reduced state have the property of donating electrons to oxygen.[5,6] Increasing the percentage of ETS carriers in the reduced states occurs through increased production of reducing equivalents in mitochondria or by decreased electron transfer capability on (or after) these carriers.[5,6] In this case, increased glycolytic flux in cells can stimulate the production of reducing equivalents, leading itself to an enhancement of ROS production.

Randle et al.[8] first suggested that fatty acids compete with glucose for substrate oxidation in isolated cardiac muscles and that increased oxidation of fatty acids would cause an increase in the intramitochondrial NADH/NAD^+ ratio. In this way, increased free fatty acids (FFA) may also lead to increased ROS production. Indeed, incubation with increased concentrations of glucose or FFA initiates the formation of ROS in muscle, adipocytes, pancreatic β cells, and other cells.[2,9–11]

The ETC is coupled to ATP synthesis through an electrochemical gradient, which includes mitochondrial membrane potential ($\Delta\psi_m$). The increased $\Delta\psi_m$ increases ATP production, decreases electron transport capability leading to a reduced state of the carriers, and dramatically increases ROS production[5,6,12] Since $\Delta\psi_m$ is used to make ATP from ADP and P_i, driven by proton movement back through the ATP synthase complex, its value depends on the ATP production rate, and, in particular, on free ADP concentration. The mitochondrial oxidative phosphorylation rate increases with increased free ADP concentration.[13] Therefore, a decrease in free-ADP concentration leads to decreased ATP production, that in turn can increase $\Delta\psi_m$ and, correspondingly, ROS production. This explains the sharp increase in ROS production with decreased ADP concentration (see Ref. 12 and Fig. 2 from Ref. 14). Both active and passive transport of cations or anions across the mitochondrial inner membrane will also influence $\Delta\psi_m$. Uncoupling agents (for example, UCPs) degrade the proton gradient across the mitochondrial inner membrane and decrease $\Delta\psi_m$, causing decreased ATP secretion, increased ADP concentration, and diminished ROS production rates (FIG. 1).[6,12,13]

Many antioxidant systems exist within cells to neutralize ROS (FIG. 1). Superoxide anions are enzymatically converted to hydrogen peroxide by a manganese-superoxide dismutase (MnSOD) within mitochondria. Hydrogen peroxide is then disposed of by the mitochondrial enzyme glutathione (GSH) peroxidase. The inner mitochondrial membrane contains vitamin E, a powerful antioxidant. The intermembrane mitochondrial space contains the superoxide dismutase isozyme and cytochrome c, which also plays a role in control of ROS concentration and defense. Catalase, on the other hand, is the major hydrogen peroxide detoxifying enzyme found exclusively in peroxisomes.[5–7]

Even in a normally functioning cell, ROS can evade these defense mechanisms, resulting in a slow accumulation of chronic damage. Modified lipids and proteins may be removed via normal turnover of the molecules. However, damage to DNA must be repaired. Both the mitochondrion and nucleus have a variety of DNA repair enzymes to correct chemical modifications to DNA resulting from oxidative damage.[5,16]

Since reactive species are produced mainly in mitochondria, these organelles are a primary target for reactive species–mediated reactions, leading to altered mitochondrial function and their disruption.[5–7] Disrupted and damaged mitochondria need to be removed and replaced, but turnover of mitochondria remains a field with numerous unanswered questions. The process of eliminating mitochondria is termed autophagy, but many aspects of this process are obscure.[6,16] Mitochondrial biogenesis is regulated by some transcriptional activators and co-activators as well as by hormones acting upstream of the transcriptional activators.[17]

Although ROS can evade cellular defense mechanisms, resulting in background levels of damage even in a normally functioning cell, oxidative stress–induced damage most likely occurs when the endogenous antioxidant network and repair systems are overwhelmed. Most authorities have implicated this aspect in the initiation of oxidative stress–related diseases.[1,3,5,6] However, the pathogenic effects that will take place depend on the kind of cells undergoing damage and the specific mechanisms of cell injury that exist for that cell type.

Examples of such oxidative stress–related diseases include (1) reperfusion injury in myocardial muscle cells, such as that which occurs after acute myocardial infarction or stroke (for detailed reviews, see Refs. 3 and 18); (2) neurodegenerative disease such as Alzheimer's disease or amyotrophic lateral sclerosis, which can specifically be linked to a mutation in the gene-coding copper-zinc superoxide dismutase (CuZnSOD);[3,19] and (3) atherosclerosis and hypertension connected with oxidative stress–related processes in endothelial cells.[3,20]

However, despite significant advances in molecular biology and genetics, the molecular mechanisms that underlie the connection between oxidative stress–related processes and diseases remain elusive. To illustrate these processes. we focus here on diabetes-associated problems connected with oxidative stress in pancreatic β cells, myocytes, and endothelial cells because they are likely to share the responsibility for the onset of type 2 diabetes mellitus (T2DM) and some of the complications seen in both type 1 diabetes mellitus (T1DM) and T2DM.

DAMAGING EFFECTS OF BOTH ROS AND ANTIOXIDANT DEFENSES IN DIABETES-ASSOCIATED CELL INJURY

Type 2 diabetes mellitus (T2DM) is a common metabolic disease characterized by elevation of the blood glucose concentration and lipid abnormalities. Insulin resistance (IRe), pancreatic β cell insufficiency in insulin production, and complications such as microvascular pathology are major features in the natural history of T2DM.[2,21,22]

Chronic exposure to elevated glucose concentrations can cause damage in different types of cells by mechanisms involving oxidative stress[1,22,23] that can play a key role in the pathogenesis of late diabetic complications.[1,2] Oxidative stress also plays an important role in development of β cell dysfunction and IRe because antioxidants, scavengers, and overexpression of antioxidant enzymes in transgenic mice have been reported to decrease their manifestations.[1,9,23–25] However, despite significant advances in molecular biology and genetics, the molecular mechanisms that underlie the onset of T2DM and the development of its complications remain elusive.

Insulin resistance seems to precede and predict the development of T2DM even under the normoglycemic conditions.[22] For example, substantial IRe also can be found in lean, normoglycemic, offspring of parents with T2DM, who have a high likelihood of developing diabetes later in life.[26] Recent studies suggest that insulin-stimulated muscle glycogen synthesis is the major metabolic pathway for disposing excess glucose in healthy adults after a meal.[22,27] Increased plasma concentration of FFA leads to intramyocellular lipid accumulation in humans, and this has also been proposed to play a critical role in the genesis of IRe and T2DM.[28] This raises questions regarding the mechanisms—What underlies the early prediabetic Ire? What is the connection between oxidative stress and the effects of lipid in the development of IRe?

Elevated glucose concentrations are also thought to alter metabolism, create oxidative stress, induce β cell damage, and/or apoptosis in the development of T2DM[1,23,24] Interestingly, compared to many other cell types, the β cell may be at high risk for oxidative damage with an increased sensitivity for apoptosis.[1,14] Furthermore, β cell function must be abnormal even before hyperglycemia develops.[22] This raises further questions concerning the particular danger of generation of reactive species in β cells versus other cell types and the mechanisms of the early prediabetic β cell dysfunction.

We will briefly discuss the hypothesis that damage induced by ROS and/or the failure of antioxidant defense, repair, and biogenesis in insulin-secreting and insulin target cells can cause the onset of T2DM (see also Refs. 4 and 14) and its complications.[2] From this hypothesis several specific therapeutic strategies can be suggested.

INSULIN RESISTANCE AS A PHYSIOLOGICAL MECHANISM FOR PROTECTION AGAINST OXIDATIVE STRESS

Although mammalian cells possess numerous antioxidant protection mechanisms, the question as to how mammalian cells can prevent ROS formation, as opposed to accelerated scavenging or other disposal systems, is unresolved.[29] However, one mechanism that may reduce formation of ROS in mammalian cells may be IRe.

Indeed, myocytes and adipocytes are paradigmatic cells for insulin-dependent glucose uptake (IDGU). Normally, insulin binding to its receptor results in receptor autophosphorylation on tyrosine residues accompanied by the tyrosine phosphorylation of insulin receptor substrates (IRSs) by the insulin receptor tyrosine kinase. This allows association of IRSs with the regulatory subunit of phosphoinositide 3-kinase (PI3K). Eventually, downstream events activate PI3K resulting in the translocation of GLUT-4–containing vesicles from their intracellular pool to the plasma membrane where they allow uptake of glucose into the cell[27,30] (FIG. 2).

Since glucose flux through glycolysis is stimulated following insulin receptor activation, this leads to increased ROS production when the electron supply to the respiratory chain is increased by glucose metabolism. Consequently, in these cells, the IDGU mechanism can cause increased ROS production following an insulin challenge in the presence of sufficient concentrations of glucose. On the other hand, IRe should lead to an inhibition of insulin action and to decreased glucose uptake. The end result is that IRe leads to decreased cellular ROS production. For this reason, IRe could be a physiological mechanism activated at the cellular level in response to activation of ROS production, leading to prevention of oxidative stress.[4]

Experimental studies support the idea that increased ROS can lead to IRe. Direct exposure of cultured L6 muscle cells to H_2O_2 has been reported to cause IRe.[31] An-

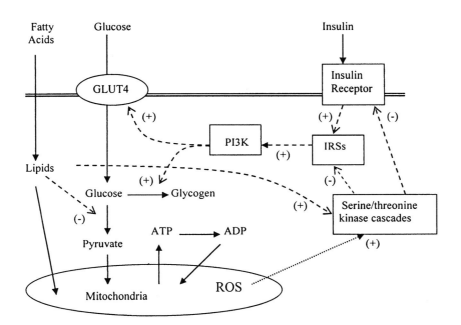

FIGURE 2. Proposed causative link between elevated mitochondrial ROS generation, oxidative stress, mitochondrial dysfunction, and ROS- and lipid-induced insulin resistance. An increase in ROS concentration leads directly to ROS-dependent IRe (see text) and to mitochondrial damage. Mitochondrial dysfunction leads to an increase in intracellular fatty acid metabolites and to lipid-induced IRe. *Solid lines* indicate flux of substrates, and *dashed lines* indicate regulating effects, where (+) represents activation and (−) represents repression.

tioxidants such as *N*-acetylcysteine and taurine prevent hyperglycemia-induced IRe *in vivo* in skeletal muscle of nondiabetic rats.[9] Likewise, an increase in UCP content can decrease IRe in some tissues including skeletal muscle.[32] Because an increase in UCP content can decrease ROS production, this is indirect evidence that ROS can induce IRe.

Muscle cells have a specific molecular mechanism to decrease glucose influx by an IRe-related pathway. *In vitro*, ROS and oxidative stress lead to the activation of multiple serine/threonine kinase signaling cascades (FIG. 2).[33] These activated kinases can act on a number of potential targets in insulin signaling pathway including the insulin receptor and the family of IRS proteins. For IRS-1 and -2, an increase in serine phosphorylation decreases the extent of tyrosine phosphorylation.[33] This reduces the association and/or activities of downstream signaling molecules, resulting in a deactivation of PI3K and to reduced IDGU. This would lead to decreased glucose uptake and to an associated decrease in ROS production. The cellular response to ROS overproduction could therefore involve IRe leading to decreased glucose uptake via deactivation of IDGU.

Increased plasma concentration of FFA and increased intramyocellular lipid content are typically associated with many IRe states, including T2DM.[28,34] Lipid-induced IRe could be partially due to FFA serving as a source of reducing equivalents for the ETC, leading to increased ROS production, because according to the Randle hypothesis, the fatty acids can compete with glucose for substrate oxidation.[8] In turn, increased FFA-induced ROS production may lead to IRe via activation by the proposed cellular ROS-dependent protection mechanism.

It was also proposed that increased delivery of FFA to muscle or a decrease in intracellular metabolism of FFA leads to an increase in intracellular fatty acid metabolites, which activates a serine/threonine kinase cascade leading to phosphorylation of serine/threonine sites on IRS-1 and IRS-2.[34,35] This in turn reduces the ability of the IRSs to activate phosphoinositide kinase (FIG. 2). As a consequence, glucose transport activity and other events downstream of insulin receptor signaling are diminished, leading to lipid-induced IRe, which is analogous to mechanisms protecting against ROS overproduction. For this reason increased FFA metabolites can amplify IRe in a similar fashion to ROS-induced IRe.

We have also already discussed how ROS-induced oxidative stress can lead to mitochondrial damage and then to decreased mitochondrial number that in turn can lead to intracellular lipid accumulation and lipid-induced IRe.[4] Hence, ROS-induced oxidative stress could be a mechanism underlying the decreased intracellular metabolism of FFA and lipid-induced IRe. This could all occur before the appearance of T2DM.

Insulin also stimulates adipocyte glucose uptake, thereby increasing the availability of glycerol-3 phosphate for triglyceride synthesis. The initial actions of insulin are similar to those in muscle, that is, activation of IRSs and downstream effects.[30]

Incubation with increased concentrations of glucose significantly increased the ROS level in mouse adipocytes.[11] Exposure of rat adipocytes in primary culture to high levels of glucose plus insulin also resulted in increased generation of ROS, which was prevented by preincubation with the antioxidant N-acetylcysteine.[36] As in other cell types, exposure of cultured 3T3-L1 adipocytes to H_2O_2 has been reported to cause IRe.[37] These data suggest that similar ROS-dependent mechanisms, as suggested for glucose-dependent IRe in muscle, can act in adipocytes.[37]

OXIDATIVE STRESS AND β CELL DYSFUNCTION

Glucose-stimulated insulin secretion (GSIS) as currently understood is summarized in (FIG. 3) (cf. Refs. 14, 21, and 38). Glucose rapidly equilibrates across the plasma membrane (via GLUT 1 and GLUT2) and is phosphorylated by a specific glucokinase with high K_m for glucose, which determines metabolic flux through glycolysis. This enables β cells to increase glucose metabolism in proportion to elevations in extracellular glucose, underlying the dependence of the β cell insulin secretory response to glucose in the physiological range. Increased glycolytic flux in β cells stimulates a steep increase in the production of reducing equivalents leading to increased ATP production in mitochondria (FIG. 3) and in an enhanced ratio of ATP to ADP in the cytoplasm. Decreased free ADP concentrations, rather than an increase in ATP, serves as the primary signal for glucose-induced block of the ATP-sensitive K^+ channels (K_{ATP}), decreasing the hyperpolarizing outward K^+ flux.[14,38] An inward cation current results in depolarization of the plasma membrane, influx

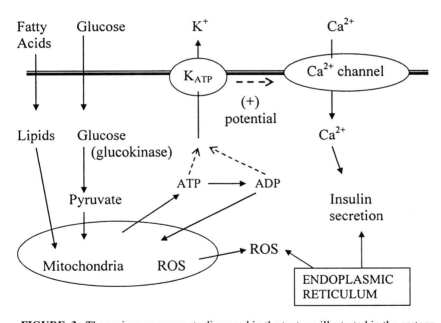

FIGURE 3. The various components discussed in the text are illustrated in the cartoon of a β cell and a mitochondrion within it. Glucose and fatty acid metabolism leads to an increase in ATP production and Ca^{2+} cytoplasmic concentrations, and to a decrease in ADP. Subsequently, closure of K_{ATP} channels depolarizes the cell membrane. This opens voltage-gated Ca^{2+} channels, raising the cytosolic Ca^{2+} concentrations, which triggers insulin exocytosis. Decreased ADP leads also to increased mitochondrial membrane potential ($\Delta\psi_m$) and a corresponding increase in ROS production. Increased insulin biosynthesis also increases ROS production in the endoplasmic reticulum. *Solid lines* indicate flux of substrates, and *dashed lines* indicate regulating effects, where (+) represents activation and (−) represents repression.

of extracellular Ca^{2+}, a sharp increase in intracellular Ca^{2+}, and activation of protein motors and kinases, which then mediate exocytosis of insulin. A considerable increase in glycolytic flux and a sharp decrease in free ADP levels are necessary for closure of K_{ATP} channels and subsequent insulin secretion following glucose challenge (for details, see reviews Refs. 14, 21, and 38).

In this case, in contrast to most other mammalian cell types, increased glucose concentration stimulates a steeply increased glycolytic flux in β cells followed by a steep stimulation in the production of reducing equivalents that can itself lead to an enhancement of ROS production. Indeed, we recently found that stimulation with 10 mM glucose (from an initial 2 mM) increased the $O_2^{\bullet}$ production rate nearly twofold in pancreatic β cells from Zucker lean control rats, confirming the possibility of increased $O_2^{\bullet}$ production with increased glucose.[10]

As was pointed out, after the addition of glucose, free ADP concentration decreases in β cells. This decrease in free [ADP] can also be directly responsible for an overproduction of ROS (see above). This idea was recently confirmed for β cells by the demonstration that ADP inhibited ROS generation in permeabilized MIN6 cells.[39] Decreased ADP concentration induced by glucose is a specific property of β cell stimulus-secretion coupling, possibly shared with other cell types that have a fuel-sensing function.

An elevation of intracellular Ca^{2+} induced by increased Ca^{2+} influx through voltage-gated Ca^{2+} channels is an integral part of the GSIS mechanism. However, further increases in intracellular Ca^{2+} can stimulate mitochondrial generation of ROS and induce apoptosis.[40] Hence, an increase in cytoplasmic Ca^{2+} concentration may share responsibility for increased oxidative stress and mediate apoptosis.

Recently, it was also found that disulfide bond formation during protein synthesis could make a significant contribution to ROS production. This ROS production occurs in endoplasmic reticulum and could be significant, especially in specialized secretory cells.[41] Since biosynthesis of every insulin molecule requires the formation of three disulfide bonds,[42] this mechanism could also be a significant source of ROS in β cells. Therefore, ROS formation should increase with increased insulin biosynthesis.

We can conclude that at least four stages of the GSIS mechanism (increased glycolytic flux, decreased ADP concentration, increased intracellular Ca^{2+} concentration, and increased insulin biosynthesis) contribute to increased ROS production and to the development of oxidative stress in pancreatic β cells. We have termed this connection the GSIS→ROS hypothesis.[14]

To complicate matters further, β cells have relatively low levels of free radical detoxifying and redox-regulating enzymes, such as superoxide dismutase, glutathione peroxidase, catalase, and thioredoxin.[1,43] The limited scavenging systems suggest that ROS concentrations in β cells may increase because of both decreased scavenging systems and ROS overproduction.

Protective effects of antioxidants, scavengers, and overexpression of antioxidant enzymes in transgenic mouse islets also indirectly suggest that ROS overproduction can lead to manifestations of oxidative stress and apoptosis in β cells.[1,23] Human pancreatic islets from T2DM patients also have functional defects and increased apoptosis, again associated with increased oxidative stress compared with controls.[44] All these data emphasize that β cells may be uniquely at high risk for oxidative damage and apoptosis.

PATHOBIOLOGY OF DIABETIC COMPLICATIONS

Hyperglycemia is likely to be a principal initiating cause of diabetic tissue damage that we see clinically as diabetic complications. Several studies and reviews have focused on oxidative stress as the mediator of the effects of hyperglycemia. It has been pointed out that the mechanism that allows an eventual reduction in the transport of glucose inside the cell following exposure to hyperglycemia (i.e., insulin resistance) is absent in cells such as endothelial and mesangial, and neurons.[2,45] (This kind of insulin-dependent mechanism was described above for muscle cells.) In this case, the overproduction of superoxide in mitochondria with increased glucose-dependent production of reducing equivalents can be a process that activates several potentially damaging pathways leading to oxidative stress.[2]

A specific molecular mechanism to explain the pathogenesis of diabetic microvascular disease was suggested, that is, that hyperglycemia-induced mitochondrial superoxide production decreases the activity of the key glycolytic enzyme glyceraldehydes-3 phosphate dehydrogenase, leading to the accumulation of upstream glycolytic intermediates—glucose, fructose-6 phosphate, and glyceraldehyde-3-phosphate.[2] The accumulation of these intermediates in turn activates four major pathways of hyperglycemic damage: the polyol pathway, formation of advanced glycation end products, changes in protein kinase C isoforms, and activation of the hexosamine pathway (FIG. 1).

We also suggest that ROS-induced mitochondrial damage and/or decreased mitochondrial number can also lead to the accumulation of glycolytic intermediates via decreased consumption of pyruvate. This mechanism of accumulation of glycolytic intermediates resembles lipid accumulation in muscle cells during decreased mitochondrial activity, as was considered above, as an explanation of ROS and lipid-induced IRe (see also Ref. 4).

A ROS mechanism for diabetes-associated macrovascular disease was also suggested. In macrovascular, but not in microvascular, disease, endothelial cells increased FFA flux from adipocytes, leading to increased fatty acid oxidation that can then lead to increased ROS production. This can lead to activation of the same damaging pathways as in microvascular endothelial cells, leading to cardiovascular disease.[2]

OVERPRODUCTION OF REACTIVE SPECIES AND/OR A FAILURE IN ANTIOXIDANT DEFENSE, REPAIR, AND BIOGENESIS MECHANISMS AS AN INITIATING FACTOR IN T2DM—AN HYPOTHESIS

Previously, we offered a hypothesis as to how ROS and lipid factors can lead to mitochondrial dysfunction and intracellular IRe in the early stages of T2DM when oxidative stress has not yet become manifest in muscle cells and adipocytes. We have suggested that these same factors can lead to β cell dysfunction and/or apoptosis since β cells can be very sensitive to oxidative stress.

However, oxidative stress–induced damage occurs only if the endogenous antioxidant network and repair system fail to provide a sufficient compensatory response. This is the essence of the hypothesis, which suggests that an imbalance in ROS production versus protection and/or defects in repair in β cells, muscle, and adipocytes can result in ROS-induced β cell dysfunction and IRe, even at apparently normal blood glucose and FFA levels. This state of affairs could ultimately lead to T2DM.[4]

After onset of T2DM, a positive feedback loop, or "vicious cycle," is engendered by hyperglycemia. This increased glucose concentration in turn causes an increased production of ROS and, therefore, increased oxidative damage, which in turn disrupts the ability of β cells to respond to elevated blood glucose and increases IRe, leading to further hyperglycemia.[46]

Several clinical studies show increased levels of oxidative stress markers in diabetic patients, when compared to healthy age-matched subjects (for details, see Refs. 1, 16, 25, and 46). Several studies have also demonstrated increased mtDNA mutations in skeletal muscle[47] and increased free radical damage to mitochondrial DNA for other tissues[2,16] in T2DM, both consistent with ROS-induced damage.

More direct evidence for our hypothesis can be found in studies examining prediabetic conditions. For example, Lee et al.[48] reported that a decrease in mtDNA content in peripheral blood leukocytes preceded development of T2DM. Expression of genes involved in oxidative phosphorylation was reduced in skeletal muscle of prediabetic (and diabetic) humans,[49] a possible consequence of decreased functional mitochondrial mass. However, further experiments are required in prediabetic conditions.

THERAPEUTIC IMPLICATIONS

The importance of ROS and oxidative stress in the etiology of T2DM suggests some novel interpretations of the impact of pharmaceutical agents used in diseases such as diabetes which may also help to prevent onset and complications. Obviously, reversal of the imbalance between ROS, antioxidant capacity, and repair and biogenesis mechanisms should improve IRe and β cell function, reduce progression to T2DM in high-risk individuals, and decrease diabetic complications. Decrements in glucose and/or lipid content in blood can lead to decreased ROS production in cells and/or to decreased lipid intracellular accumulation, an indirect approach to improving ROS imbalance. Several pharmaceutical agents that decrease glucose and lipid concentrations in blood, such as metformin, thiazolidinediones, and statins, are in this category.[50,51]

Direct actions that may have therapeutic potential in the treatment of diabetes, its complications, (and those of other ROS-associated diseases), include strategies to decrease mitochondrial radical production and/or to increase antioxidant capacity to avoid cell injury. This could be achieved, for example, by the use of antioxidants or by decreasing the mitochondrial membrane potential.[6,7] In animal models of diabetes, antioxidants, especially lipoic acid, decrease some of the manifestations of IRe[25,52] and can preserve β cell function and mass.[1,23,24] The protective effects of antioxidants on induced IRe could relate to their ability to neutralize ROS in muscle[53] and in β cells.[23,24] However, despite evidence on the damaging consequences of oxidative stress and its role in experimental diabetes, large-scale clinical trials with classic antioxidants failed to demonstrate any benefit for diabetic patients.[1,25,54]

One possible explanation for such spectacular failures is that conventional antioxidants temporarily neutralize reactive oxygen molecules on a one-for-one molar basis, while production of reactive species is a continuous process and results in accumulation of oxidative damage. Therefore, cellular damage might not be prevented or repaired by the late addition of antioxidants.[2,55] At the present time, α-lipoic acid, which has an indirect antioxidant action, is only considered likely to be safe, and while controversial it may be effective in some aspects of treatment of T2DM.[1,25,54] For this reason, the development of special antioxidants targeted to mitochondria and/or a catalytic antioxidant, such as a SOD/catalase mimetic, which works continuously like the enzymes, could be required.[7]

Thiazolidinediones and metformin are now the most highly used agents for treatment of T2DM, although the detailed mechanisms of their action remain under investigation. Some of their beneficial effect may be due in part to the antioxidative effect of these drugs because these agents may decrease oxidative stress in pancreatic islets from T2DM diabetic patients[44] and from obese diabetic db/db mice.[56] Indeed, both the thiazolidinediones and metformin, can inhibit respiratory complex I, reducing cell respiration.[57,58] Metformin could also directly scavenge ROS.[59] If their action can be explained by a decrease of electron supply to the ETC by an inhibition of respiratory complex I, then it should lead to decreased ROS production.

Another mechanism is also possible. In light of the preceding discussion, mitochondrial impairment is likely to play an important role in the development of ROS-induced T2DM. Accordingly, treatments that increase mitochondrial biogenesis could be extremely effective. Although the mechanisms of mitochondrial biogenesis remain unclear, several approaches could be pursued. For example, thiazolidinediones target IRe in skeletal muscle and adipocytes and function as selective and potent agonists of the transcription factor PPAR-γ,[27,50] which can induce mitochondrial biogenesis.[60] In one experiment, troglitazone prevented the occurrence of abnormal mitochondria in the ZDF rat model.[61] Therefore, we suggest that thiazolidinediones can decrease IRe in muscle, adipocytes, and possibly repair β cells in part by activating mitochondrial repair or biogenesis. However, a great deal is still to be learned about mitochondrial biogenesis and its regulation.

CONCLUSION

There is emerging evidence that the generation of reactive species, and particularly ROS, is a major factor in the onset and the development of T2DM and other diseases. Oxidative stress clearly plays a role in this pathway. However, the details of the pathogenic mechanisms remain unclear in most such diseases.

ROS can induce inactivation of the signaling pathway between the insulin receptor and the glucose transporter system leading to the onset of IRe in T2DM. In addition, ROS-induced mitochondrial dysfunction can lead to increased intracellular lipid content leading to lipid-dependent IRe. Comparing metabolic pathways of GSIS and ROS production suggests that secretagogues causing increased insulin secretion by the GSIS mechanism can also lead to increased ROS production. This should lead to activation of oxidative stress concomitantly with stimulation of the GSIS mechanism.

Thus, the essence of our specific hypothesis for the onset of T2DM is that an imbalance in ROS production versus protection and/or defects in repair in β cells, muscle, and adipocytes can result in ROS-induced β cell dysfunction and IRe even at apparently normal blood glucose and FFA levels. A corollary of this hypothesis is that appropriate therapeutic strategies should be directed toward reducing IRe and improving β cell function without an increase in ROS production or concentration. Such considerations also will be critical in therapeutic strategies for other diseases resulting from oxidative stress–induced cell injury.

ACKNOWLEDGMENTS

This work was partially supported by grants from the National Institutes of Health (DK44840, DK20595, and DK48494).

REFERENCES

1. EVANS, J.L., I.D. GOLDFINE, B.A. MADDUX & G.M. GRODSKY. 2002. Oxidative stress and stress-activated signaling pathways: a unifying hypothesis of type 2 diabetes. Endocr. Rev. **23:** 599–622.
2. BROWNLEE, M. 2005. The pathobiology of diabetic complications: a unifying mechanism. Diabetes **54:** 1615–1625.
3. DROGE, W. 2002. Free radicals in the physiological control of cell function. Physiol. Rev. **82:** 47–95.
4. FRIDLYAND, L.E. & L.H. PHILIPSON. 2005. Reactive species and early manifestation of insulin resistance in type 2 diabetes. Diabetes, Obesity and Metabolism. doi:10.1111/j.1463-1326.2005.00496.x
5. TURRENS, J.F. 2003. Mitochondrial formation of reactive oxygen species. J. Physiol. **552:** 335–344.
6. SKULACHEV, V.P. 1999. Mitochondrial physiology and pathology; concepts of programmed death of organelles, cells and organisms. Mol. Aspects Med. **20:** 139–184.
7. GREEN, K., M.D. BRAND & M.P. MURPHY. 2004. Prevention of mitochondrial oxidative damage as a therapeutic strategy in diabetes. Diabetes **53:** S110–118.
8. RANDLE, P.J., P.B. GARLAND, C.N. HALES & E.A. NEWSHOLME. 1963. The glucose fatty acid cycle: its role in insulin sensitivity and the metabolic disturbances of diabetes mellitus. Lancet **1:** 785–789.
9. HABER, C.A., T.K. LAM, Z. YU, et al. 2003. N-acetylcysteine and taurine prevent hyperglycemia-induced insulin resistance in vivo: possible role of oxidative stress. Am. J. Physiol. Endocrinol. Metab. **285:** E744–753.
10. BINDOKAS, V.P., A. KUZNETSOV, S. SREENAN, et al. 2003. Visualizing superoxide production in normal and diabetic rat islets of Langerhans. J. Biol. Chem. **278:** 9796–9801.
11. TALIOR, I., M. YARKONI, N. BASHAN & H. ELDAR-FINKELMAN. 2003. Increased glucose uptake promotes oxidative stress and PKC-delta activation in adipocytes of obese, insulin-resistant mice. Am. J. Physiol. Endocrinol. Metab. **285:** E295–E302.
12. KORSHUNOV, S.S., V.P. SKULACHEV & A.A. STARKOV. 1997. High protonic potential actuates a mechanism of production of reactive oxygen species in mitochondria. FEBS Lett. **416:** 15–18.
13. JENESON, J.A., R.W. WISEMAN, H.V. WESTERHOFF & M.J. KUSHMERICK. 1996. The signal transduction function for oxidative phosphorylation is at least second order in ADP. J. Biol. Chem. **271:** 27995–27998.
14. FRIDLYAND, L.E. & L.H. PHILIPSON. 2004. Does the glucose-dependent insulin secretion mechanism itself cause oxidative stress in pancreatic beta-cells. Diabetes **53:** 1942–1948.

15. SALEH, M.C., M.B. WHEELER & C.B. CHAN. 2002. Uncoupling protein-2: evidence for its function as a metabolic regulator. Diabetologia **45:** 174–187.
16. EVANS, M.D., M. DIZDAROGLU & M.S. COOKE. 2004. Oxidative DNA damage and disease: induction, repair and significance. Mutat. Res. **567:** 1–61.
17. GOFFART, S. & R.J. WIESNER. 2003. Regulation and co-ordination of nuclear gene expression during mitochondrial biogenesis. Exp. Physiol. **88:** 33–40.
18. OTANI, H. 2004. Reactive oxygen species as mediators of signal transduction in ischemic preconditioning. Antioxid. Redox Signal. **6:** 449–469.
19. VALENTINE, J.S., P.A. DOUCETTE & S.Z. POTTER. 2005. Copper-zinc superoxide dismutase and amyotrophic lateral sclerosis. Annu. Rev. Biochem. **74:** 563–593.
20. LANE DUVALL, W. 2005. Endothelial dysfunction and antioxidants. Mt. Sinai J. Med. **72:** 71–80.
21. BELL, G.I. & K.S. POLONSKY. 2001. Diabetes mellitus and genetically programmed defects in beta-cell function. Nature **414:** 788–791.
22. KAHN, S.E. 2003. The relative contributions of insulin resistance and beta-cell dysfunction to the pathophysiology of type 2 diabetes. Diabetologia **46:** 3–19.
23. KAJIMOTO, Y. & H. KANETO. 2004. Role of oxidative stress in pancreatic beta-cell dysfunction. Ann. N.Y. Acad. Sci. **1011:** 168–176.
24. ROBERTSON, R.P. 2004. Chronic oxidative stress as a central mechanism for glucose toxicity in pancreatic islet beta cells in diabetes. J. Biol. Chem. **279:** 42351–42354.
25. ROSEN, P., P.P. NAWROTH, G. KING, et al. 2001. The role of oxidative stress in the onset and progression of diabetes and its complications: a summary of a Congress Series sponsored by UNESCO-MCBN, the American Diabetes Association and the German Diabetes Society. Diabetes Metab. Res. Rev. **17:** 189–212.
26. PETERSEN, K.F., S. DUFOUR, D. BEFROY, et al. 2004. Impaired mitochondrial activity in the insulin-resistant offspring of patients with type 2 diabetes. N. Engl. J. Med. **350:** 664–671.
27. PETERSEN, K.F. & G.I. SHULMAN. 2002. Pathogenesis of skeletal muscle insulin resistance in type 2 diabetes mellitus. Am. J. Cardiol. **90:** 11G–18G.
28. MCGARRY, J.D. 2002. Banting lecture 2001: dysregulation of fatty acid metabolism in the etiology of type 2 diabetes. Diabetes **51:** 7–18.
29. SKULACHEV, V.P. 1997. Membrane-linked systems preventing superoxide formation. Biosci. Rep. **17:** 347–366.
30. WATSON, R.T., M. KANZAKI & J.E. PESSIN. 2004. Regulated membrane trafficking of the insulin-responsive glucose transporter 4 in adipocytes. Endocr. Rev. **25:** 177–204.
31. BLAIR, A.S., E. HAJDUCH, G.J. LITHERLAND & H.S. HUNDAL. 1999. Regulation of glucose transport and glycogen synthesis in L6 muscle cells during oxidative stress: evidence for cross-talk between the insulin and SAPK2/p38 mitogen-activated protein kinase signaling pathways. J. Biol. Chem. **274:** 36293–32699.
32. HAN, D.H., L.A. NOLTE, J.S. JU, et al. 2004. UCP-mediated energy depletion in skeletal muscle increases glucose transport despite lipid accumulation and mitochondrial dysfunction. Am. J. Physiol. Endocrinol. Metab. **286:** E347–353.
33. EVANS, J.L., I.D. GOLDFINE, B.A. MADDUX & G.M. GRODSKY. 2003. Are oxidative stress-activated signaling pathways mediators of insulin resistance and beta-cell dysfunction? Diabetes **52:** 1–8.
34. BODEN, G. & G.I. SHULMAN. 2002. Free fatty acids in obesity and type 2 diabetes: defining their role in the development of insulin resistance and beta-cell dysfunction. Eur. J. Clin. Invest. **32:** 14–23.
35. YU, C., Y. CHEN, G.W. CLINE, et al. 2002. Mechanism by which fatty acids inhibit insulin activation of insulin receptor substrate-1 (IRS-1)-associated phosphatidylinositol 3-kinase activity in muscle. J. Biol. Chem. **277:** 50230–50236.
36. LU, B., D. ENNIS, R. LAI, et al. 2001. Enhanced sensitivity of insulin-resistant adipocytes to vanadate is associated with oxidative stress and decreased reduction of vanadate (+5) to vanadyl (+4). J. Biol. Chem. **276:** 35589–35598.
37. TIROSH, A., R. POTASHNIK, N. BASHAN & A. RUDICH. 1999. Oxidative stress disrupts insulin-induced cellular redistribution of insulin receptor substrate-1 and phosphatidylinositol 3-kinase in 3T3-L1 adipocytes. A putative cellular mechanism for

impaired protein kinase B activation and GLUT4 translocation. J. Biol. Chem. **274:** 10595–10602.
38. RUTTER, G.A. 2001. Nutrient-secretion coupling in the pancreatic islet beta-cell: recent advances. Mol. Aspects Med. **22:** 247–284.
39. KOSHKIN, V., X. WANG, P.E. SCHERER, et al. 2003. Mitochondrial functional state in clonal pancreatic beta-cells exposed to free fatty acids. J. Biol. Chem. **278:** 19709–19715.
40. BROOKES, P.S., Y. YOON & J.L. ROBOTHAM. 2004. Calcium, ATP, and ROS: a mitochondrial love-hate triangle. Am. J. Physiol. Cell Physiol. **287:** C817–833.
41. TU, B.P. & J.S. WEISSMAN. 2004. Oxidative protein folding in eukaryotes: mechanisms and consequences. J. Cell Biol. **164:** 341–346.
42. DEREWENDA, U. & G.G. DODSON. 1993. The structure and sequence of insulin. *In* Molecular Structures in Biology. R. Diamond, T.F. Koetzle, K. Prout & J.S. Richardson, Eds.: 217–230. Oxford University Press. Oxford, UK.
43. TIEDGE, M., S. LORTZ, J. DRINKGERN & S. LENZEN. 1997. Relation between antioxidant enzyme gene expression and antioxidative defense status of insulin-producing cells. Diabetes **46:** 1733–1742.
44. MARCHETTI, P., S. DEL GUERRA, L. MARSELLI, et al. 2004. Pancreatic islets from type 2 diabetic patients have functional defects and increased apoptosis that are ameliorated by metformin. J. Clin. Endocrinol. Metab. **89:** 5535–5541.
45. BAYNES, J.W. & S.R. THORPE. 1999. Role of oxidative stress in diabetic complications: a new perspective on an old paradigm. Diabetes **48:** 1–9.
46. WEST, I.C. 2000. Radicals and oxidative stress in diabetes. Diabetic Med. **17:** 171–180.
47. LIANG, P., V. HUGHES & N.K. FUKAGWA. 1997. Increased prevalence of mitochondrial DNA deletions in skeletal muscle of older individuals with impaired glucose tolerance. Diabetes **46:** 920–923.
48. LEE, H.K., J.H. SONG, C.S. SHIN, et al. 1998. Decreased mitochondrial DNA content in peripheral blood precedes the development of non-insulin-dependent diabetes mellitus. Diabetes Res. Clin. Pract. **42:** 161–167.
49. PATTI, M.E., A.J. BUTTE, S. CRUNKHORN, et al. 2003. Coordinated reduction of genes of oxidative metabolism in humans with insulin resistance and diabetes: potential role of PGC1 and NRF1. Proc. Natl. Acad. Sci. USA **100:** 8466–8471.
50. HENRY, R.R. 2003. Insulin resistance: from predisposing factor to therapeutic target in type 2 diabetes. Clin. Ther. **25:** B47–B63.
51. LEBOVITZ, H.E. & M.A. BANERJI. 2004. Treatment of insulin resistance in diabetes mellitus. Eur. J. Pharmacol. **490:** 135–146.
52. MIDAOUI, A.E., A. ELIMADI, L. WU, et al. 2003. Lipoic acid prevents hypertension, hyperglycemia, and the increase in heart mitochondrial superoxide production. Am. J. Hypertens. **16:** 173–179.
53. MADDUX, B.A., S. SEE, J.C. LAWRENCE, JR., et al. 2001. Protection against oxidative stress-induced insulin resistance in rat L6 muscle cells by micromolar concentrations of α-lipoic acid. Diabetes **50:** 404–410.
54. JOHANSEN, J.S., A.K. HARRIS, D.J. RYCHLY & A. ERGUL. 2005. Oxidative stress and the use of antioxidants in diabetes: linking basic science to clinical practice. Cardiovasc. Diabetol. **4:** 5.
55. KOHEN, R. & A. NYSKA. 2002. Oxidation of biological systems: oxidative stress phenomena, antioxidants, redox reactions, and methods for their quantification. Toxicol. Pathol. **30:** 620–650.
56. ISHIDA, H., M. TAKIZAWA, S. OZAWA, et al. 2004. Pioglitazone improves insulin secretory capacity and prevents the loss of beta-cell mass in obese diabetic db/db mice: possible protection of beta cells from oxidative stress. Metabolism **53:** 488–494.
57. OWEN, M.R., E. DORAN & A.P. HALESTRAP. 2000. Evidence that metformin exerts its anti-diabetic effects through inhibition of complex 1 of the mitochondrial respiratory chain. Biochem. J. **348:** 607–614.
58. BRUNMAIR, B., K. STANIEK. F. GRAS, et al. 2004. Thiazolidinediones, like metformin, inhibit respiratory complex I: a common mechanism contributing to their anti-diabetic actions? Diabetes **53:** 1052–1059.

59. BONNEFONT-ROUSSELOT, D., B. RAJI, S. WALRAND, *et al.* 2003. An intracellular modulation of free radical production could contribute to the beneficial effects of metformin towards oxidative stress. Metabolism **52:** 586–589.
60. WILSON-FRITCH, L., A. BURKAR, G. BELL, *et al.* 2003. Mitochondrial biogenesis and remodeling during adipogenesis and in response to the insulin sensitizer rosiglitazone. Mol. Cell. Biol. **23:** 1085–1094.
61. HIGA, M., Y.T. ZHOU, M. RAVAZZOLA, *et al.* 1999. Troglitazone prevents mitochondrial alterations, beta cell destruction, and diabetes in obese prediabetic rats. Proc. Natl. Acad. Sci. USA **96:** 11513–11518.

The Mechanisms of Cell Membrane Repair
A Tutorial Guide to Key Experiments

RICHARD A. STEINHARDT

Department of Molecular Cell Biology, University of California, Berkeley, Berkeley, California 94720-3200, USA

ABSTRACT: The best way to approach a new area is to study closely a sample of the key papers, and spread out from there. In this tutorial paper I present my personal selection of papers introducing concepts in the study of the mechanisms of cell membrane repair. For a more comprehensive review up to 2003, I refer the student to McNeil and Steinhardt (2003).

KEYWORDS: cell membrane repair; exocytosis; sea urchin egg; 3T3 fibroblasts; membrane vesicles; cell membrane puncture

BACKGROUND

As recently as 1993, a summary of the field of membrane disruption and repair had only one short paragraph on postulated mechanisms of plasma membrane repair, and no references that addressed how the bilayer might reseal.[1] It was known that cells could recover from serious breaks in the cell membranes, whether artificially or naturally occurring. For very large disruptions, roles for the cytoskeleton were proposed in plugging gaps or alternately forming a contractile ring and helping close the wound. For a comprehensive review up to 2003, see McNeil and Steinhardt.[2]

The basic underlying assumption found in every textbook was that the cell membrane would spontaneously reseal if broken because lipid bilayers naturally sought the lowest energy configuration where the hydrophobic portions of the bilayer faced each other.[3]

In spite of this universal assumption, a few investigations had been undertaken, not so much as to challenge the assumption, but to add additional detail. Two basic ideas were that the cytoskeleton might be used to bring ruptured cell membranes together and that proteases might play a role in removing structures that might impair membrane–membrane contact.[4] One thing emerged clearly from these studies was that calcium ions were required and that magnesium ions inhibited cell membrane repair, observations made much earlier by De Mello.[5]

For artificial liposomes it was indeed true that calcium would facilitate fusion. This was explained as the action of divalents in screening the charge at the hydrophilic surfaces of these phospholipid vesicles.[6] However, for liposome fusion, mag-

---

Address for correspondence: Richard A. Steinhardt, Department of Molecular Cell Biology, University of California, Berkeley, CA 94720-3200. Voice: 510-520-1073.
ricksteinhardt@berkeley.edu

nesium was almost as good as calcium.[7,8] But to the contrary, in real nucleated cells, magnesium *is* a strong antagonist of membrane fusion. Further, possession of even a small degree of biological intuition would lead one to suspect that calcium would always have specific molecular targets and would never be acting simply as a divalent cation.

THE ADVENT OF THE MODERN THEORY: EXOCYTOTIC CELL MEMBRANE REPAIR

To systematically examine possible targets for calcium, it is necessary to be able to observe the rate of the cell membrane resealing process. Starting in 1992, studies of this type were undertaken in our laboratory that led to a new hypothesis for the mechanism of cell membrane repair.[9] We observed the rate of cell membrane repair by quantitatively following the loss of intracellular dye after micropuncture.

FIGURE 1 illustrates how the cessation of dye loss after micropuncture was used to determine the time to complete the resealing of the cell membrane. One can clearly see the rise in intracellular Ca^{2+} following the puncture, and that when the membrane resealed, dye loss ceased. Intracellular Ca^{2+} remains somewhat higher for a variable period after micropuncture and was never by itself used to determine the time to complete the resealing process. Intracellular Ca^{2+} levels were only used to confirm a successful wound or to confirm a failure to complete resealing and eventual cell death.

A second wound reseals faster than a first. The calcium influx at a first wound invokes pathways that speed up the healing process for a second wound. These are discussed later. Once repair could be monitored, a systematic search for likely calcium targets was undertaken.

Three model systems, the unfertilized sea urchin egg, the activated sea urchin embryo, and cultured 3T3 fibroblasts, were used to quantify the requirement for calcium and the antagonistic action of magnesium. The first two model systems, the unfertilized sea urchin egg and the activated sea urchin embryo, offer a special advantageous contrast. The unfertilized egg has thousands of docked secretory vesicles (the cortical granules), while the activated embryo has very few, since they have been discharged during activation by an intracellular calcium rise.[11,12]

The threshold value of calcium required for resealing depends on whether there are docked vesicles already in place and the threshold is sharply raised if magnesium is present. In the absence of magnesium and with docked vesicles, the calcium threshold is on the order of 0.3 mM. Without docked vesicles the threshold is 1.3 mm. 3T3 fibroblasts in the absence of magnesium have a threshold of 0.3–0.4 mM calcium, similar to that of sea urchin eggs with docked vesicles.

The large effect of docked vesicles on the calcium threshold suggested to us that delivery and docking of vesicles for exocytosis was involved in resealing, and that the low threshold for unactivated eggs could be explained by the successful exocytosis of pre-docked vesicles triggered by calcium influx at a wound site. This was confirmed by direct microscopic observation of cortical granule discharge at the threshold concentration of calcium for membrane resealing.[9]

For the embryo, which lacks docked secretory vesicles, it was reasonable to suspect a role in resealing for CaM kinase (Ca^{2+}-calmodulin-dependent kinase). In

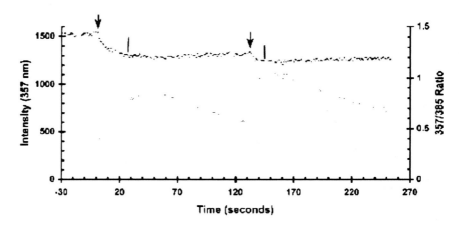

FIGURE 1. 3T3 cells exhibit facilitated membrane resealing. A typical record of a cell micropunctured by a 0.3-s jab of a microneedle (*first arrow*). The membrane wound causes a loss in fura-2 dye, detected as a drop in intensity at the Ca^{2+}-insensitive 357-nm excitation (*top trace*), and a rise in intracellular Ca^{2+} activity, detected as an increase in the emission ratio of intracellular fura-2 dye excited at 357 and 385 nm (*bottom trace*). Resealing of the membrane causes a cessation of dye loss (*bar*). In this example, the duration of dye loss after the initial wound was 27 s. The cell wounded a second time (*second arrow*) at the same site required only 16 s to reseal. The faster resealing to a second wounding at the same or different site is called a facilitated or potentiated response, respectively. (Modified from Shen and Steinhardt.[10])

neurons CaM kinase is postulated to be involved in freeing vesicles from the actin cytoskeleton so that they can be delivered for docking at sites for exocytosis.[13,14]

A specific auto-inhibitory peptide to CaM kinase, which had already shown to be effective in sea urchins, knocked out resealing in activated sea urchin embryos, and had no effect on unfertilized sea urchin eggs with already docked vesicles. The control peptide had no effect. A similar result was obtained with 3T3 fibroblasts with roughly three-quarters of cells failing to reseal when CaM kinase was inhibited. The implication was, if vesicles were not yet docked, CaM kinase was needed, possibly to free them for delivery to the docking sites for exocytosis.

Could we see other evidence that delivery of vesicles might be required? Was kinesin function essential to membrane repair? Fortunately two antibodies to conventional kinesin had been characterized in sea urchins, one that blocked function (SUK-4) and another that bound but did not block function (SUK-2). SUK-4 blocked resealing in activated embryos that lacked pre-docked vesicles, and blocked resealing in 80% of 3T3 fibroblasts. SUK-2, which does not inhibit the transport function of kinesin, had no effect. The findings of inhibition of resealing in the majority of fibroblasts by CaM kinase and kinesin reagents implied that additional delivery and docking of vesicles was usually required for successful resealing of a micropuncture in cultured cells.

The implication that exocytosis was essential for membrane repair was supported by microinjection of neurotoxins from *Clostridium botulinum*. The botulinum neurotoxins (BNTs) are proteases that inhibit exocytosis by cleaving SNARE proteins, which are required for exocytosis.[15] BNT-B cleaves synaptobrevin, and BNT-A cleaves SNAP-25.

In embryos, which lack the docked vesicles of the unfertilized egg, exocytosis and membrane resealing were inhibited by microinjection of BNT-B or BNT-A. The action of BNT-B was shown to be specific by delaying its effect with co-injection of excess cleavage site peptide. Similar neurotoxin inhibitions of membrane repair were observed in fibroblasts.[9,25]

DIRECT CORRELATION OF EXOCYTOTIC EVENTS WITH CELL MEMBRANE REPAIR

Using the inflow of the fluorescent dye FM1-43 into the exocytotic pocket, individual exocytotic events were observed and correlated with membrane resealing in sea urchin eggs and embryos.[16,17] Upon wounding by a laser beam, both eggs and embryos showed a rapid burst of localized Ca^{2+}-regulated exocytosis. The rate of exocytosis was correlated quantitatively with successfully resealing. In embryos, whose activated surfaces must first dock vesicles before fusion, exocytosis and membrane resealing were inhibited by neurotoxins that selectively cleave synaptobrevin, SNAP-25, or syntaxin, all SNARE complex proteins. The specificity of each neurotoxin inhibition was confirmed by the corresponding specificity of delays in the block by co-injection of the corresponding cleavage site peptides. Specificity was also confirmed by later experiments that directly demonstrated that sea urchin eggs contain the SNARE protein targets with matching protease cleavage site sequences that are sensitive to the neurotoxins.[18,19]

The unique ability to directly observe individual exocytotic events, combined with the fact that the unfertilized egg has thousands of pre-docked cortical vesicles that the activated embryo lacks, make the sea urchin egg and embryo ideal preparations for studies of exocytosis and membrane resealing. The cortical granules are the pool of vesicles used for plasma membrane resealing in the unfertilized egg. In an unfertilized egg, each exocytotic figure observed after wounding corresponds exactly to a previously docked cortical granule. The ability to reseal can be correlated with the rate of exocytosis, and therefore with the exocytosis of these identified vesicles. An unresolved question was whether unfertilized eggs used the same exocytotic mechanism to reseal as did embryos, since reagents that blocked resealing in embryos and other cells did not affect unfertilized eggs, whose cortical vesicles are already docked at the plasma membrane. This question could be approached experimentally. A study by others had shown that high osmolarity caused by extracellular 0.8 M stachyose in sea water could undock the cortical granules in unfertilized sea urchin eggs and that these undocked vesicles were unable to exocytose upon calcium ionophore stimulation.[20]

Since this undocking method was fully reversible when the stachyose was diluted, one could compare docking status with resealing ability in unfertilized eggs and also transiently expose the otherwise inaccessible docking/fusion machinery of cortical vesicles to neurotoxins.[16] In eggs, whose cortical granules are already docked, these vesicles could be reversibly undocked with externally applied stachyose. Then the undocked cortical vesicles could be re-docked with dilution of stachyose and were able to exocytose when exposed to calcium.

When cortical vesicles were transiently undocked and exposed to tetanus toxin or to botulinum neurotoxin C1, these vesicles were no longer competent for exocytosis or resealing. Cortical granule secretory vesicles transiently undocked in the presence of tetanus toxin were subsequently fusion-incompetent and could not exocytose even though they retained their ability to visually re-dock when the stachyose was diluted. We concluded that addition of internal membranes by exocytosis of cortical granules was required for resealing, and that a SNARE-like complex played necessary roles in vesicle docking and fusion for the repair of disrupted plasma membrane.

In this example, the exocytosis of a single identified intracellular compartment, the cortical granule secretory vesicle, was proved to be necessary and sufficient to complete membrane repair after wounding.

The identity of the vesicles that arrived and exocytosed near a wound site in an activated embryo surface was unknown, but there too it was possible to show specific inhibition of exocytosis by neurotoxin proteases and to correlate the degree and rate of exocytosis with the ability to successfully reseal a wound.

Miyake and McNeil completed a similar study with 3T3 cells.[17] First, 3T3 endothelial cells were exposed to FM 1-43, and endocytosed this lipophilic plasma membrane dye from the solution. When the FM 1-43 was removed from the outside solution, the remaining dye was trapped in intracellular vesicles. While the fluorescence of the population was continuously monitored in a fluorometer, the labeled cells were wounded by forcing them through a syringe. Exocytosis of the labeled vesicles freed the dye from the membrane, and this free dye was no longer fluorescent. The calcium dependence of both exocytosis and membrane repair was closely correlated. Using electron micrographs, Miyake and McNeil were able to observe that these larger wounds caused a massive fusion of intracellular vesicles during the resealing process. This was an additional step in membrane repair for wounds that removed large areas of surface membrane. These early clues to additional steps for very large membrane disruptions gave rise later to the patch hypothesis (see below).

THE MACHINERY FOR DELIVERY OF VESICLES USED IN REPAIR

Kinesin and myosin had been proposed to transport intracellular organelles and vesicles to the cell periphery in several cell systems. However, there had been little direct observation of the role of these motor proteins in the delivery of vesicles during regulated exocytosis in intact cells. Using a confocal microscope, we triggered local bursts of Ca^{2+}-regulated exocytosis by wounding the cell membrane and visualized the resulting individual exocytotic events in real time. Different temporal phases of the exocytosis burst were distinguished by their sensitivities to reagents targeting different motor proteins. The function blocking kinesin antibody SUK-4 as well as the stalk-tail fragment of kinesin heavy chain specifically inhibited a slow phase, while butanedione monoxime, a myosin ATPase inhibitor, inhibited both the slow and fast phases. The blockage of Ca^{2+}/calmodulin-dependent protein kinase II with its autoinhibitory peptide also inhibited the slow and fast phases, consistent with disruption of a myosin-actin–dependent step of vesicle recruitment. Membrane resealing in wounded cells was also inhibited by these reagents.

These direct observations of individual vesicles provided evidence that in intact living cells, kinesin and myosin motors may mediate two sequential transport steps that recruit vesicles to the release sites of Ca^{2+}-regulated exocytosis, although the identity of the responsible myosin isoform was not yet known. These results demonstrated the existence of three semi-stable vesicular pools along this regulated membrane trafficking pathway, a fast pool of vesicles already docked, a second pool dependent on myosin function, and the late slow pool dependent on kinesin function.[21]

It became possible to identify the myosin candidates for resealing in 3T3 cells by using antisense sequences for specific isoforms of non-muscle myosins. The functions of non-muscle myosin IIA and IIB in this exocytotic process of membrane repair were studied using the antisense technique. Knockdown of myosin IIB suppressed wound-induced exocytosis and the membrane resealing process. Knockdown of myosin IIA did not suppress exocytosis at an initial wound and had no inhibitory effect on the resealing at initial wounds, but did inhibit the facilitated rate of resealing normally found at repeated wounds made at the same site. COS-7 cells, which lack myosin IIA, did not show the facilitated response of membrane resealing to a repeated wound. S91 melanoma cells, a mutant cell line lacking myosin Va, showed normal membrane resealing and normal facilitated responses. We concluded that myosin IIB was required for exocytosis, and therefore cell membrane repair itself, and that myosin IIA was required in facilitation of cell membrane repair at repeated wounds. Myosin IIB was primarily located in the sub-plasmalemma cortex and myosin IIA was concentrated at the trans-Golgi network, consistent with their distinct roles in vesicle trafficking in cell membrane repair.[22]

THE ROLE FOR EXOCYTOSIS: DIRECT OBSERVATIONS OF MEMBRANE TENSION[23]

It was known that the rate of resealing in artificial liposomes was dependent on the applied tension.[24] We hypothesized that the requirement for Ca^{2+}-dependent exocytosis in cell-membrane repair was to provide an adequate lowering of membrane tension to permit membrane resealing. We used laser tweezers to form membrane tethers and measured the force of those tethers to estimate the membrane tension of Swiss 3T3 fibroblasts after membrane disruption and during resealing. These measurements showed that, for fibroblasts wounded in normal Ca^{2+} Ringer's solution, the membrane tension decreased dramatically after the wounding and resealing coincided with a decrease of approximately 60% of control tether force values (FIG. 2).[23]

However, the tension did not decrease if cells were wounded in a low Ca^{2+} Ringer's solution, which inhibited both membrane resealing and FM 1-43–monitored exocytosis. When cells were wounded twice in normal Ca^{2+} Ringer's solution, decreases in tension at the second wound were 2.3 times faster than at the first wound, correlating well with twofold faster resealing rates for repeated wounds. The facilitated resealing to a second wound was dependent on new vesicles, which were generated via a protein kinase C (PKC)-dependent and brefeldin A (BFA)-sensitive process from the Golgi. Therefore, tension decreases at second wounds were slowed or inhibited by PKC inhibitor or BFA. Lowering membrane tension by cytochalasin D treatment could substitute for exocytosis and could restore membrane resealing in low Ca^{2+} Ringer's solution. Tension could also be lowered with poloxamer 188 (Pluronic F68), and this surfactant promoted membrane repair even if the normal mechanisms were blocked.[25]

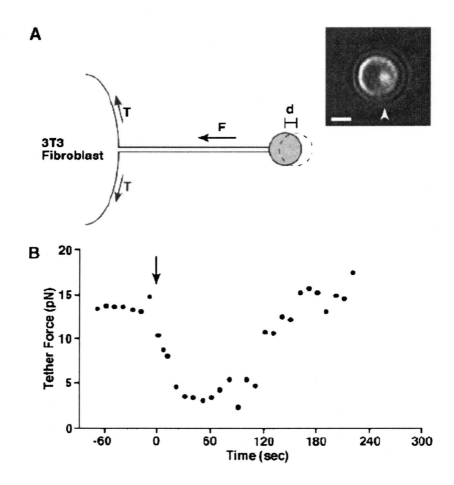

FIGURE 2. (A) Schematic drawing of a tether force measurement from a 3T3 fibroblast. An IgG-coated bead, held by the laser tweezers was attached to the plasma membrane, and a membrane tether was formed by moving the cell to one side. Tether force (F) can be estimated from the displacement of the bead (d) from the center of laser trap. T signifies the apparent membrane tension. The bright fluorescent spot (*arrowhead*) in the bead image shows the center of the laser trap. The distance between the fluorescent spot, which marks the center of the laser trap, and the geometric center of the spherical bead indicates the distance of bead displacement. Tether force squared is proportional to the apparent membrane tension. Bar = 1 μm. (B) Tether force changes after wounding in 1.8 mM Ca^{2+} Ringer's solution. The cell membrane was cut by laser scissors at 0 s (*arrow*). (Modified from Togo *et al.*[23])

ADDITIONAL MECHANISMS USED AT HIGHER DIMENSIONS OF CELL MEMBRANE DISRUPTION

Cells, embryos and eggs can also heal much larger membrane disruptions. When a 40 by 10 μm surface patch was torn off a sea urchin egg, entry of high molecular weight fluorescein-labeled saccharides from the medium was not detected by confocal microscopy. Moreover, only a brief (5–10 s) rise in cytosolic Ca^{2+} was detected at the wound site. Intracellular membranes are the primary source of the membrane recruited for this massive resealing event. When fluorescein-labeled saccharides in sea water were injected deep into the cells, a vesicle formed immediately, entrapping within its confines most of the label. Electron microscopy confirmed that the barrier delimiting the injected sea water was a membrane bilayer. The threshold for vesicle formation was 3 mM Ca^{2+}, 10× higher than the level required for resealing of wounds of several microns. When egg cytoplasm stratified by centrifugation was exposed to sea water, only the yolk platelet–rich domain formed a membrane, suggesting that the yolk platelet was a critical element in this response and that the ER was not required. McNeil *et al.* proposed that plasma membrane disruption evokes Ca^{2+}-regulated vesicle–vesicle (including endocytic compartments, but excluding ER) fusion reactions. In the resealing of large wounds, this cytoplasmic fusion reaction forms a replacement bilayer patch. Then the patch has to be added to the disrupted surface bilayer by exocytotic fusion events.[26]

Three predictions of this "patch" hypothesis were later confirmed by McNeil *et al.*[27] First, they showed that surface markers for plasma membrane protein and lipid were initially absent over disruption sites after resealing was complete. Second, they demonstrated that resealing capacity is strongly dependent upon local availability of fusion-competent cytoplasmic organelles, specifically the yolk granules. Lastly, they demonstrated that the yolk granules were capable of rapid ($t_{1/2} < 1$ s), Ca^{2+}-regulated (high threshold) fusion capable of erecting large (>1000 mm^2), continuous membrane boundaries. They proposed that production of patch vesicles for resealing may proceed by an "emergency" Ca^{2+}-dependent fusion mechanism. Obviously, for these large wounds in which the entire membrane was removed over large areas there would be no membrane to exocytose into and an entirely new one would necessarily be created first by additional mechanisms.

This work has been followed recently by a study on yolk granule–tethering proteins by McNeil and McNeil.[28]

What happens when there are too few intracellular vesicles to fuse to create a large enough area of new membrane? If too few vesicles are present to seal a transected cell, additional vesicles are created in a massive endocytosis that precedes the SNARE-mediated fusion mechanism.[29,30]

MODIFICATION OF THE RESPONSE WITH REPEATED WOUNDS

Wounding a cell a second time at the same spot evokes a more rapid membrane resealing response (FIG. 1). The facilitated response of resealing can be inhibited by either low external Ca^{2+} concentration or the specific protein kinase C (PKC) inhibitors, bisindolylmaleimide I (BIS) and Go-6976. In addition, pre-activation of PKC by phorbol ester facilitated the resealing of the initial first wound. BIS and Go-6976 suppressed the resealing rate enhancement of phorbol ester. Exocytosis-dependent fluorescent dye loss from a FM1-43 pre-labeled endocytotic compartment (destaining) was used to investigate the relationship between exocytosis, resealing, and the facilitation of resealing. Exocytosis of endocytotic compartments near the wounding site was correlated with successful resealing. The FM 1-43 destaining did not occur when exocytosis and resealing were inhibited by low external Ca^{2+} concentration or by injected tetanus toxin. When the dye-loaded cells were wounded twice, FM1-43 destaining at the second wound was less than at the first wound. Less destaining was also observed in cells pre-treated with phorbol ester, suggesting that newly formed vesicles, which were FM1-43 unlabeled, were exocytosed in the resealing of repeated woundings.

Facilitation was also blocked by brefeldin A (BFA), a fungal metabolite that inhibits vesicle formation at the Golgi apparatus. Lowering the temperature below 20°C also blocked facilitation, as expected from a block of Golgi function at that temperature. BFA had no effect on the resealing rate of an initial wound. The facilitation of the resealing by phorbol ester was blocked by pre-treatment with BFA. We suggested that, at first wounding, the cell used an endocytotic compartment to add membrane necessary for resealing. At a second wounding, PKC, activated by Ca^{2+} entry at the first wound, stimulated vesicle formation (unlabeled) from the TGN and Golgi apparatus, resulting in more rapid resealing of the second membrane disruption. Tatsuru Togo (personal communication) has directly confirmed that lipid is delivered to the site of first wounds, as expected, and is delivered in a PKC-dependent matter. However, BFA and lower temperatures could have pleiotropic effects disrupting most endosomes and trafficking aside from their dispersal and ablation of Golgi function. An alternative mechanism would be mobilization of a store of vesicles that was previously stable and inactive and that therefore had not been labeled by the pretreatment with FM1-43.

Since vesicle pools were implicated in both membrane resealing and facilitation of membrane resealing, we reasoned that artificial decreases in membrane surface tension would have the same result. Decreases in surface tension induced by the addition of a surfactant poloxamer 188 (Pluronic F68) or cytochalasin D facilitated resealing at first wounding. Furthermore, Poloxamer 188 restored resealing even when exocytosis and resealing was blocked by tetanus toxin.[25] These results were the first clue that membrane resealing required a decrease in surface tension and under natural conditions this would be provided by Ca^{2+}-dependent exocytosis of new membrane near the site of disruption (see above).

Although facilitated resealing of repeated wounds at the same site could be followed only on a short time scale for the obvious reason of changes in cell shape and position, repeated wounding at different sites can be studied over long periods. It was possible demonstrate that repeated membrane disruption at different sites led to long-term potentiation of Ca^{2+}-regulated exocytosis in 3T3 fibroblasts, which was closely correlated with faster membrane resealing rates. In this case the additional exocytosis was seen by FM1-43, indicating mobilization of an existing store of pre-labeled vesicles. This potentiation of exocytosis at different sites was cAMP-dependent protein kinase A–dependent in the early stages (minutes); required protein synthesis in the intermediate term (hours); and depended on the activation of cAMP response element–binding protein (CREB) for the long term (24 hours). We were able to demonstrate that wounding cells activated CREB within 3.5 hours. In all three phases, the increase in the amount of exocytosis was correlated with an increase in the rate of membrane resealing. However, a brief treatment with forskolin, which is effective for short-term potentiation and which could also activate CREB, was not sufficient to induce long-term potentiation of resealing. These results imply that long-term potentiation by CREB required activation by another, cAMP-independent pathway.[31] This long-term potentiation of exocytosis and membrane repair was subsequently proved to be dependent on cAMP-response element-mediated transcription via a PKC and p38 MAPK-dependent pathway.[32]

THE PUZZLE OF WHICH INTRACELLULAR COMPARTMENTS ARE USED IN EXOCYTOTIC CELL MEMBRANE REPAIR

We have already seen evidence that the cortical granules of unfertilized sea urchin eggs were responsible for membrane repair in that system. There was genetic evidence that a peroxisomal vesicle was required for cell membrane repair in the fungi *Neurospora*.[33]

Yolk platelets had been shown to be crucial in the formation of large membrane patches for large disruptions in eggs. There was also evidence implicating the endosomal pathway as well as new vesicles from the Golgi in the case of repeated wounds at the same site.[17,23,25,31] A person with biological intuition would suspect that any major source of Ca^{2+}-sensitive intracellular organelles would be fused with the plasma membrane in the vicinity of the massive Ca^{2+} influx at a wound site. Any new student in this area, however, would soon be aware of the claims by one group that lysosomes and lysosomes alone account for cell membrane repair. Yolk granules were considered to be honorary lysosomes in this context. These claims were based on several suggestive lines of evidence. For example, lysosomes were calcium-sensitive and could exocytose their contents when exposed to calcium ionophore in Ringer's solution. The SNARE protein synaptotagmin VII (Syt VII) was located to lysosomes (but not exclusively). Anti-Syt VII antibodies, a recombinant C2A Syt VII peptide, and antibodies to the cytoplasmic domain of lysosomal protein Lamp-1, all inhibited lysosomal exocytosis and membrane resealing.[34] Finally a Syt VII knockout mouse model was claimed to be impaired in membrane repair.[35] However, there were complications with this evidence. Inhibition by syt VII C2 fragments may not implicate particular syt isoforms in normal function. For example, recombinant fragments of syt VII C2 domains are potent inhibitors of PC12 neurosecretion, yet PC12 cells express syt VII at very low levels and Syt IX is implicated in secretion there.[36,37] This was confirmed by a recent gene-silencing approach that suggested that syt IX, not syt VII, is indispensable for PC12 exocytosis.[38] Studies had also suggested that syt VII C2 domains are nonselective because of binding with a variety of other molecules, including oligomerization with other subclasses of syts and competition with native syts for binding to effectors, such as SNARE proteins and phosphatidylinositol 4,5-bisphosphate.[37] The C2B peptide region, a calcium-sensing domain of syt VII was essential to exocytosis, yet it had no effect on lysosomal exocytosis.[39,40]

When the syt VII C2B peptide was introduced into fibroblasts, it was the most effective at blocking cell membrane resealing. The fact that a potent inhibitor of membrane resealing had no inhibitory effect on lysosomal exocytosis casts doubt on the theory of exclusive use of lysosomes in membrane repair.[10] There was a further unexplained complication in that the absence of syt VII in the knockout mouse cell actually accelerated observed exocytosis of lysosomes.[41] Further, a compound, vacuolin-1, could block lysosomal exocytosis and not inhibit membrane resealing.[42] The vacuolin-1 result was challenged.[43] In their attempt to repeat the original vacuolin experiment, Huynh and Andrews omitted BSA during the incubation with ionomycin, a detail in methods that had been mistakenly left out of the original publication and that has since been corrected in an erratum.[42] When the experiments are repeated under the original conditions with BSA or serum present, vacuolin-1 does completely block Lamp-1 surface expression (lysosomal exocytosis) without having any effect on membrane resealing (McNeil and Kirchausen, personal communication). This seems to conclusively eliminate the hypothesis that lysosomes are the exclusive source of intracellular membrane for cell membrane repair. The fact that BSA or serum could make such a difference in outcome underlines the many complications inherent in intracellular vesicular trafficking. Attempts to identify specific intracellular compartments in membrane repair will continue to be difficult, given the fluid nature and rapid interchange between compartments.

REMOVING CONFUSION: WOUNDING METHODS AND LEVELS OF DISRUPTION

The major source of confusion in the field of cell membrane repair results from comparing apples with oranges. As we have seen, there are several different levels of mechanisms that come into play, depending on the degree to which cell membranes are disrupted. At the smallest dimensions, well below one micron, lipid flow is unimpaired by the tension of the cytoskeleton and membrane repair is independent of calcium and exocytosis. These size wounds are typical in the electroporation of cells and are well known to reseal without calcium.[44–46] Turning up the voltage can result in larger wounds, but the experimenter would be well advised to check whether the electroporated cells actually do require calcium to reseal.

At dimensions near one micron and above, calcium-dependent exocytosis is required. When a much larger area of membrane is removed, an additional mechanism is needed to create a patch of new membrane by a massive fusion of intracellular vesicles. If too few vesicles are present to seal a cell transsection, additional vesicles are created in a massive endocytosis that precedes the fusion mechanism.

With this in mind it would be better to consistently make wounds on the same dimension by the same method when comparing mechanisms. Additional confusions can result otherwise. Adding ionophore is actually quite different than wounding cells. Both increase intracellular Ca^{2+}, but the levels attained and the distribution is quite different. Dropping beads on cells will wound them, but this method wounds them repeatedly and different mechanisms come into play in resealing first and second wounds. For example, Shen et al.,[10] using micropuncture, were able to see that C2A syt VII peptides did not inhibit resealing to an initial wound, a finding that was missed by Reddy et al.,[34] who used beads. Although it was known that C2B syt VII peptide would not block lysosomal exocytosis, for some reason Reddy el al.[34] refrained from testing that peptide on membrane repair altogether. Look for this sort of discrepancy when reading and new questions will open up to you.

ACKNOWLEDGMENTS

This research was supported by the National Institutes of Health.

REFERENCES

1. McNeil, P.L. 1993. Cellular and molecular adaptations to injurious mechanical stress. Trends Cell Biol. **3**: 302–307.
2. McNeil, P.L. & R.A. Steinhardt. 2003. Plasma membrane disruption: repair, prevention, adaptation. Annu. Rev. Cell Dev. Biol. **19**: 697–731.
3. Alberts, B., D. Bray, J. Lewis, et al. 2004. In Essential Cell Biology: 370. Garland. New York & London.
4. Xie, X. & J.N. Barrett. 1991. Membrane resealing in cultured rat septal neurons after neurite transection: evidence for enhancement by Ca^{2+}-triggered protease activity and cytoskeletal disassembly. J. Neurosci. **11**: 3257–3267.
5. De Mello, W.C. 1973. Membrane sealing in frog skeletal-muscle fibers. Proc. Nat. Acad. Sci. USA **70**: 982–984.
6. Papahadjopoulos, D., W. Vail, C. Newton, et al. 1977. Studies on membrane fusion. III. The role of calcium-induced phase changes. Biochim. Biophys. Acta **465**: 579–598.
7. Ohki, S. & O. Zschornig. 1993. Ion-induced fusion of phosphatidic acid vesicles and correlation between surface hydrophobicity and membrane fusion Chem. Phys. Lipids **65**: 193–204.
8. Leventis, R., J. Gagne, N. Fuller, et al. 1986. Divalent cation induced fusion and lipid lateral segregation in phosphatidylcholine-phosphatidic acid vesicles. Biochemistry **25**: 6978–6987.
9. Steinhardt, R.A, G. Bi & J.M. Alderton. 1994. Cell membrane resealing by a vesicular mechanism similar to neurotransmitter release. Science **263**: 390–393.
10. Shen, S.S., W.C. Tucker, E.R. Chapman & R.A. Steinhardt. 2005. Molecular regulation of membrane resealing in 3T3 fibroblasts. J. Biol. Chem. **280**: 1652–1660.
11. Steinhardt, R.A. & D. Epel. 1974. Activation of sea-urchin eggs by a calcium ionophore. Proc. Natl. Acad. Sci. USA **71**: 1915–1919.
12. Steinhardt, R., R. Zucker & G. Schatten. 1977. Intracellular calcium release at fertilization in the sea urchin egg. Dev. Biol. **58**: 185–196.
13. Llinas, R., J.A. Gruner, M. Sugimori, et al. 1991. Regulation by synapsin I and Ca(2+)-calmodulin-dependent protein kinase II of the transmitter release in squid giant synapse. J. Physiol. **436**: 257–282.
14. Benfenati, F., F. Valtorta, J.L. Rubenstein, et al. 1992. Synaptic vesicle-associated Ca2+/calmodulin-dependent protein kinase II is a binding protein for synapsin I. Nature **359**: 417–420.
15. Schiavo, G., F. Benfenati, B. Poulain, et al. 1992. Tetanus and botulinum-B neurotoxins block neurotransmitter release by proteolytic cleavage of synaptobrevin. Nature **359**: 832–835.
16. Bi, G.Q., J.M. Alderton & R.A. Steinhardt. 1995. Calcium-regulated exocytosis is required for cell membrane resealing. J. Cell. Biol. **131**: 1747–1758.
17. Miyake, K. & P.L. McNeil. 1995. Vesicle accumulation and exocytosis at sites of plasma membrane disruption. J. Cell Biol. **131**: 1737–1745.
18. Conner, S., D. Leaf & G. Wessel. 1997. Members of the SNARE hypothesis are associated with cortical granule exocytosis in the sea urchin egg. Mol. Reprod. Dev. **48**: 106–118.
19. Avery, J., A. Hodel & M. Whitaker. 1997. In vitro exocytosis in sea urchin eggs requires a synaptobrevin-related protein. J. Cell Sci. **110**: 1555–1561.
20. Chandler, D.E., M. Whitaker & J. Zimmerberg. 1989. High molecular weight polymers block cortical granule exocytosis in sea urchin eggs at the level of granule matrix disassembly. J. Cell Biol. **109**: 1269–1278.
21. Bi, G.Q., R.L. Morris, G. Liao, et al. 1997. Kinesin- and myosin-driven steps of vesicle recruitment for Ca2+-regulated exocytosis. J. Cell Biol. **138**: 999–1008.

22. TOGO, T. & R.A. STEINHARDT. 2004. Nonmuscle myosin IIA and IIB have distinct functions in the exocytosis-dependent process of cell membrane repair. Mol. Biol. Cell **15:** 688–695.
23. TOGO, T., T.B. KRASIEVA & R.A. STEINHARDT. 2000. A decrease in membrane tension precedes successful cell-membrane repair. Mol. Biol Cell **11:** 4339–4346.
24. ZHELEV, D.V., & D. NEEDHAM. 1993. Tension-stabilized pores in giant vesicles: determination of pore size and pore line tension. Biochim. Biophys.Acta **1147:** 89–104.
25. TOGO, T., J.M. ALDERTON, G.Q. BI & R.A. STEINHARDT. 1999. The mechanism of facilitated cell membrane resealing. J. Cell Sci. **112:** 719–731.
26. TERASAKI, M., K. MIYAKE & P.L. MCNEIL. 1997. Large plasma membrane disruptions are rapidly resealed by Ca2+-dependent vesicle-vesicle fusion events. J. Cell Biol. **139:** 63–74.
27. MCNEIL, P.L., S.S. VOGEL, K. MIYAKE & M. TERASAKI. 2000. Patching plasma membrane disruptions with cytoplasmic membrane. J. Cell Sci. **113:** 1891–902.
28. MCNEIL, A. & P.L. MCNEIL. 2005. Yolk granule tethering: a role in cell resealing and identification of several protein components. J. Cell Sci. **118:** 4701–4708.
29. EDDLEMAN, C.S., M.L. BALLINGER, M.E. SMYERS, *et al.* 1998. Endocytotic formation of vesicles and other membranous structures induced by Ca2+ and axolemmal injury. J. Neurosci. **18:** 4029–4041.
30. DETRAIT, E., C.S. EDDLEMAN, S. YOO, *et al.* 2000. Axolemmal repair requires proteins that mediate synaptic vesicle fusion. J. Neurobiol. **44:** 382–291.
31. TOGO, T., J.M. ALDERTON & R.A. STEINHARDT. 2003. Long-term potentiation of exocytosis and cell membrane repair in fibroblasts. Mol. Biol. Cell **14:** 93–106.
32. TOGO, T. 2004. Long-term potentiation of wound-induced exocytosis and plasma membrane repair is dependent on cAMP-response element-mediated transcription via a protein kinase C- and p38 MAPK-dependent pathway. J. Biol. Chem. **279:** 44996–45003.
33. JEDD, G. & N.H. CHUA. 2000. A new self-assembled peroxisomal vesicle required for efficient resealing of the plasma membrane. Nat. Cell Biol. **2:** 226–231.
34. REDDY, A., E.V. CALER & N.W. ANDREWS. 2001. Plasma membrane repair is mediated by Ca(2+)-regulated exocytosis of lysosomes. Cell **106:** 157–169.
35. CHAKRABARTI, S., K.S. KOBAYASHI, R.A. FLAVELL, *et al.* 2003. Impaired membrane resealing and autoimmune myositis in synaptotagmin VII-deficient mice. J. Cell Biol. **162:** 543–549.
36. ZHANG, X., J. M. KIM-MILLER, M. FUKUDA, *et al.* 2002. Ca2+-dependent synaptotagmin binding to SNAP-25 is essential for Ca2+-triggered exocytosis. Neuron **34:** 599–611.
37. TUCKER, W.C., J.M. EDWARDSON, J. BAI, *et al.* 2003. Identification of synaptotagmin effectors via acute inhibition of secretion from cracked PC12 cells. J. Cell Biol. **162:** 199–209.
38. FUKUDA, M. 2004. RNA interference-mediated silencing of synaptotagmin IX, but not synaptotagmin I, inhibits dense-core vesicle exocytosis in PC12 cells. Biochem. J. **380:** 875–879.
39. DESAI, R.C., B. VYAS, C.A. EARLES, *et al.* 2000. The C2B domain of synaptotagmin is a Ca(2+)-sensing module essential for exocytosis. J. Cell Biol. **150:** 1125–1136.
40. MARTINEZ, I., S. CHAKRABARTI, T. HELLEVIK, *et al.* 2000. Synaptotagmin VII regulates Ca(2+)-dependent exocytosis of lysosomes in fibroblasts. J. Cell Biol. **148:** 1141–1149.
41. JAISWAL, J.K., S. CHAKRABARTI, N.W. ANDREWS & S.M. SIMON. 2004. Synaptotagmin VII restricts fusion pore expansion during lysosomal exocytosis. PLoS Biol. 2, E233.
42. CERNY, J., Y. FENG, A. YU, *et al.* 2004. The small chemical vacuolin-1 inhibits Ca(2+)-dependent lysosomal exocytosis but not cell resealing. EMBO Rep. **5:** 883–888; erratum in EMBO Rep. **6:** 898 (2005).
43. HUYNH, C. & N.W. ANDREWS. 2005. The small chemical vacuolin-1 alters the morphology of lysosomes without inhibiting Ca2+-regulated exocytosis. EMBO Rep. **6:** 843–847.
44. MULLER, K.J., M. HORBASCHEK, K. LUCAS, *et al.* 2003. Electrotransfection of anchorage-dependent mammalian cells. Exp. Cell Res. **288:** 344–353.

45. DJUZENOVA, C.S., U. ZIMMERMANN, H. FRANK, et al. 1996. Effect of medium conductivity and composition on the uptake of propidium iodide into electropermeabilized myeloma cells. Biochim. Biophys. Acta **1284:** 143–152.
46. SHIRAKASHI, R., C.M. KOSTNER, K.J. MULLER, et al. 2002. Intracellular delivery of trehalose into mammalian cells by electropermeabilization. J. Membr. Biol. **189:** 45–54.

The Role of Ca^{2+} in Muscle Cell Damage

HANNE GISSEL

Institute of Physiology and Biophysics, University of Aarhus, DK-8000 Århus C, Denmark

ABSTRACT: Skeletal muscle is the largest single organ of the body. Skeletal muscle damage may lead to loss of muscle function, and widespread muscle damage may have serious systemic implications due to leakage of intracellular constituents to the circulation. Ca^{2+} acts as a second messenger in all muscle and may activate a whole range of processes ranging from activation of contraction to degradation of the muscle cell. It is therefore of vital importance for the muscle cell to control [Ca^{2+}] in the cytoplasm ([Ca^{2+}]$_c$). If the permeability of the sarcolemma for Ca^{2+} is increased, the muscle cell may suffer Ca^{2+} overload, defined as an inability to control [Ca^{2+}]$_c$. This could lead to the activation of calpains, resulting in proteolysis of cellular constituents, activation of phospholipase A_2 (PLA$_2$), affecting membrane integrity, an increased production of reactive oxygen species (ROS), causing lipid peroxidation, and possibly mitochondrial Ca^{2+} overload, all of which may further worsen the damage in a self-reinforcing process. An increased influx of Ca^{2+} leading to Ca^{2+} overload in muscle may occur in a range of situations such as exercise, mechanical and electrical trauma, prolonged ischemia, Duchenne muscular dystrophy, and cachexia. Counteractions include membrane stabilizing agents, Ca^{2+} channel blockers, calpain inhibitors, PLA$_2$ inhibitors, and ROS scavengers.

KEYWORDS: calpain; phospholipase A_2; reactive oxygen species; exercise; Duchenne muscular dystrophy; surfactant; skeletal muscle

INTRODUCTION

Skeletal muscle is the largest single organ of the body, constituting about 40% of the mass of the average human body. Skeletal muscle is formed of long multinucleate, cylindrical cells called muscle fibers. Muscle damage may lead to loss of function, and widespread damage to muscle may have serious systemic implications due to leakage of intracellular constituents to the circulation.

Ca^{2+} plays a very important role in skeletal muscle. All muscles use Ca^{2+} as their main regulatory and signaling molecule, and the function of all muscle types is controlled by Ca^{2+} as a second messenger.

In this review I will focus on the role of Ca^{2+} in the healthy muscle cell and in the development of skeletal muscle damage. I will present a model placing an increased influx of Ca^{2+} across the cellular membrane as the initiating factor leading to skele-

Address for correspondence: Hanne Gissel, Institute of Physiology and Biophysics, University of Aarhus, Ole Worms Alle 1160, DK-8000 Århus C, Denmark. Voice: +45-8942-2820; fax: +45-8612-9065.

HGH@fi.au.dk

tal muscle damage and propose a range of clinical situations where this may be the case. Finally I will evaluate possible countermeasures to prevent or reduce skeletal muscle damage.

CA^{2+} IN THE HEALTHY MUSCLE CELL

The concentration of free Ca^{2+} in the cytosol ($[Ca^{2+}]_c$) is maintained at ~50 nM in the resting state.[1] This is in sharp contrast to the concentration of Ca^{2+} in the extracellular fluid ($[Ca^{2+}]_o$), which is around 1 mM, creating an enormous chemical gradient for Ca^{2+} across the cellular membrane. Despite a very low permeability of the sarcolemma for Ca^{2+} there is continuous passive diffusion of Ca^{2+} into the muscle cells owing to the large electrochemical gradient across the membrane.[2] In order to maintain the low resting $[Ca^{2+}]_c$ the muscle cell must actively remove Ca^{2+} from the cytosol by the expenditure of ATP using the Ca^{2+}-ATPases in the sarcoplasmic reticulum (SR) and sarcolemmal/T-tubular membranes.[2–4] During a contraction relaxation cycle there is a large flux of Ca^{2+} ions from the SR to the contractile filaments and back, causing relatively large (up to 100-fold locally) but short-lived changes in $[Ca^{2+}]_c$. Excitation is also associated with a small influx of Ca^{2+} across the cellular membrane (FIG. 1).[5]

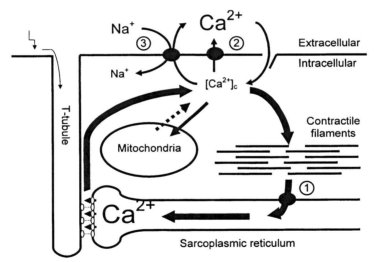

FIGURE 1. Schematic diagram of a skeletal muscle cell showing the fluxes of Ca^{2+} during excitation. As the action potential moves along the cellular membrane and down into the T-tubules, the electric signal is transferred to the SR, where it is converted to a chemical signal by the release of Ca^{2+}. The increase in $[Ca^{2+}]_c$ activates contraction. The Ca^{2+} released from the SR is quickly reaccumulated into the SR by Ca^{2+}-ATPases in the SR membrane and contraction is stopped. Thus during contraction there is a large flux of Ca^{2+} from the SR to the contractile filaments and back. The major storage compartment for Ca^{2+} is the SR, although the mitochondria may also participate in Ca^{2+} regulation. Ca^{2+} influx can also occur through Ca^{2+}-conducting ion channels and Ca^{2+} is exported from the cell via the Ca^{2+}-ATPase in the cellular/T-tubular membrane or via Na^+/Ca^{2+} exchange. (1) SR Ca^{2+}-ATPase; (2) SL Ca^{2+}-ATPase; (3) Na^+/Ca^{2+} exchanger.

In the resting cell, approximately 5% of the total energy turnover is dedicated to Ca^{2+} cycling, whereas during contraction Ca^{2+} cycling may consume 20–50% of the total energy turnover.[6]

Transport Mechanisms

The skeletal muscle cell has several ways of clearing Ca^{2+} from the cytoplasm. The most important is the SR Ca^{2+}-ATPase, which is responsible for reuptake of Ca^{2+} into the SR. It has a high affinity for Ca^{2+} ($K_m < 0.5$ µM) on the cytoplasmic side and a maximum velocity of 6.4 and 2.4 µmol/g fiber protein/min in fast and slow single human fibers, respectively.[7] The sarcolemmal and T-tubular Ca^{2+}-ATPases (SL Ca^{2+}-ATPase) actively export Ca^{2+} from the muscle cell. These are high-affinity ($K_m = 0.5$ µM) but low-capacity systems (12–30 nmol Ca^{2+}/mg protein/min in rabbit white muscle)[2,3] that are active in regulating $[Ca^{2+}]_c$ especially under resting conditions.[2,8]

A second export mechanism is Na^+-Ca^{2+} exchange. Na^+-Ca^{2+} exchange has been widely studied in cardiac muscle, where it plays a major role in Ca^{2+} regulation. In skeletal muscle the capacity and quantitative importance of Na^+-Ca^{2+} exchange is more controversial. The existence of Na^+-Ca^{2+} exchange in mammalian skeletal muscle has been documented in several studies.[9,10] It is a low-affinity but high capacity system. In sarcolemmal vesicles the transport rate was 5–10-fold lower and K_m for Ca^{2+} was an order of magnitude higher in Na^+/Ca^{2+} exchange compared to Ca^{2+}-ATPase transport.[3,11] Studies on frog muscle and single mouse fibers showed that Na^+/Ca^{2+} exchange activity was low during normal resting conditions and only becomes measurable when $[Ca^{2+}]_c$ is increased substantially above 80 nM.[10] Thus it was suggested that Na^+-Ca^{2+} exchange may be active in Ca^{2+} removal at high $[Ca^{2+}]_c$, as during fatiguing stimulation,[10] and may participate in the long-term regulation of myoplasmic Ca^{2+}.[9]

Internal Storage of Ca^{2+}

The muscle cell is capable of storing large amounts of Ca^{2+}. Two organelles are involved in Ca^{2+} storage: the sarcoplasmic reticulum (SR) and the mitochondrion.

The Sarcoplasmic Reticulum (SR)

The major storage compartment of the muscle cell is the SR, which is traditionally considered the organelle responsible for the ambiguous regulation of Ca^{2+}. Because of the presence of the Ca^{2+}-binding protein calsequestrin, the buffer capacity of the SR for Ca^{2+} is large. Calsequestrin is capable of binding 40–50 moles of Ca^{2+}/mol.[1] The total concentration of Ca^{2+} within the SR at rest has been reported to be 11 and 21 mM for slow-twitch and fast-twitch fibers, respectively, whereas the concentration of free Ca^{2+} is kept around 1 mM.[12] Studies have shown that at resting $[Ca^{2+}]_c$ the SR of slow-twitch fibers is saturated with Ca^{2+}, whereas the SR of fast twitch fibers is only one-third saturated.[12] This means that the buffering capacity of the SR in slow-twitch fibers is smaller than that of the fast-twitch fibers.

Mitochondria

As early as the 1960s, it was discovered that the mitochondria were capable of taking up large amounts of Ca^{2+}.[13] In mitochondria Ca^{2+} uptake occurs via the Ca^{2+}

uniporter located in the inner membrane of the mitochondrion. The K_m for the uniporter is between 10 and 100 µM.[14] Since global increases of $[Ca^{2+}]_c$ to this range are usually only seen under pathologic conditions it was seriously questioned whether Ca^{2+} was taken up by the mitochondria under physiological conditions. Recently the mitochondria have reemerged as active players in the regulation of Ca^{2+}. On the basis of 3-D fluorescence imaging, it has been proposed that the mitochondria are located close to Ca^{2+}-release channels of the SR, and upon activation the mitochondria sense highly localized increases in $[Ca^{2+}]$ (20–30 µM) as it is being released. The high local $[Ca^{2+}]$ enables the uniporter to take up Ca^{2+}.[15] Recently it was shown *in vivo* in mouse skeletal muscle that mitochondria take up Ca^{2+} during contraction and release it during relaxation. The mitochondrial Ca^{2+} uptake was delayed by a few milliseconds compared with the cytosolic rise and occurred both during a single twitch and upon tetanic contraction.[16] Thus, it has become apparent that the mitochondria may play an important role in the regulation of Ca^{2+} in skeletal muscle.

EFFECTS OF CA^{2+} OVERLOAD

Ca^{2+} overload, defined as an excessive accumulation of Ca^{2+} in the muscle cell, resulting in an inability of the muscle cell to control $[Ca^{2+}]_c$, is a serious condition involved in a wide range of situations leading to muscle cell damage.

As mentioned previously, Ca^{2+} is an important messenger in skeletal muscle, capable of activating a wide range of processes. Therefore, it is vital to the muscle cell to tightly control $[Ca^{2+}]_c$, a function performed by the combined action of the SR, the mitochondria, and the SL Ca^{2+}-ATPase. In addition fast muscle fibers from lower vertebrates and small mammals contain high amounts of the cytosolic Ca^{2+}-binding protein parvalbumin.[17] In humans parvalbumin is only present in the intrafusal fibers and is therefore not important in Ca^{2+} regulation in skeletal muscle.[18]

If the permeability of the sarcolemma for Ca^{2+} is increased (e.g., due to damage or opening of Ca^{2+}-conducting channels), a larger influx of Ca^{2+} will occur, increasing the Ca^{2+} load on the cell. Because of the large storage capacity of the SR and the mitochondria this may not necessarily present a problem for the cell, as the extra Ca^{2+} can be stored. However, if the increased influx of Ca^{2+} persists, Ca^{2+} overload of the cell may result.

Increases in $[Ca^{2+}]_c$ may result in activation of Ca^{2+}-activated neutral proteases (calpains), activation of PLA_2, an increased production of ROS, and/or mitochondrial Ca^{2+} overload. Each of these will be discussed in the following sections.

Calpain

It has become increasingly clear over the past years that calpains play an important regulatory role in cellular functions. Three types of calpains are of interest in skeletal muscle: calpain 1 (µ-calpain), calpain 2 (m-calpain), and calpain 3 (p94). The latter is almost exclusively expressed in skeletal muscle[19] and the amount of mRNA for this calpain is 10 times higher than mRNA for calpains 1 and 2.[20]

The three calpains differ in their sensitivity to Ca^{2+}. The threshold for initial activation is lowest for calpain 3 (approximately 0.5 µM),[21] while for calpain 1 it is

3 µM and for calpain 2 it is 400 µM. However, when the protein is autolyzed, proteolytic activity occurs at [Ca^{2+}] of 0.5–2 µM for calpain 1 and 50–150 µM for calpain 2.[19] Recent evidence suggests that autolysis is necessary for activation of calpain 3.[22]

Calpains 1 and 3 are expected to be activated at physiological [Ca^{2+}] and are thought to be important in the regulatory function of the cell. Calpain 2, with the higher Ca^{2+} requirement, is thought to be involved in fiber degeneration.[23]

Calpain cleaves a variety of protein substrates including cytoskeletal, myofibrillar, and membrane proteins. In particular titin (α-connectin), a large elastic protein that tethers the thick filament at the center of the sarcomere, by anchoring it to the Z-line, is readily proteolyzed by calpain.[19] Calpain is closely associated with the I and the Z band regions and calpain-mediated degradation is thought to contribute to the changes in muscle structure (Z-line streaming) and function (EC uncoupling) that occur immediately after exercise.[24–26] Increases in calpain activity in muscle has been observed during prolonged running,[26,27] following chronic low-frequency stimulation,[23] in sepsis[28] and in muscle from *mdx* mice (the murine model for Duchennes muscular dystrophy).[29]

PLA_2

PLA_2 may be activated by elevated [Ca^{2+}]$_c$.[30] The activated PLA_2 will attack mitochondrial and other membrane phospholipids, giving rise to lysophospholipids and free fatty acids. Lysophospholipids will disrupt membrane lipid organization and free fatty acids may have a detergent action, causing membrane damage.[31] Among the free fatty acids liberated is arachidonic acid, which may promote ROS production by the mitochondria (see the next section). There is evidence that increased [Ca^{2+}]$_c$ activates PLA_2, based on the observation that enzyme efflux from experimentally damaged muscles (suffering Ca^{2+} overload) can be blocked by PLA_2 inhibitors.[31,32] NDGA (a lipo-oxygenase inhibitor) completely inhibited CK efflux from skeletal muscle following treatment with the Ca^{2+} ionophore A23187 or DNP (dinitrophenol, poisoning of the mitochondria) showing that PLA_2 activation, and in particular the lipooxygenase pathway, affects cellular membrane integrity.[33]

Reactive Oxygen Species

The mitochodria of skeletal muscle cells continuously generate ROS. Increased Ca^{2+} load on the muscle cell may increase superoxide anion synthesis. Enzymatic synthesis by xanthine oxidase is stimulated by increased [Ca^{2+}]$_c$.[34] Furthermore, it has been shown that mitochondrial H_2O_2 production is stimulated by physiological levels of ADP and calcium, possibly via a PLA_2-mediated pathway.[35]

Oxidative stress may disrupt the normal Ca^{2+} handling kinetics in muscle cells by disturbing SR function[36] and may be a causative factor in opening of the mitochondrial permeability transition pore (mPTP), especially in combination with high mitochondrial Ca^{2+} content ([Ca^{2+}]$_m$) (see the next section).[37,38] Oxidative stress leads to lipid peroxidation, affecting membrane structures in the cell, which has been shown following exhaustive exercise in both human and animal studies.[39–41]

Mitochondrial Ca^{2+} Overload

In the healthy cell, mitochondria may assist in refilling SR Ca^{2+} stores by interaction with store-operated Ca^{2+} channels;[42] furthermore the mitochondria may act as Ca^{2+} buffers, rapidly taking up surplus Ca^{2+} and slowly releasing it again to avoid detrimental oscillations in $[Ca^{2+}]_c$. If the muscle cell faces increased influx of Ca^{2+} from the outside, the mitochondria will accumulate Ca^{2+}.

It is by now well established that small increases in mitochondrial Ca^{2+} content ($[Ca^{2+}]_m$) stimulate ATP synthesis. Uptake of Ca^{2+} during activity may therefore serve as an important signal to ensure that energy production matches energy expenditure. However, larger accumulations may have detrimental effects. As mentioned above, increased Ca^{2+} load of the mitochondria will increase ROS production, and with it, the risk of increasing nonspecific inner membrane permeabilization.[43] Increases in $[Ca^{2+}]_m$ also increase the probability of opening of the permeabilization transition (mPTP) pore.[37] The mPTP is a large conductance pore spanning both the inner and outer membrane of the mitochondria. The pore opens most clearly under specific and usually pathologic conditions. Opening leads to depolarization of the mitochondrial membrane potential, effectively shutting down ATP production and resulting in a massive efflux of Ca^{2+} from the mitochondria, causing further rises in $[Ca^{2+}]_c$ and initiating a vicious cycle leading to apoptosis or necrosis.[37,38]

The Vicious Cycle

The vicious cycle model describes how the above-mentioned factors act in concert, creating a self-reinforcing cycle that leads to muscle cell necrosis/apoptosis (FIG. 2).

An increase in Ca^{2+} influx across the cellular membrane is observed in many situations (some of which I will discuss in detail later). Influx of Ca^{2+} will lead to local increases in $[Ca^{2+}]$ in the subsarcolemmal compartments, which may lead to local activation of calpain. Some of the Ca^{2+} will be exported again by the SL Ca^{2+}-ATPase and some will be taken up by the SR via the SR Ca^{2+}-ATPase and stored. The SR is capable of storing quite large quantities of Ca^{2+}. However, buffer capacity differs between different fiber types. The SR of fast-twitch fibers is capable of increasing their Ca^{2+} content threefold, whereas the SR of slow-twitch fibers is almost saturated at resting $[Ca^{2+}]_c$. Thus in the fast-twitch fibers, the SR represents a large buffer capacity, whereas slow-twitch fibers most likely rely on their more abundant mitochondria for Ca^{2+} buffering. In addition fast fibers from lower vertebrates and small mammals contain high concentrations of parvalbumin,[1] which may act as a significant buffer for Ca^{2+}.

The mitochondria will participate actively in the Ca^{2+} handling, particularly at the time of SR saturation. Mitochondria are capable of storing large amounts of Ca^{2+}; however, if the storage capacity is exceeded, detrimental consequences may occur. Increased Ca^{2+} load leads to increased production of ROS, causing peroxidation of membrane lipids, and possibly decreases in ATP synthesis. Mitochondrial Ca^{2+} overload also increases the risk of opening of the mPTP, which will ultimately lead to apoptosis or necrosis. Local increases in $[Ca^{2+}]_c$ may lead to PLA_2 activation, resulting in degradation of the cellular membrane as well as the membranes of organelles. This leads to an increased production of arachidonic acid leading to fur-

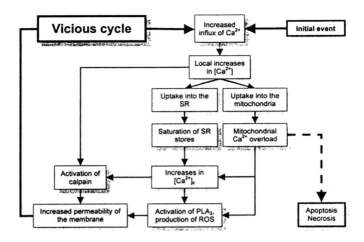

FIGURE 2. The "vicious cycle" model of how an initial event or condition leading to an increased influx of Ca^{2+} may result in muscle cell apoptosis or necrosis.

ther increases in production of ROS in the mitochondria. Finally, activation of calpain is expected, leading to degradation of cytoskeletal, myofibrillar, and membrane proteins. These processes ultimately lead to an increased permeability of the cellular membrane, causing further influx of Ca^{2+}, and the vicious cycle is activated.

The muscle cell has a high capacity for repairing damages to the membrane,[44] and therefore if the degradation process is stopped in time, the vicious cycle may be halted and the muscle cell may recover. However, a "point of no return" exists, beyond which it is not possible for the cell to regain control of cellular Ca^{2+}, eventually leading to the destruction of the cell.

CA^{2+} OVERLOAD SCENARIOS

Ca^{2+} overload occurs in a wide range of situations, and even though many of these situations seems to differ, a common denominator across the following examples is an initial increase in Ca^{2+} influx leading to muscle Ca^{2+} overload and muscle damage.

One of the serious consequences of muscle Ca^{2+} overload is rhabdomyolysis, a condition that is characterized by muscle tissue breakdown, resulting in the release of cytosolic components into the blood circulation. The degree of rhabdomyolysis ranges from an asymptomatic illness with elevation in the plasma CK level to a life-threatening condition associated with extreme elevations in plasma CK, electrolyte imbalances, acute renal failure, and disseminated intravascular coagulation.[45] Hypocalcemia is frequently observed during rhabdomyolysis owing to massive influx of Ca^{2+} into the muscle tissue. Administration of Ca^{2+} in this situation should be very carefully considered as clinical investigations have shown that it may worsen the damage considerably.[46]

Exercise

It is well documented in both human and animal studies that prolonged or unaccustomed excercise, and in particular lengthening (eccentric) exercise, may lead to muscle cell damage.[47–49] This damage may be ultrastructural alterations leading to loss of force or loss of sarcolemmal integrity leading to loss of intracellular enzymes, such as lactate dehydrogenase (LDH) or CK. Although not necessarily a pathologic phenomenon, it is a frequently occurring consequence of certain types of contractile activity, and might serve as an adaptive response to exercise. However, if the exercise is excessive, rhabdomyolysis may occur.

Much attention has been diverted to the milder forms of muscle damage—in particular, delayed onset muscle damage (DOMS), since it is of great importance for muscle performance in athletes. A model for the development of DOMS was proposed by Armstrong in 1990.[49] The model describes four stages in exercise-induced muscle injury: (1) the initial stage; (2) the autogenic stage, or the "Ca^{2+} overload" stage; (3) the phagocytic stage; and (4) the regenerative stage. The autogenic stage is characterized by Ca^{2+} overload. This is supported by evidence that exercise leads to an increased activation of calpain[26] as well as PLA_2,[50] both of which are activated by increased $[Ca^{2+}]_c$. Activation of calpain and PLA_2 may then further worsen the damage as described above in the section on the vicious cycle. The increased production of ROS may also have deleterious effects on muscle cell integrity.

There has been much debate on the nature of the initial event triggering DOMS. Some believe that mechanical damage to the fiber initiates exercise-induced muscle damage.[48] This may very well be the case during lengthening contractions, where the strain on the muscle fiber is high. Our lab has shown that excitation of rat skeletal muscle both *in situ* and *ex vivo* (isometric and shortening contractions) is associated with an increased influx of Ca^{2+} (up to 34-fold increase within the first 30 sec of stimulation) and that prolonged stimulation may lead to accumulation of Ca^{2+} in the muscle cells and loss of membrane integrity.[5,51,52] Similar, but smaller accumulations of Ca^{2+} in muscles from athletes after prolonged running has also been reported.[47,53] We suggest that the excitation-induced influx of Ca^{2+} may act as or contribute to the initiating factor leading to muscle damage during prolonged or otherwise excessive exercise.

Trauma

Trauma affects muscle membrane integrity directly. In this section I will discuss mechanical and electrical trauma both leading to an increased influx of Ca^{2+} although by two very different mechanisms.

Mechanical trauma causes direct injury to the cellular membrane causing Ca^{2+} and Na^+ to flood the injured tissue. Often circulation is compromised, causing ischemia of the tissue, and the damage process may accelerate in the following reperfusion period. The effect is a massive Ca^{2+} overload of the muscle cells, resulting in necrosis of many fibers.[45]

Electrical trauma due to high-voltage electrical injury or lightening may lead to electroporation of cellular membranes, rendering them permeable to ions and molecules. This will result in an increased influx of Ca^{2+}.[54] If resealing does not occur rapidly enough, cellular Ca^{2+} homeostasis will be lost, leading to destruction of the muscle cell. (Electric force–mediated membrane disruption will be reviewed elsewhere in this issue by Raphael Lee.)

Ischemia

Prolonged ischemia and reperfusion results in tissue damage and necrosis in many organs.

Prolonged ischemia is associated with hypoxia. Results from this laboratory as well as others have shown that anoxia increases the permeability of the membrane for Ca^{2+} and may lead to a loss of cellular integrity.[55–57] The mechanism behind the increased influx of Ca^{2+} has not been identified, but if anoxia was combined with electrical stimulation, possibly depleting cellular ATP stores, the stimulation-induced accumulation of Ca^{2+} was markedly increased and so was the release of LDH.[55] During prolonged ischemia/anoxia, ATP levels would be expected to be low, possibly compromising the Ca^{2+} handling ability of the muscle cell. In line with this increasing baseline $[Ca^{2+}]_c$ was observed during fatiguing stimulation under hypoxic conditions.[58]

In vivo reperfusion may be associated with an increased generation of ROS from circulating neutrophils, causing further damage to the cellular membrane.[59]

DMD

The human Duchenne muscular dystrophy (DMD) is the best known of the muscular dystrophies, since it is one of the most frequent genetic human diseases.[1] It is characterized by muscle fiber necrosis and progressive muscle wasting and weakness. Ca^{2+} seems to be critically involved in the pathology of DMD. The primary defect in DMD is the lack of dystrophin, a subsarcolemmal cytoskeletal protein. In normal muscle dystrophin links the cytoskeleton, via a complex of membrane proteins, to laminin in the extracellular matrix mechanically stabilizing the cellular membrane.[60] Lack of dystrophin renders the sarcolemma more vulnerable to mechanical damage. The number of transient disruptions of the sarcolemma was found to be sixfold higher in muscles from exercised *mdx* mice compared to controls.[61] These transient disruptions may give rise to Ca^{2+} entry, leading to local activation of calpain. Studies have shown that $[Ca^{2+}]$ in the subsarcolemmal compartments was 4.5-fold higher in *mdx* myotubes following depolarization compared to controls.[62] Furthermore an increased calpain activity has been observed in muscle from *mdx* mice.[29] It was suggested that the activation of calpain would lead to proteolysis of the Ca^{2+} leak channels (possibly TRP channels[63]), causing an increased activity. This in turn would further increase Ca^{2+} influx and Ca^{2+}-activated proteolysis.[64] An increased Ca^{2+} leak channel activity has been observed in *mdx* fibers and this was thought to be the mechanism behind the greater levels of Ca^{2+} influx, leading to increased $[Ca^{2+}]_c$ and increased proteolysis.[65] In addition to the increased calpain activity an increase in PLA_2 activity was shown in DMD patients.[66]

Muscle Cachexia

Muscle cachexia is a common metabolic response to several disease states seen in surgical patients including sepsis, severe injury, renal failure, and cancer.[67] Although not identical, the intracellular mechanisms and molecular regulation of muscle atrophy are similar in many of these conditions. The early phase of muscle cachexia may be beneficial, providing amino acids for gluconeogenesis and acute-phase protein synthesis. However, prolonged and severe cachexia has significant deleterious consequences.[67]

It is quite likely that early cachexia is associated with an increased influx of Ca^{2+}. Sepsis has been shown to increase ^{45}Ca uptake and Ca^{2+} content in isolated rat skeletal muscle, which was associated with an increased protein breakdown.[68,69] Furthermore, muscle cachexia induced by severe injury, sepsis, and cancer is associated with increased gene expression and activity of calpain 1, 2, and 3.[67] It has been proposed that early cachexia is associated with a calcium-calpain–dependent degradation of Z-disks and release of myofilaments from the myofibrils. This is followed by ubiquitination of the myofilaments and a subsequent degradation by the proteasome.[67] Supported for this was provided by morphologic evidence showing disintegration of the Z-disk and release of myofilaments from myofibrils in cachectic muscle from septic rats. Treatment with dantrolene (inhibiting the release of calcium from intracellular stores) prevented the sepsis-induced release of myofilaments.[70]

In addition, mitochondria from septic animals demonstrate substantially higher (PLA_2-mediated) H_2O_2 formation compared to controls,[35] increasing the oxidative stress on the muscle fiber.

COUNTERACTING MEASURES

Inhibiting Influx of Ca^{2+}

One obvious way of preventing Ca^{2+}-induced muscle damage is to inhibit or reduce the increase in $[Ca^{2+}]_c$. This can be done by resealing of the cellular membrane or reducing Ca^{2+} flux through Ca^{2+} channels in the sarcolemma and SR membrane.

Resealing

A lot of research has gone into developing agents that assist in resealing or stabilization of the membrane. Muscle damage can be considered as a continuous process of membrane degradation and repair. Assisting membrane repair may stop the vicious cycle, resulting in survival of the muscle fiber. The use of synthetic surfactants to stabilize membranes and thus restore cellular integrity has yielded promising results both *in vitro* and *in vivo*.[71] (Membrane-sealing therapy will be reviewed elsewhere in this volume by Lee and Lee.)

Ca^{2+} Channel Blockers

A certain influx of Ca^{2+} occurs via the L-type Ca^{2+} channels in the outer membrane and in the T-tubules when the muscle is excited. If the muscle cell experiences an increased Ca^{2+} load, blocking any further influx may allow the muscle cell to maintain control over Ca^{2+}. Studies using the Ca^{2+} channel blockers have showed a reduction in the exercise-induced damage both in human and animal studies.[72,73]

The ryanodine receptor Ca^{2+}-release channels in the SR likewise release Ca^{2+} into the cytoplasm following an action potential. Blocking of this (by, for example, dantrolene) has proven efficient during malignant hyperthermia and sepsis.[70]

Inhibiting the Actions of Ca^{2+}

A second approach to prevent or reduce muscle damage is to prevent the consequences of the increased $[Ca^{2+}]_c$. This may be done by blocking the actions of calpain, PLA_2, and ROS.

Calpain Inhibitors

Much discussion regarding the role of calpain in muscle damage has come from the conflicting evidence that an increase in $[Ca^{2+}]_c$ results in cellular damage, but in some cases this could not be prevented by the use of calpain inhibitors. This discrepancy may be the result of the different types of calpain, all with different characteristics. The calpain inhibitor calpastatin only inhibits calpains 1 and 2, with little effect on calpain 3. Leupeptin on the other hand inhibits all three calpains, but calpain 3 needs to undergo autolysis before leupeptin can bind to the active site.[22] Ca^{2+} causes autolytic activation of single calpain 3 molecules, and these can then proteolytically activate other nearby calpain 3 molecules in a strongly self-reinforcing activation cascade.[74] Since leupeptin cannot bind before calpain 3 is autolyzed it may not be possible for leupeptin to prevent proteolytic activity of calpain 3 when $[Ca^{2+}]$ is appreciably above the threshold level for calpain activation.[25] However, if the activating $[Ca^{2+}]$ is relatively low (e.g., ~2 μM) the actions of calpain may be blocked by leupeptin.[25] This may be the case in DMD/*mdx*, where it was recently shown that muscle fiber degeneration in *mdx* mice was delayed by intramuscular administration of leupeptin.[29]

PLA_2 Inhibitors

Inhibition of PLA_2 has been found to reduce enzyme release (e.g., CK and LDH) from skeletal muscle,[31,32] although PLA_2 inhibitors failed to protect against myofibrillar damage.[33] Thus blocking PLA_2 activity may reduce further damage to the membrane, but does not provide protection for structural damage within the muscle fiber.

ROS Scavengers

There is little doubt that ROS are involved in the development of skeletal muscle damage over a range of situations. Production of ROS is dramatically accelerated during contractile activity, mainly owing to the increased oxygen consumption.[34] Many studies have shown that an increased Ca^{2+} load also leads to an increased production of ROS. In addition, as mentioned above (Section 3.3), the actions of ROS may result in an increased Ca^{2+} load on the muscle cell.

The effect of antioxidants in reducing muscle damage after exercise has been moderate, with some studies showing no effect and others showing small benefits. Recent studies have suggested that xanthine oxidase–mediated oxidative stress leads to useful cellular adaptions to exercise and that the practice of taking antioxidants prior to exercise may have to be re-evaluated.[75]

Nevertheless antioxidant treatment may prove to be beneficial in other situations (e.g., sepsis), where a pathologic increase in ROS production leads muscle cell damage.

CONCLUDING REMARKS

There is good reason to believe that Ca^{2+} plays an important role in the development of many types of muscle damage. Situations of an increased influx of Ca^{2+} may initiate a vicious cycle, leading to muscle damage and death. Therefore, inhibiting the actions of Ca^{2+} may be important in preventing or reducing skeletal muscle damage. This may involve the use of a combination of compounds, as in combining membrane-stabilizing agents with antioxidants and calpain and PLA_2 inhibitors.

ACKNOWLEDGMENTS

This study was supported by the Danish Medical Research Council (Grant 22-02-0523). Thanks are extended to Dr. William McDonald for useful comments on the manuscript.

REFERENCES

1. BERCHTOLD, M.W., H. BRINKMEIER & M. MUNTENER. 2000. Calcium ion in skeletal muscle: its crucial role for muscle function, plasticity, and disease. Physiol. Rev. **80:** 1215–1265.
2. HIDALGO, C., M.E. GONZALEZ & A.M. GARCIA. 1986. Calcium transport in transverse tubules isolated from rabbit skeletal muscle. Biochim. Biophys. Acta **854:** 279–286.
3. MICHALAK, M., K. FAMULSKI & E. CARAFOLI. 1984. The Ca^{2+}-pumping ATPase in skeletal muscle sarcolemma: calmodulin dependence, regulation by cAMP-dependent phosphorylation, and purification. J. Biol. Chem. **259:** 15540–15547.
4. CARAFOLI, E. 1987. Intracellular calcium homeostasis. Annu. Rev. Biochem. **56:** 395–433.
5. GISSEL, H. & T. CLAUSEN. 2000. Excitation-induced Ca^{2+} influx in rat soleus and EDL muscle: mechanisms and effects on cellular integrity. Am. J. Physiol. Regul. Integr. Comp. Physiol. **279:** R917–R924.
6. CLAUSEN, T., C. VAN HARDEVELD & M.E. EVERTS. 1991. Significance of cation transport in control of energy metabolism and thermogenesis. Physiol. Rev. **71:** 733–774.
7. SALVIATI, G., R. BETTO, B.D. DANIELI et al. 1984. Myofibrillar-protein isoforms and sarcoplasmic-reticulum Ca^{2+}-transport activity of single human muscle fibres. Biochem. J. **224:** 215–225.
8. MONTEITH, G.R. & B.D. ROUFOGALIS. 1995. The plasma membrane calcium pump: a physiological perspective on its regulation. Cell Calcium **18:** 459–470.
9. DEVAL, E., G. RAYMOND & C. COGNARD. 2002. Na^+-Ca^{2+} exchange activity in rat skeletal myotubes: effect of lithium ions. Cell Calcium **31:** 37–44.
10. BALNAVE, C.D. & D.G. ALLEN. 1998. Evidence for Na^+/Ca^{2+} exchange in intact single skeletal muscle fibers from the mouse. Am. J. Physiol. **274:** C940–946.
11. MICKELSON, J.R., T.M. BEAUDRY & C.F. LOUIS. 1985. Regulation of skeletal muscle sarcolemmal ATP-dependent calcium transport by calmodulin and cAMP-dependent protein kinase. Arch. Biochem. Biophys. **242:** 127–136.
12. FRYER, M.W. & D.G. STEPHENSON. 1996. Total and sarcoplasmic reticulum calcium contents of skinned fibres from rat skeletal muscle. J. Physiol. Lond. **493:** 357–370.
13. DE LUCA, H.F. & G.W. ENGSTROM. 1961. Calcium uptake by rat kidney mitochondria. Proc. Natl. Acad. Sci. USA **47:** 1744–1750.
14. GUNTER, T.E. & D.R. PFEIFFER. 1990. Mechanisms by which mitochondria transport calcium. Am. J. Physiol. **258:** C755–C786.
15. RIZZUTO, R., P. PINTON, M. BRINI, et al. 1999. Mitochondria as biosensors of calcium microdomains. Cell Calcium **26:** 193–199.
16. RUDOLF, R., M. MONGILLO, P. J. MAGALHAES, et al. 2004. In vivo monitoring of Ca^{2+} uptake into mitochondria of mouse skeletal muscle during contraction. J. Cell Biol. **166:** 527–536.

17. GAILLY, P. 2002. New aspects of calcium signaling in skeletal muscle cells: implications in Duchenne muscular dystrophy. Biochim. Biophys. Acta **1600:** 38–44.
18. FOHR, U. G., B. R. WEBER, M. MUNTENER, et al. 1993. Human alpha and beta parvalbumins: structure and tissue-specific expression. Eur. J. Biochem. **215:** 719–727.
19. GOLL, D.E., V.F. THOMPSON, H. LI et al. 2003. The calpain system. Physiol. Rev. **83:** 731–801.
20. KINBARA, K., H. SORIMACHI, S. ISHIURA, et al. 1997. Muscle-specific calpain, p94, interacts with the extreme C-terminal region of connectin, a unique region flanked by two immunoglobulin C2 motifs. Arch. Biochem. Biophys. **342:** 99–107.
21. BRANCA, D., A. GUGLIUCCI, D. BANO, et al. 1999. Expression, partial purification and functional properties of the muscle-specific calpain isoform p94. Eur. J. Biochem. **265:** 839–846.
22. DIAZ, B.G., T. MOLDOVEANU, M.J. KUIPER et al. 2004. Insertion sequence 1 of muscle-specific calpain, p94, acts as an internal propeptide. J. Biol. Chem. **279:** 27656–27666.
23. SULTAN, K.R., B.T. DITTRICH, E. LEISNER, et al. 2001. Fiber type-specific expression of major proteolytic systems in fast- to slow-transforming rabbit muscle. Am. J. Physiol. Cell Physiol. **280:** C239–C247.
24. BELCASTRO, A.N., L.D. SHEWCHUK & D.A. RAJ. 1998. Exercise-induced muscle injury: a calpain hypothesis. Mol. Cell Biochem. **179:** 135–145.
25. VERBURG, E., R.M. MURPHY, D.G. STEPHENSON, et al. 2005. Disruption of excitation-contraction coupling and titin by endogenous Ca^{2+}-activated proteases in toad muscle fibres. J. Physiol. [online] **564:** 775–790.
26. BELCASTRO, A.N. 1993. Skeletal muscle calcium-activated neutral protease (calpain) with exercise. J. Appl. Physiol. **74:** 1381–1386.
27. ARTHUR, G.D., T.S. BOOKER & A.N. BELCASTRO. 1999. Exercise promotes a subcellular redistribution of calcium-stimulated protease activity in striated muscle. Can. J. Physiol. Pharmacol. **77:** 42–47.
28. WEI, W., M.U. FAREED, A. EVENSON, et al. 2005. Sepsis stimulates calpain activity in skeletal muscle by decreasing calpastatin activity but does not activate caspase-3. Am. J. Physiol. Regul. Integr. Comp. Physiol. **288:** R580–R590.
29. BADALAMENTE, M.A. & A. STRACHER. 2000. Delay of muscle degeneration and necrosis in mdx mice by calpain inhibition. Muscle Nerve **23:** 106–111.
30. BRUTON, J.D., J. LANNERGREN & H. WESTERBLAD. 1998. Mechanisms underlying the slow recovery of force after fatigue: importance of intracellular calcium. Acta Physiol. Scand. **162:** 285–293.
31. JACKSON, M.J., D.A. JONES & R.H. EDWARDS. 1984. Experimental skeletal muscle damage: the nature of the calcium-activated degenerative processes. Eur. J. Clin. Invest. **14:** 369–374.
32. SANDERCOCK, D.A. & M.A. MITCHELL. 2003. Myopathy in broiler chickens: a role for Ca(2+)-activated phospholipase A2? Poult. Sci. **82:** 1307–1312.
33. DUNCAN, C.J. & M.J. JACKSON. 1987. Different mechanisms mediate structural changes and intracellular enzyme efflux following damage to skeletal muscle. J. Cell Sci. **87:** 183–188.
34. REID, M.B. 2001. Invited review: redox modulation of skeletal muscle contraction: what we know and what we don't. J. Appl. Physiol. **90:** 724–731.
35. NETHERY, D., L.A. CALLAHAN, D. STOFAN, et al. 2000. PLA_2 dependence of diaphragm mitochondrial formation of reactive oxygen species. J. Appl. Physiol. **89:** 72–80.
36. XU, K.Y., J.L. ZWEIER & L.C. BECKER. 1997. Hydroxyl radical inhibits sarcoplasmic reticulum Ca(2+)-ATPase function by direct attack on the ATP binding site. Circ. Res. **80:** 76–81.
37. DUCHEN, M. R. 2004. Mitochondria in health and disease: perspectives on a new mitochondrial biology. Mol. Asp. Med. **25:** 365–451.
38. CROMPTON, M., S. VIRJI, V. DOYLE et al. 1999. The mitochondrial permeability transition pore. Biochem. Soc. Symp. **66:** 167–179.
39. DAVIES, K.J., A.T. QUINTANILHA, G.A. BROOKS et al. 1982. Free radicals and tissue damage produced by exercise. Biochem. Biophys. Res. Commun. **107:** 1198–1205.

40. RAJGURU, S.U., G.S. YEARGANS & N.W. SEIDLER. 1994. Exercise causes oxidative damage to rat skeletal muscle microsomes while increasing cellular sulfhydryls. Life Sci. **54:** 149–157.
41. MCANULTY, S.R., L.S. MCANULTY, D.C. NIEMAN et al. 2003. Influence of carbohydrate ingestion on oxidative stress and plasma antioxidant potential following a 3 h run. Free Radic. Res. **37:** 835–840.
42. PAREKH, A.B. & J.W. PUTNEY, JR. 2005. Store-operated calcium channels. Physiol. Rev. **85:** 757–810.
43. GRIJALBA, M.T., A.E. VERCESI & S. SCHREIER. 1999. Ca^{2+}-induced increased lipid packing and domain formation in submitochondrial particles: a possible early step in the mechanism of Ca^{2+}-stimulated generation of reactive oxygen species by the respiratory chain. Biochemistry **38:** 13279–13287.
44. MCNEIL, P.L. & R.A. STEINHARDT. 1997. Loss, restoration, and maintenance of plasma membrane integrity. J. Cell Biol. **137:** 1–4.
45. HUERTA-ALARDIN, A., J. VARON & P. MARIK. 2005. Bench-to-bedside review: rhabdomyolysis—an overview for clinicians. Crit. Care **9:** 158–169.
46. THYSSEN, E.P., S.H. HOU, J.C. ALVERDY, et al. 1990. Temporary loss of limb function secondary to soft tissue calcification in a patient with rhabdomyolysis-induced acute renal failure. Am. J. Kidney Dis. **16:** 491–494.
47. OVERGAARD, K., T. LINDSTROM, T. INGEMANN-HANSEN, et al. 2002. Membrane leakage and increased content of Na(+)-K(+) pumps and Ca(2+) in human muscle after a 100-km run. J. Appl. Physiol. **92:** 1891–1898.
48. FRIDEN, J. & R.L. LIEBER. 2001. Eccentric exercise-induced injuries to contractile and cytoskeletal muscle fibre components. Acta Physiol. Scand. **171:** 321–326.
49. ARMSTRONG, R.B. 1990. Initial events in exercise-induced muscular injury. Med. Sci. Sports Exerc. **22:** 429–435.
50. FEDERSPIL, G., B. BAGGIO, P.C. DE, et al. 1987. Effect of prolonged physical exercise on muscular phospholipase A2 activity in rats. Diabetes Metab. **13:** 171–175.
51. EVERTS, M.E., T. LØMO & T. CLAUSEN. 1993. Changes in K^+, Na^+ and calcium contents during in vivo stimulation of rat skeletal muscle. Acta Physiol. Scand. **147:** 357–368.
52. GISSEL, H. & T. CLAUSEN. 1999. Excitation-induced Ca^{2+} uptake in rat skeletal muscle. Am. J. Physiol. **276:** R331–R339.
53. OVERGAARD, K., A. FREDSTED, A. HYLDAL, et al. 2004. Effects of running distance and training on Ca^{2+} content and damage in human muscle. Med. Sci. Sports Exerc. **36:** 821–829.
54. GISSEL, H. & T. CLAUSEN. 2003. Ca^{2+} uptake and cellular integrity in rat EDL muscle exposed to electrostimulation, electroporation, or A23187. Am. J. Physiol. Regul. Integr. Comp. Physiol. **285:** R132–R142.
55. FREDSTED, A., U.R. MIKKELSEN, H. GISSEL, et al. 2005. Anoxia induces Ca^{2+} influx and loss of cell membrane integrity in rat EDL muscle. Exp. Physiol. In press.
56. JONES, D.A., M.J. JACKSON, G. MCPHAIL, et al. 1984. Experimental mouse muscle damage: the importance of external calcium. Clin. Sci. **66:** 317–322.
57. LAMBERT, I.H., J.H. NIELSEN, H.J. ANDERSEN, et al. 2001. Cellular model for induction of drip loss in meat. J. Agric. Food Chem. **49:** 4876–4883.
58. STARY, C.M. & M.C. HOGAN. 2000. Impairment of Ca^{2+} release in single *Xenopus* muscle fibers fatigued at varied extracellular PO_2. J. Appl. Physiol. **88:** 1743–1748.
59. RUBIN, B.B., A. ROMASCHIN, P.M. WALKER, et al. 1996. Mechanisms of postischemic injury in skeletal muscle: intervention strategies. J. Appl. Physiol. **80:** 369–387.
60. LAPIDOS, K.A., R. KAKKAR & E.M. MCNALLY. 2004. The Dystrophin Glycoprotein complex: signaling strength and integrity for the sarcolemma. Circ. Res. **94:** 1023–1031.
61. CLARKE, M.S., R. KHAKEE & P.L. MCNEIL. 1993. Loss of cytoplasmic basic fibroblast growth factor from physiologically wounded myofibers of normal and dystrophic muscle. J. Cell Sci. **106:** 121–133.
62. BASSET, O., F.X. BOITTIN, O.M. DORCHIES, et al. 2004. Involvement of inositol 1,4,5-trisphosphate in nicotinic calcium responses in dystrophic myotubes assessed by near-plasma membrane calcium measurement. J. Biol. Chem. **279:** 47092–47100.

63. VANDEBROUCK, C., D. MARTIN, M.C.-V. SCHOOR, et al. 2002. Involvement of TRPC in the abnormal calcium influx observed in dystrophic (mdx) mouse skeletal muscle fibers. J. Cell Biol. **158:** 1089–1096.
64. TURNER, P. R., R. SCHULTZ, B. GANGULY, et al. 1993. Proteolysis results in altered leak channel kinetics and elevated free calcium in mdx muscle. J. Membr. Biol. **133:** 243–251.
65. HOPF, F.W., P.R. TURNER, W.F. DENETCLAW, JR., et al. 1996. A critical evaluation of resting intracellular free calcium regulation in dystrophic mdx muscle. Am. J. Physiol. **271:** C1325–1339.
66. LINDAHL, M., E. BACKMAN, K. G. HENRIKSSON et al. 1995. Phospholipase A2 activity in dystrophinopathies. Neuromuscular Disorders **5:** 193–199.
67. HASSELGREN, P.O. & J.E. FISCHER. 2001. Muscle cachexia: current concepts of intracellular mechanisms and molecular regulation. Ann. Surg. **233:** 9–17.
68. BHATTACHARYYA, J., K.D. THOMPSON & M.M. SAYEED. 1993. Skeletal muscle Ca^{2+} flux and catabolic response during sepsis. Am. J. Physiol. Regul. Integr. Comp. Physiol. **265:** R487–R493.
69. BENSON, D.W., P. HASSELGREN, D.T. HIYAMA, et al. 1989. Effect of sepsis on calcium uptake and content in skeletal muscle and regulation in vitro by calcium of total myofibrilar protein breakdown in control and septic muscle: results from a preliminary study. Surgery **106:** 87–93.
70. WILLIAMS, A. B., G. M. DECOURTEN-MYERS, J. E. FISCHER, et al. 1999. Sepsis stimulates release of myofilaments in skeletal muscle by a calcium-dependent mechanism. FASEB J. **13:** 1435–1443.
71. LEE, R.C., J. HANNIG, K. L. MATTHEWS, et al. 1999. Pharmaceutical therapies for sealing of permeabilized cell membranes in electrical injuries. Ann. N. Y. Acad. Sci. **888:** 266–273.
72. BEATON, L.J., M.A. TARNOPOLSKY & S.M. PHILLIPS. 2002. Contraction-induced muscle damage in humans following calcium channel blocker administration. J. Physiol. **544:** 849–859.
73. DUARTE, J.A., J.M. SOARES & H.J. APPELL. 1992. Nifedipine diminishes exercise-induced muscle damage in mouse. Int. J. Sports Med. **13:** 274–277.
74. TAVEAU, M., N. BOURG, G. SILLON, et al. 2003. Calpain 3 is activated through autolysis within the active site and lyses sarcomeric and sarcolemmal components. Mol. Cell. Biol. **23:** 9127–9135.
75. GOMEZ-CABRERA, M.C., C. BORRAS, F.V. PALLARDO, et al. 2005. Decreasing xanthine oxidase mediated oxidative stress prevents useful cellular adaptions to exercise in rats. J. Physiol. **567:** 113–120.

Protein Denaturation and Aggregation

Cellular Responses to Denatured and Aggregated Proteins

STEPHEN C. MEREDITH

Departments of Pathology, and Biochemistry and Molecular Biology, University of Chicago, Chicago, Illinois 60637, USA

ABSTRACT: Protein aggregation is a prominent feature of many neurodegenerative diseases, such as Alzheimer's, Huntington's, and Parkinson's diseases, as well as spongiform encephalopathies and systemic amyloidoses. These diseases are sometimes called protein misfolding diseases, but the latter term begs the question of what is the "folded" state of proteins for which normal structure and function are unknown. Amyloid consists of linear, unbranched protein or peptide fibrils of ≈100 Å diameter. These fibrils are composed of a wide variety of proteins that have no sequence homology, and no similarity in three-dimensional structures—and yet, as fibrils, they share a common secondary structure, the β-sheet. Because of the prominence of amyloid deposits in many of these diseases, much effort has gone into elucidation of fibril structure. Recent advances in solid-state NMR spectroscopy and other biophysical techniques have led to the partial elucidation of fibril structure. Surprisingly at the time, for β-amyloid, a set of 39–43-amino-acid peptides believed to play a pathogenic role in Alzheimer's disease, the β-sheets are parallel with all amino acids of the sheets in-register. Since the time of those observations, however, it has become clear that there is no universal structure for amyloid fibrils. While many of the amyloid fibrils described thus far have a parallel β-sheet structure, some have antiparallel β-sheets, and other, more subtle structural differences among amyloids exist as well. Amyloids demonstrate *conformational plasticity*, the ability to adopt more than one stable tertiary fold. Conformational plasticity could account for "strain" differences in prions, and for the fact that a single polypeptide can form different fibril types with conformational differences at the atomic level.
More recent data now indicate that the fibrils may not be the most potent or proximate mediators of cyto- and neurotoxicity. This damage is not confined to cell death, but also includes more subtle forms of damage, such as disruption of synaptic plasticity in the central nervous system. Rather than fibrils, prefibrillar aggregates, variously called "micelles," "protofibrils," or ADDLs (β-amyloid-derived diffusible ligands in the case of β-amyloid) may be the more proximate mediators of cell damage. These are soluble oligomers of aggregating peptides or proteins, but their structure is very challenging to study, because they are generally difficult to obtain in large enough quantities for high-resolution structural techniques, and they are temporally unstable, rapidly changing into more mature, and eventually fibrillar forms. Consequently, the mechanisms by which they disrupt cellular function are also not well understood. Nevertheless, three broad, overlapping, nonexclusive sets of mechanisms

Address for correspondence: Stephen C. Meredith, Department of Pathology, University of Chicago, 5841 S. Maryland Avenue, MC 6079, Chicago IL 60637. Voice: 773-702-1267; fax: 773-834-5251.
scmeredi@uchicago.edu

have been proposed as responsible for the cellular damage caused by soluble, oligomeric protein aggregates. These are: (1) disruption of cell membranes and their functions [e.g., by inserting into membranes and disrupting normal ion gradients]; (2) inactivation of normally folded, functional proteins [e.g., by sequestering or localizing transcription factors to the wrong cellular compartment]; and (3) "gumming up the works," by binding to and inactivating components of the quality-control system of cells, such as the proteasome or chaperone proteins.

KEYWORDS: protein aggregation; protein aggregation diseases; protein misfolding diseases; conformational diseases; amyloid; amyloidosis; neurodegenerative diseases; Alzheimer's disease; Parkinson's disease; Huntington's disease; neurotoxicity; cytotoxicity; solid state NMR; β-sheets; β-amyloid; α-synuclein; transthyretin; amyloid A; huntingtin; prions

HISTORY AND DEFINITIONS

In 1857, Rudolf Virchow used the term "amyloid" to describe an amorphous, eosinophilic extracellular deposit occurring in corpora amylacea of the nervous system. He was not, however, the first person to use that term: it was coined by Matthias Schleiden, a German botanist, in 1838.[1] Virchow was unimpressed with reports of the high nitrogen content of this material, and denied that it was proteinaceous, because of its having tinctorial properties or starches ("amylins"): the corpora amylacea were found to stain with iodine. Although he was wrong in believing that amyloid was composed of carbohydrates, he also believed that the data were inadequate to resolve the question of amyloid's composition, though as it turns out, its tinctorial resemblance to starch was more than coincidental. Routine histologic sections stained with hematoxylin and eosin ("H&E") show amyloid to be intensely eosinophilic. For diagnostic purposes, pathologists stain tissues with Congo red, a symmetric diazo compound that stains amyloid pink, and demonstrates apple-green *birefringence* when the tissue is observed through crossed polarizing filters. The birefringence of Congo red dye bound to amyloid reflects the fact that amyloid fibrils are ordered—but less so than crystals, for which reason amyloid fibrils are sometimes referred to as paracrystalline. Another stain is thioflavin T, which fluoresces when it binds to amyloid. Somewhat surprisingly, the exact chemical mechanism by which amyloid induces thioflavin fluorescence is still not well understood. Electron microscopy shows that amyloid is composed of linear, unbranched rod-shaped fibrils of ≈100-Å diameter (see review by Cohen *et al.*[2]). Amyloids sometimes appear to be made up of twisting duplexes, but this appearance is variable, even for amyloids derived from a single protein.

The chemical composition of amyloids was described in the 1960s and 1970s. Surprisingly (at that time), despite the fact that all amyloids look pretty much alike at light and electron microscopic levels, the insoluble fibrils of amyloids were found to be composed of a wide variety of proteins with little or no sequence homology. Indeed, the same is true of these proteins even at the level of tertiary fold, and it is now widely held that many proteins, perhaps all proteins, are capable of forming amyloid.

At the same time that the chemical nature of amyloid was being defined,[3,4] several physical techniques indicated that the amyloid fibrils are composed of β-sheets.[5] The structures of the monomeric forms of many amyloid-forming proteins are known. The limited water-solubility of most amyloid-forming proteins, however, and the paracrystalline state of amyloid fibrils themselves precluded the use of most high-resolution structural techniques. Through the use of X-ray fiber diffraction, FTIR, and other physical techniques, it was discovered in the 1980s that most or all amyloid fibrils had a common secondary structure, the β-sheet.[5–8] From X-ray fiber diffraction, the "cross β-fibril" structure was observed for many amyloids. This term, which is used widely, though sometimes inaccurately, refers to a meridional reflection at 5 Å, corresponding to hydrogen-bonded β-strands in line with longitudinal fiber axis, and an equatorial reflection at 10-Å reflection, corresponding to lamination of β-sheets at right angles to the long axis of the fibril. The β-sheets interact with each other through side chains, so the exact size of the equatorial reflection is a function of the average sizes of the side chains, and can range from 7.5 Å for some silks with mainly Gly and Ser side chains, to 12 Å for laminated β-sheets rich in aromatic and other bulky amino acids. These techniques do not have sufficient resolution to determine the detailed relationship among peptide chains in the fibril. For historical reasons, most of the early models of amyloid fibrils based on the observation of the cross-β structure assumed that the β-sheets were antiparallel. Most recently, however, solid-state NMR (discussed in more detail below) has shown a parallel β-sheet structure for several β-amyloid peptides, and a recent X-ray crystallographic study of a short poly Gln/Asn amyloid suggests that the parallel β-sheet may be the most common structure for amyloids in general. Nevertheless, despite impressive strides in investigating the structure of amyloid fibrils, the structure of amyloid is not known in much detail.

THE CHEMICAL NATURE OF AMYLOIDS

Systemic Amyloidosis Syndromes

Amyloid fibrils can be isolated by as an insoluble remnant after differential sieving, centrifugation, and extraction of lipids.[9–11] Further purification takes advantage of the fact that amyloid fibrils are protease-resistant and insoluble, and thus can be from most other proteins by extraction and proteolysis of non-amyloid proteins. Such procedures allow the isolation of fibrils of the relevant protein or peptide at purities of ≈90–95%. TABLE 1 shows a classification of fibril-associated diseases and the relevant fibril-forming proteins.

In addition to the fibril-forming protein, most amyloid fibrils formed *in vivo* also contain protein P—also called amyloid P, somewhat improperly, since this protein does not itself form amyloid fibrils, but is a practically invariant non-fibrillar component of amyloid fibrils *in vivo*. Protein P is a member of the pentraxin protein family, which shares a pentameric disc-like protein structure with a five-fold axis of rotational symmetry around the minor axis of the disk.[12,13] This family also includes C-reactive protein.[14] Amyloid P binds a variety of ligands, including DNA and amyloid fibrils, in a calcium-dependent manner, with each subunit having a ligand-binding site, all positioned on the B face of the molecule. The physiological function of protein P is not known. It is not necessary for amyloid fibril formation, and indeed, it probably coats the exterior of fibrils and limits their lateral growth. No naturally occurring deficiency or polymorphism of amyloid P has been described, and it is highly conserved in evolution, both of which suggests a necessary but probably redundant physiological function. Amyloid P knockout mice are viable and apparently

TABLE 1. Classification of fibril-associated diseases and the relevant fibril-forming proteins

Clinicopathologic Category	Diseases	Precursor Protein
Systemic (Generalized) Amyloidosis		
Immune dyscrasias	Multiple myeloma Waldenström's macroglobulineamia	Immunoglobulin light chains, disproportionately λ
Reactive systemic amyloidosis	Tuberculosis, rheumatoid arthritis, other chronic inflammatory conditions	Amyloid A (N-terminal fragment of serum amyloid A
Hemodialysis-associated amyloidosis	Chronic hemodialysis	β_2-microglobulin
Hereditary amyloidosis		
Polyneuropathy	Familial Mediteranean fever	
Familial amyloidoses	Various organ failure	Mutant transthyretin, mutant apolipoproteins, many others
Localized Amyloidoses		
Senile cardiac amyloidosis	Usually asymptomatic	Transthyretin (wild-type)
Senile and presenile cerebral amyloidosis	Alzheimer's disease	β-amyloid, derived from β-amyloid precursor protein
Medullary thyroid carcinoma		Calcitonin
Islets of Langerhans	Type II diabetes mellitus	Islet amyloid precursor protein
Isolated atrial amyloidosis		Atrial natriuretic protein
"Conformational Diseases"		
Spongiform encephalopathies	Creutzfeld–Jakob disease Fatal familial insomnia Gerstmann	Human prion protein in "scrapie" form
Huntington's disease	Huntington's disease	Huntingtin (polyglutamine repeats)
Pediatric cirrhosis	α_1-antitrypsin deficiency, PiZZ type	Mutant α_1-antitrypsin

healthy, but have alterations in plasma clearance of DNA and chromatin. Like C-reactive protein, amyloid P may be a part of innate immunity and confer some degree of resistance to a wide variety of infections. The physiological function of its amyloid-binding properties is not known. When bound to amyloid, amyloid P protein is highly resistant to proteases, as a result of which, fibrils—already resistant to proteases in the absence of amyloid P— are even more protease resistant. Indeed, amyloid P knockout accelerates amyloid fibril clearance. Whatever its role in amyloid formation, amyloid P has become useful as the basis of a diagnostic test for amyloidosis, whole-body ^{123}I-labelled protein P (or serum amyloid P) scintigraphy. Because the protein binds to amyloid fibrils, ^{123}I-labelled protein is used to image amyloid load in patients, essentially regardless of the chemical nature of the particular amyloid.[15]

The following is a brief survey of several notable amyloids associated with human (and in some cases, non-human) diseases:

Serum Amyloid A (SAA)

Serum amyloid A is an acute-phase reactant of unknown physiological function.[16] It has properties of an apolipoprotein,[17] and indeed, when it is made in large quantities during the acute phase, it displaces apolipoprotein A-I from HDL, and apolipoprotein A-I is a negative acute-phase reactant. SAA is a polymorphic protein with three isoforms in humans (SAA1, SAA2 and SAA4, plus one pseudogene, SAA3). Amyloid fibrils are composed of SAA1 and SAA2, both of which are acute-phase reactants (whereas SAA4 is constitutively expressed), and there are a number of common allelic variants of each of these proteins.[18-21] SAA1 and SAA2 have 104 amino acids (the N-terminal Arg is often removed post-translationally). The tertiary structure of SAA is not known, but the N-terminal domain is necessary for both lipid binding and amyloid formation.

The concentration of SAA in plasma rises in the acute phase from 1 to 1000 mg/L, which suggests an important role for this protein in inflammation. A number of such roles have been proposed, most of which involve modulation or inhibition of inflammation (e.g., inhibition of antibody production,[22] platelet aggregation,[23] the oxidative burst reaction in neutrophils,[24] induction of collagenase,[25] and chemotaxis for neutrophils and monocytes[26]).

Secondary structural algorithms predict an α-helical structure, and this is indeed the structure of the lipid-bound form of the intact protein. Amyloid, however, is formed not from the intact protein, but from an N-terminal proteolytic fragment,[21,27] and the site of cleavage is critical for amyloid formation: a 76-amino-acid fragment (residues 1–76) generated by cathepsin G forms amyloid, while a 79-amino-acid fragment (residues 1–79) made by cathepsin B generates a non-amyloidogenic peptide. These fragments appear to undergo an α-helix $\rightarrow$ β-sheet transition, which may be inhibited by the C-terminal domain of the intact protein.

Clinically, SAA deposition is the classic secondary amyloidosis, and is associated with chronic inflammation in conditions such as osteomyelitis, rheumatoid arthritis, tuberculosis and chronic cholecystitis, in which secretion of SAA and other acute phase reactants persists over prolonged times. This fact, however, begs the question of why only a small minority of patients with chronic inflammatory conditions develops SAA amyloidosis.

Transthyretin (TTR)

Transthyretin (previously called "prealbumin") is an abundant plasma protein (0.2–0.25 mg/mL) that binds a number of ligands, including thyroxine and retinol binding protein.[28] It is encoded on chromosome 18 and is made in the liver and the choroid plexus. It is also one of the most abundant proteins in CSF (0.02–0.04 mg/mL), and its synthesis in the choroid plexus is believed to facilitate transport of thyroid hormone in CSF.

Wild-type transthyretin is a 129-amino-acid protein that circulates mainly as a homotetramer. The wild-type protein is associated with senile cardiac amyloidosis,[29] a condition that is of no clinical significance in most individuals, but which nevertheless demonstrates the amyloidogenic potential of even the wild-type protein. Of greater clinical significance are point mutations in TTR that lead to earlier onset of TTR deposition in the heart, and clinically significant cardiomyopathy. In addition, familial amyloid polyneuropathy results from any of more than 50 different point mutations in TTR, due to deposition of the amyloid in peripheral and autonomic nerves. The inheritance of these forms of familial amyloidosis is autosomal dominant, and most patients with this disease are heterozygotes.[30]

The structure of normal TTR has been solved by X-ray crystallography.[31,28] It consists of a β-sheet sandwich, in which one β-sheet (D-A-G-H β-strands) is stacked upon another β-sheet (C-B-E-F β-strands) of the same molecule. Although the main form of the wild-type protein is tetrameric, the protein undergoes a monomer–dimer–tetramer equilibrium. In the dimeric form of TTR, there are interactions between the H strands of identical subunits. The contacts between strands in the tetramer are a bit more complex, and are briefly described below (see also Koo et al.[32]).

Aggregation of TTR mutants has been shown to result from aberrations not of the tertiary fold of these proteins, but of their conformational dynamics.[33-37] The dynamics of TTR structure and their role in aggregation have been described in elegant work by Kelly and colleagues.[38] By way of illustration, consider two particularly amyloidogenic mutants, Val30Met and Leu55Pro point mutations. The X-ray crystallographic structures of these mutant proteins are almost identical to that of the wild-type protein. The differences were quite subtle: in the immediate vicinity of the mutation, there was slight increase in the sheet-to-sheet separation in the monomer. The main differences occur in the conformational dynamics of the proteins. Some of the mutants are thermodynamically less stable than the wild-type protein; other point mutants are not less stable than wild-type protein thermodynamically (free energy difference between folded and denatured forms), but altered conformational dynamics and denaturation kinetics allow the protein to enter an amyloidogenic state more frequently or rapidly than the wild-type. In either case, a structural element of the protein, particularly the loop containing the C and D β-strands (e.g., in the Leu55Pro mutant), in effect, partially denatures, and forms a flexible loop that becomes available for intermolecular, rather than intramolecular contacts. Thus, the mutant proteins populate an amyloidogenic form of the protein more readily than does the wild-type protein. *In vitro*, TTR proteins can be induced to form amyloids in a mildly acidic pH range of the lysosome (i.e., 4–5.5). The essence of the proposed pathway for amyloid formation, shown below, is that denaturation and fibrillogenesis are competing processes, where the latter is favored in the mutant proteins and the former is favored in the wild-type proteins. A TTR bearing point mutation populates the partially denatured state (called the A state in the Diagram 1 below) more often than does the wild-type protein, and hence is more susceptible to forming fibrils.

Structural studies of TTR have led to novel approaches to the therapy of familial amyloid polyneuropathy and other amyloid-related diseases associated with point mutations in TTR.[40-43] As described above, the tetramer of TTR is a β-sheet–rich

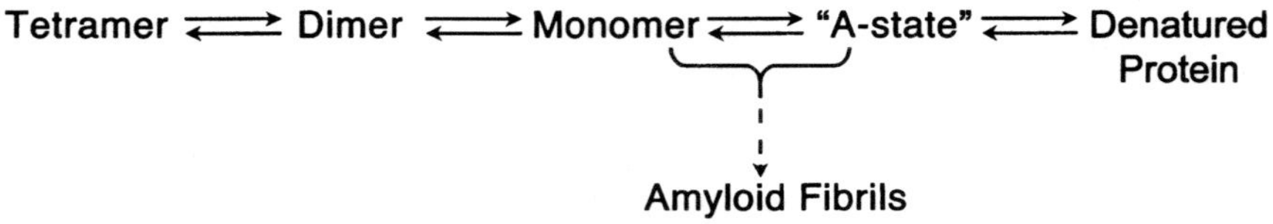

Tetramer ⇌ Dimer ⇌ Monomer ⇌ "A-state" ⇌ Denatured Protein
                                ↓
                        Amyloid Fibrils

DIAGRAM 1.

structure. It exhibits 2,2,2 symmetry, and thus consists of two pairs of dimers (A-B and C-D). Within each of the dimers, the two monomer subunits associate along the H β-strands, which terminate one of the two β-sheets. The two dimers are related to one another through a two-fold axis of rotational symmetry, the crystallographic C_2 symmetry axis. Two binding sites for thyroxine and other small ligands occur at the interface between the two dimers, one between the A and C subunits, and one between the B and D subunits. TTR is not the main thyroxine binding protein in plasma, and in the circulation, 90% of the binding sites for thyroxine in TTR are vacant. In CSF, however, TTR is the main thyroxine transporter, and 75% of these sites are occupied. Binding of thyroxine and other small ligands stabilizes the tetrameric form of the monomer–dimer–tetramer equilibrium, and, indeed, in patients with fibrillogenic point mutants of TTR, fibrils tend not to form in brain tissue. This is consonant with the observation that a monomeric form of TTR is the fibril-forming species. In theory, one could inhibit TTR fibril formation by saturating the protein with thyroxine, were it not for the toxicity of the high levels of thyroxine that would be needed. Accordingly, Kelly and colleagues have sought and developed other high-affinity, but hormonally inactive ligands for the thyroxine binding sites of TTR to stabilize the tetrameric form of the protein and prevent its dissociation into monomers, the precursor of the fibril. These ligands including several with a bisarylamine scaffold that includes a nonsteroidal anti-inflammatory agent, flufenamic acid,[40,41] which served as a starting point for further rational structure-based drug design of fibril inhibitors.[44]

β2-microglobulin (β2M)

β2-microglobulin is a normal component of the class I MHC complex that is also secreted as a plasma protein. It is sufficiently small (12 kDa) that it is normally cleared from plasma by excretion into urine, but in patients receiving chronic hemodialysis, it is not cleared into the urine and is not cleared by hemodialysis procedures, and is retained in plasma.[45] In patients on hemodialysis for more than 10 years, dialysis-related amyloidosis is a common and serious complication. β2M is deposited in connective tissues, especially synovium, tendons, and bones, where it appears to bind to collagen. In addition to its clinical importance, the small size of β2M makes it a suitable subject for detailed studies of protein folding. Structurally (FIG. 1), β2M is a mainly β-sheet protein, containing a sandwich of two sheets, one with four β-strands (A, B, D, and E) and one with three β-strands (C, F, and G).[46]

β2M fibrils contain many other constituents in addition to β2M itself, including glycosaminoglycans, apolipoprotein E, α2-macroglobulin and other protease inhibitors, and serum amyloid P.[47] While some of these components may serve to limit or regulate fibril growth, some of them are probably essential for fibril growth *in vivo*. *In vivo*, β2M fibrils form first in cartilage, and glycosaminoglycans and proteoglycans have been implicated in seeding the growth of β2M fibrils. It may be that under conditions of physiological pH, homogeneous nucleation of β2M fibrils is kinetically hindered, but heterogeneous nucleation on a connective tissue (e.g., cartilage) surface can occur.

β2M readily forms fibrils *in vitro* under acidic conditions, in the presence or absence of seed fibrils. In contrast, at pH closer to physiological pH, unseeded solutions of β2M yield fibrils only at low yields and inconsistently. Goto and coworkers[48] observed that repeated self-seeding at pH 7.0 of β2M amyloid fibrils prepared originally at pH 2.5 results in the maturation and stabilization of fibrils to pH 7.0, which contrasts with ordinary self-propagation of amyloid fibrils. To investigate this phenome-

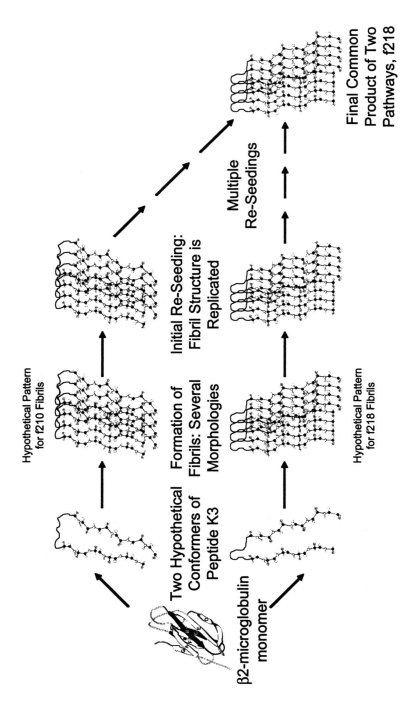

FIGURE 1. Yamaguchi et al.[48] observed that the peptide K3 (shown in black, sequence –SNFLNCYVSGFHPSDIEVDLLK–) residues 20–41 from β2-microglobulin was able to form two distinct types of fibrils in 20% (v/v) TFE and 10 mM HCl at 25°C. Both types of fibrils could be replicated by seeding, indicating that there was template-dependent propagation of a fibril's conformation. Repeated self-seeding, however, led to the conversion of one fibril type (f218) into the other, apparently more mature fibril type (f210). These observations were explicable entirely on the basis of competitive propagation of two fibrils.

non further, these investigators studied an amyloidogenic fragment Ser20–Lys41 (K3). This peptide formed two distinct types of fibrils (termed f210 and f218) when incubated in 20% (v/v) 2,2,2-trifluoroethanol and 10 mM HCl. The two fibril types differed by ultrastructural appearance and β-sheet content (CD and FTIR). Both fibril types could be reproduced by self-seeding, but upon repeated self-seeding, f218 fibrils were gradually transformed into f210 fibrils, indicating conformational maturation. The maturation phenomenon could be explained in terms of kinetic competition in the propagation of the two fibril types. This type of "strain phenomenon" in amyloid fibrils is reminiscent of prion fibrils. In the latter case, however, the barrier for converting one fibril type to another is sufficiently high that in any given patient with a prion disease, the strain of the prion—and with that, the disease phenotype—remains constant. In the case of β2M, however, conformational maturation of amyloid fibrils may occur, yielding a common final amyloid fibril structure.

Not to Slight Others...

The above cases were chosen for particular didactic points, but one would be remiss not to mention the following clinically and scientifically important proteins that give rise to amyloid fibrils and clinically can cause amyloidosis.

1. Immunoglobulin light chains: In B lymphocyte malignancies, especially multiple myeloma and Waldenström's macroglobulinemia, immunoglobulin light chain production is dysregulated (i.e., the normal coordination with heavy chain production is lost). Light chain monomers or, more commonly, dimers, are secreted, and a low percentage of these are amyloidogenic. The "rules" governing which light chains will form amyloid are under investigation. A disproportionate percentage (75%) of amyloid-forming light chains are type VI λ light chains (75% of normal light chains and light chains secreted by myelomas are κ chains). Often, the intact light chain is not amyloidogenic, but an N-terminal fragment of the light chain is. The same is true for other forms of amyloid (e.g., amyloid A).

2. Apolipoprotein mutants: A number of mutant proteins—many of them derived from apolipoproteins—can form amyloid. There are rare, amyloidogenic mutant forms of apolipoprotein: A-I, apolipoprotein A-II, and apolipoprotein E, among others. These often are associated with peripheral neuropathy.

3. Calcitonin: Among the peptide hormones that frequently form amyloid in patients is calcitonin, which is made by medullary carcinomas of the thyroid. This peptide hormone routinely gets deposited in extracellular matrix of medullary thyroid carcinomas.

4. Amylin, also called *islet amyloid polypeptide (IAPP)*: This peptide is made in the β-cells of the pancreatic islets. Normally, its synthesis is coordinated with that of insulin, but in many patients with diabetes, especially type II diabetes, the synthesis of IAPP increases as that of insulin decreases. In most patients with advanced type II diabetes, IAPP amyloid is seen in the islets.

5. Insulin: Human insulin tends to form fibrils, a significant concern to pharmaceutical companies making preparations of insulin, and to designers and manufacturers of insulin pumps. In addition, the degu, a Chilean relative of the guinea pig, is very prone to develop diabetes, and its islets become hyalinized and replaced by fibrillar material. Surprisingly, however, the fibrillar material is insulin, not IAPP.

6. And still others: Atrial natriuretic peptide can sometimes be deposited as amyloid in an isolated atrial form of amyloidosis. Some gelsolin fragments can make

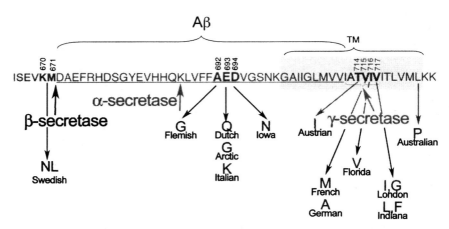

FIGURE 2. Structure of Aβ peptides, showing cleavage sites for α-, β-, and γ-secretases. The main product of the action of the β- and γ- secretases is Aβ(1–40); Aβ(1–42) is normally ≤5% of the product of this reaction, but is disproportionately deposited in neuritic plaques in Alzheimer's disease, Also shown are several mutation sites associated with early-onset, familial Alzheimer's disease, within and outside of the Aβ peptide itself. Many of these sites are either adjacent to the cleavage sites for the β- and γ-secretases, or adjacent to the bend region of Aβ(1–40), which is the likely focus of the conformational plasticity of Aβ peptides.

amyloid and are associated with a form of familial hereditary polyneuropathy. In addition, mutants or fragments of lysozyme, fibrinogen, and cystatin C, among many others, are also rare causes of amyloidosis.

Amyloid and Neurodegenerative Diseases

β-Amyloid Peptide (Aβ) and Alzheimer's Disease

a. Sequence and Origin of β-Amyloid Peptides: β–amyloid (Aβ) is a group of peptides of 39-43 amino acids, of which the most abundant ones in CSF are 40- and 42-amino-acids long.[49,50] The sequence of Aβ(1–42) is:

$$NH_2-DAEFRHDSGY^{10}\ EVHHQKLVFF^{20}\ AEDVGSNKSA^{30}\ IIGLMVGGVV^{40}\ IA-COOH$$

Aβ (FIG. 2) is derived from a larger precursor protein, the β-amyloid precursor protein (β-APP), a large-membrane glycoprotein with a single transmembrane domain. Most of the protein, including the N-terminus, is extracellular. It has some domains that are homologous with Kunitz protease inhibitors. The normal function of this protein is not known. β–amyloid is derived from the sequential action of two proteolytic activities, called secretases (FIG. 3). The intracellular itinerary of β-APP is quite complex and the site(s) at which it is cleaved are controversial. The β-secretase, called BACE1, is a transmembrane aspartyl protease that cleaves within the ectodomain of β-APP, leaving a C-terminal fragment that still contains the single transmembrane domain, and becomes a substrate for the γ-secretase.[51,52] Another protease, α-secretase, cleaves the ectodomain at a site closer to the transmembrane domain of the β-APP, and also creates a substrate for the γ-secretase.[53] Whereas the

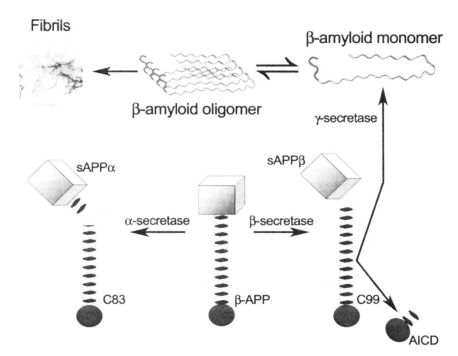

FIGURE 3. Processing of the β-amyloid precursor protein (β-APP). β-APP is a single-pass transmembrane protein, of which most of the mass is lumenal/extracellular. Cleavage by the α-secretase yields a soluble N-terminal fragment, sAPPα. The portion of the protein that remains in the membrane, C83, is a substrate for the γ-secretase, but this yields a non-pathogenic peptide. Cleavage of β-APP by the β-secretase yields an N-terminal fragment, sAPPβ, and a membrane-bound C-terminal fragment, C99. C99 is also a substrate for γ-secretase, which upon cleavage by that enzyme yields β-amyloid peptides (exact length varying from ≈38–43, main products of 40 and 42 residues), and the APP intracellular domain (AICD). Once liberated from β-APP, β-amyloid can undergo its aggregation pathway, first to soluble oligomers, and finally to fibrils. Recent evidence indicates that some soluble oligomeric forms of β-amyloid are more cytotoxic than the final fibril product of the aggregation pathway.

β-sectretase leads to production (through the γ-secretase) of an amyloidogenic protein, the product of sequential action of the α- and γ-secretases, called P3 peptide, is nonamyloidogenic.

The identity of the γ-secretase is debated, though there is growing evidence indicating that this activity may reside in the protein *presenilin*. γ-secretase is a high molecular weight, membrane-bound aspartyl protease containing at least four polypeptides: presenilins, nicastrin, APH-1 and PEN-2.[54–57] These four proteins are sufficient to reconstitute γ-secretase activity in yeast, an organism that does not possess its own γ-secretase activity.[58] In addition to its role in generating β-amyloid, the γ-secretase may be involved in several signal transduction pathways that are important for maintaining synaptic function. At present, γ-secretase is known to have at least 16 transmembrane substrate targets.[59,60] It is believed to cleave β-APP and its other targets close to the membrane–cytosol interface, although the exact location of this interface is not known for β-APP or its other targets. Presenilin1 was the first member of this complex to be described, and knockout of this protein leads to decreased production of Aβ. Presenilin1 has seven transmembrane domains, and the two catalytic Asp residues of the γ-secretase may reside in helices VI and VII; in any case, these residues are required for production of Aβ from β-APP, and action of this protease on other substrates such as Notch-1 and N-cadherin. The hypothesis that presenilin1 was the catalytic subunit of γ-secretase was greatly strengthened by the finding that aspartyl protease transition state analogue inhibitors bind directly to presenilin1.[61,62] If presenilin1 is indeed the catalytic subunit of the γ-secretase, however, it is an aspartyl protease with no homology to previously described enzymes of this class. The other proteins of the complex, nicastrin, APH-1, and PEN2, are believed to be necessary cofactors for the enzyme. A current model of γ-secretase is that nicastrin serves as a scaffold for APH-1 in the endoplasmic reticulum. This complex then recruits presenilin1,[63, 64] which undergoes internal proteolysis and becomes catalytically active only when PEN-2 joins the complex.[65,66]

Cleavage of APP by the γ-secretase yields a series of peptides with a hydrophobic C-terminal domain of variable length, and, accordingly, a set of peptides in which the propensity to aggregate is proportionate to the length of the hydrophobic domain. The most abundant form of Aβ is Aβ(1–40), and this accounts for 90–95% of the Aβ peptides in normal individuals. The second most abundant of the Aβ peptides is Aβ(1–42). Although there is an inexact relationship between cerebral Aβ load and neurodegeneration, the latter is believed to be especially neurotoxic, and accumulates disproportionately compared with Aβ(1–40) in the brains of patients with Alzheimer's disease. The physiological function of Aβ peptides, like that of its precursor, is not well understood.

b. Structure of β-Amyloid Fibrils: There are two pathognomonic lesions in Alzheimer's disease brains. Neuritic plaques are extracellular lesions of which the most abundant protein is β-amyloid, which forms a core of the plaque. β-amyloid also gets deposited in blood vessels of the brain, and is associated with cerebral amyloid angiopathy and cerebral hemorrhages. Neurofibrillary tangles are aggregates within degenerating neurons; the main protein component of tangles is a hyperphosphorylated form of the intermediate filament protein, tau (τ).

The structure of β-amyloid fibrils has been studied by many physical techniques including X-ray diffraction, small angle X-ray and neutron scattering, and most notably, solid-state NMR (see review by Tycko[67]; see also Stejskal and Memory,[68] and Duer[69]). Solid-state NMR is especially suited to this type of structural problem, since, as mentioned earlier, the amyloid fibril is refractory to other high-resolution structural approaches, including X-ray crystallography and solution NMR spectroscopy. As with other NMR techniques, the basis of solid-state NMR is the absorption of radiofrequency electromagnetic radiation at the resonant frequency by an electron in an atom with a spin ½ nucleus (^1H, ^{13}C, ^{15}N, etc.). This absorption spectrum is sensitive to neighboring groups, other spins, solvent conditions, other influences, and these environmental effects are used to obtain structural data. The fundamental problem in solid-state NMR, however, is the extreme broadening of spectral lines. This is also seen, of course, in solution NMR with macromolecules as a function of increasing molecular weight: lines broaden increasingly due to chemical shift anisotropy (CSA) and other effects as rotational tumbling slows. While this remains a serious problem when working with proteins in solution, it renders solid-state

NMR spectroscopy impossible without adaptations to deal with line widths that often exceed 20 kHz. Fortunately, it has been found that line broadening from CSA can be greatly reduced by spinning samples at high speeds, and that the line width is proportionate to a term $(3\cos^2\theta - 1)$, where θ is the angle between the rotational axis of the spinning sample and the applied magnetic field. Setting $(3\cos^2\theta - 1)$ to 0 yields a value of ≈ 54.73561, called the magic angle, at which line widths are minimal. Most of the solid-state NMR experiments performed on β-amyloid and other fibrils are dipolar recoupling experiments. Dipolar couplings are typically lost in magic angle spinning, but can be selectively restored through the use of complex pulse sequences, which at the same time preserve the spectral resolution of magic angle spinning (MAS) methods. Among the pulse sequences that have been most useful for solid-state NMR of protein fibrils are homonuclear techniques such as DRAWS (Dipolar Recoupling in a Windowless Sequence) and fp-RFDR-CT (finite pulse, constant-time, radiofrequency driven recoupling), and heteronuclear techniques such as REDOR (Rotational Echo Double Resonance) spectroscopy. These techniques allow the determination of interatomic distances (up to 7.5 Å) between spin ½ nuclei with great precision (0.2 Å for distances up to 6Å). In addition, techniques are being developed for estimating torsional angles (φ and ψ) in appropriately labeled samples.

Solid-state NMR measurements have demonstrated that fibrils made from Aβ(10–35) and Aβ(1–40) have parallel β-sheet structure, in which amino acids are aligned in exact register.[70–75] This arrangement allows the β-amyloid molecules to sequester the hydrophobic C-termini away from the water solvent. Indeed, short, non-amphiphilic peptide fragments of Aβ(1–40), such as Aβ(16–22) or Aβ(34–42), can form antiparallel β-sheets,[76,77] but the orientation of the β-sheet can be flipped to parallel and in-register by a simple covalent modification (α-amino octanoylation) that renders the peptide amphiphilic.[78] These results also showed that there is no universal structure of amyloid fibrils.

In addition to the orientation of peptide chains within a single β-sheet, the parallel β-sheets of Aβ(1–40) laminate to roughly a 4–6-fold thickness to produce the dimensions of the fibrils—typically a diameter of 10 nm in a fibril with electron microscopic appearance of a twisted duplex. The exact laminated arrangement of β-sheets within the fibril is not known, however, and is probably more complex than simple lamination of continuous planar β-sheets. One recent model, proposed by Petkova et al.,[79–81] is that the protofilament (defined as the fibril with minimum dimensions and minimum MPL that can be observed experimentally) consists of two stacks of β-amyloid molecules, each containing two β-sheets. The two β-strands formed by each β-amyloid molecule are connected by a more flexible bend region. This model has important mechanistic implications about the way β-amyloid molecules aggregate, even before they form fibrils. The β-strands associate mainly through hydrophobic interactions and also by a salt bridge between Asp23 and Lys 28. This model was based on constraints from solid-state NMR data and mass-per-length data from electron microscopy. It also maximizes hydrophobic interactions in the context of the in-register, parallel β-sheet, and also avoids unfavorable electrostatic interactions. This model is also consistent with observations of Kheterpal et al.,[82,83] who measured H/D exchange for Aβ(1–40) fibrils, and showed that between 48% and 55% of backbone amide protons are highly protected from exchange.

A critical feature of the model is a non-β-sheet segment (residues 24–29) interposed between two β-sheet regions. This region, sometimes referred to as a "bend region" differs from a β-turn because the interactions are between side chains, not backbone atoms. One of these interactions is the formation of a salt bridge between Asp23 and Lys28, which has been demonstrated by REDOR solid-state NMR data.[79,84] Formation of this salt bridge and its burial within a hydrophobic pocket necessarily requires temporary desolvation of two charged amino acids; hence it may be a rate-limiting step in fibril formation. Sciarretta *et al.*[80] showed that substitution of a pre-formed lactam between the side chains of these two residues eliminates the lag period and accelerates fibril formation by 1000-fold, compared to unmodified Aβ(1–40).

c. *Kinetics of β-Amyloid Fibril Formation:* FIGURE 4A shows a typical time course of Aβ fibril formation, followed by thioflavin T binding and fluorescence. Both thioflavins[85] and Congo red,[86] a diazo dye, bind to most, but not all amyloid fibrils, suggesting that they associate specifically with β-sheet structures. Surprisingly, however, the exact site to which these dyes bind is not known. Nevertheless, this kinetic profile is typical: after a lag period in which thioflavin T fluorescence barely increases, there is a period of relatively rapidly increasing fluorescence, during which fibril formation proceeds rapidly. The lag period is often attributed to slow formation of a "seed" or "nucleus": an organized (i.e., structured), multimolecular core. Formation of the nucleus is slow because it requires high-order oligomerization; furthermore, not all of the collisions among monomeric subunits are productive of the seed nucleus. In addition, in the case of Aβ and other fibril-forming peptides, the monomeric subunit has little defined structure, but the fibril is structured, with a

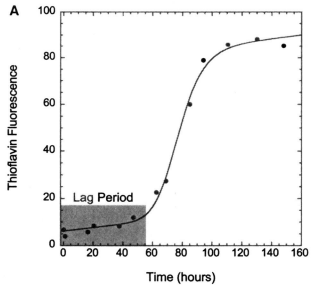

FIGURE 4A. Typical time course of fibrillogenesis. In this case, a solution of β-amyloid is incubated at 37°C in phosphate buffer (pH 7.40), and fibril formation is monitored discontinuously by thioflavin fluorescence. A long lag period occurs in which thioflavin fluorescence is fairly constant; this is followed by a relatively rapid increase in thioflavin fluorescence. During the lag period, both oligomerization and conformational changes occur. In the case of β-amyloid, this includes both formation of β-sheet structure, and the formation of a salt bridge (between Asp23 and Lys28) within a hydrophobic pocket formed by the two β-sheet domains.

high β-sheet content. Presumably, the conversion from monomer entails conformational changes, and it is also possible that some of these are slow or rate-limiting steps in fibril formation.

FIGURE 4B represents a minimalist scheme for fibril formation. Upon dissolving Aβ in buffer, even before there is a discernible increase in thioflavin or Congo Red dye binding, the peptide self-associates into large ($\approx$30–50 molecules) oligomers, and some of these show properties of micelles, such as diphenylhexatriene binding and fluorescence.[78,87–89] It is debated whether such micelle-like aggregates are on- or off-pathway for fibril formation,[90,91] and whether (or to what extent) the population of "micelles" overlaps with that of the seed nuclei. In addition, using size exclusion chromatography, one can identify smaller oligomers (the exact sizes and concentrations depend on buffer and other conditions) containing $\approx$6 molecules.[92–94] In the latter stages of fibril formation, the kinetics of aggregation are complicated by any physical disruption that tends to create new growth sites from linear fibrils (e.g., sonication, pipetting, and so forth). Finally, as shown by the PICUP (photo-inducible crosslinking of unmodified proteins) technique,[95,96] distinct pathways of oligomerization exist for Aβ(1–40) and Aβ(1–42),[97] and there is no reason to assume that such distinctions are unique to Aβ peptides. Beyond the intellectual satisfaction of understanding the ins and outs of this complicated pathway, there is the practical issue of which of these many intermediates is (are) neurotoxic. As discussed below, this issue is far from settled.

Polyglutamine Peptides and Polyglutamine Expansion Diseases

Huntington's disease is the best known of many hereditary neurodegenerative diseases linked to expansion of polyglutamine (polyQ) tracts; among the others are spinocerebellar ataxia type I (Kennedy's disease), dentatorubral–pallidoluysian atrophy, and Machado–Joseph disease.[98–100]

In Huntington's disease, there is selective loss of neurons in the striatum and cortex, which leads to movement disorders (chorea), dementia, and eventually, death. The gene for this disease is called huntingtin, a 348-kDa (3140-amino acid) protein, encoded on 4p16.3. As with other polyglutamine expansion diseases, there is a polymorphic CAG repeat, leading to expression of polyQ in huntingtin. The CAG repeats of huntingtin are encoded in exon 1 of the gene (for review, see Gusella and MacDonald[101] Ross,[100] and MacDonald et al.[102]).

In the case of Huntington's disease, there is a remarkably sharp boundary between diseased and disease-free: disease occurs only when the polyQ tract is greater than 40 residues long, and never when it is less than 35 residues long. In addition, with increasing length of the polyQ tract, there is progressively younger age of onset and more severe disease. Huntington's disease and other polyQ diseases also show a genetic phenomenon, *anticipation*: symptoms of the disease typically appear after reproductive age, and in those progeny who inherit this disease (with autosomal dominant inheritance) the disease appears at an early age and has a more severe phenotype than in the previous generation; this is associated with increased length of the expanded polyQ tract in progeny. Similar phenomena occur in other polyQ expansion diseases. For example, in spinocerebellar ataxia type I, disease occurs with polyQ tracts of > 40 residues, while non-affected individuals have polyQ tracts of <31 residues. In model organisms such as *S. cerevisiae*[103] and *C. elegans,*[104] similar

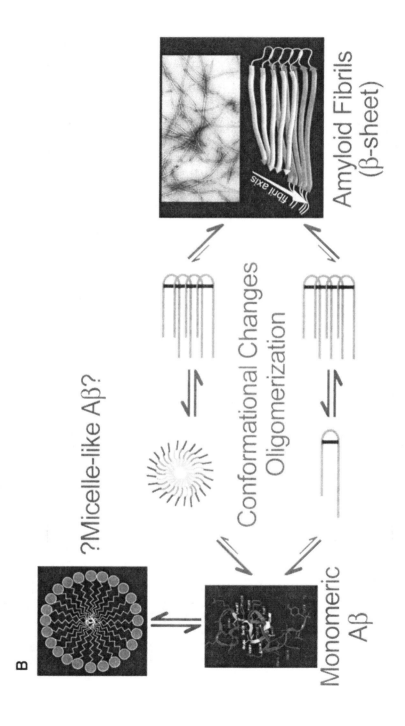

FIGURE 4B. *See opposite page for legend.*

threshholds have been seen for both aggregation and dysfunction (in the case of *C. elegans*, lower rates of motility and pharyngeal pumping).

PolyQ forms insoluble, granular nuclear and cytoplasmic deposits.[105,106] In transgenic mice, the appearance of these deposits precedes neural death.[107–110] The structure of these fibrillar aggregates is not known in detail. Several structures have recently been proposed, however. On the basis of X-ray diffraction data of a 19-amino acid peptide, $D_2Q_{15}K_2$, Perutz *et al.* proposed that fibrils of this peptide are hollow, water-filled nanotubes (external diameter, 3.1 nm; internal diameter, 1.2 nm) with a helical pitch of 20 residue per turn, and successive turns hydrogen-bonded through both backbone and side-chain amides.[111,112] The authors argued, furthermore, that this structure, with variations, was a general structure for amyloid fibrils made by other proteins, including β-amyloid, α-synuclein, and transthyretin. More recently, however, it has been shown that the X-ray diffraction data on which this model was based admits of an alternative interpretation: Sikorski and Atkins[113] proposed that the peptides adopt a once-folded hairpin conformation of stacked β-sheets that form cross-β crystallites. Another structure was obtained from microcrystals of a seven-residue peptide (GNNQQNY) from the yeast prion protein Sup35. These were of sufficient size, order, and stability for X-ray diffraction analysis, which showed that the peptide makes a double β-sheet, with each sheet formed from parallel segments stacked in-register.[114] Another structure, proposed from modeling simulations of 20-, 40- and 80-residue polyQ peptides, is the α-sheet. The α-extended chain conformation was originally proposed by Pauling and Corey[115] as consisting of alternation of residues in the helical $α_R$ and $α_L$ conformations. The α-sheet is formed by hydrogen bonding between adjacent strands in the α-extended chain conformation. As a result of their simulations, the authors propose that fibril formation by polyQ peptides occurs through an α-sheet structure.[116] Finally, the β-helix, observed in a crystal structure of pectate lyase[117,118] has been proposed on the basis of molecular dynamic simulations for aggregates of polyglutamine peptides, as well as other amyloids.[119]

α-Synuclein and Parkinson's Disease

The discussion of α-synuclein requires some definitions. At least 8 genetic loci for Parkinson's disease are known,[120] each given a PARK designation (α-synuclein, for example, is encoded by the *SNCA* gene at the PARK1 locus). Clinically, parkinsonism is defined as a triad of resting tremor, rigidity (and associated postural insta-

---

FIGURE 4B. Minimal scheme of fibrillogenesis. Monomers and fibrils are relatively highly populated forms of the peptide. Oligomers can, in theory, be on- or off-pathway for fibril formation. β-amyloid and other fibril-forming proteins and peptides exhibit micelle-like behavior, as shown by a critical concentration above which the peptide binds hydrophobic dyes such as diphenylhexatriene. It is not known whether such micelle-like aggregates are on- or off-pathway (i.e., their relationship to seed nuclei is not known). The most neuro/cytotoxic species of aggregating proteins and peptides is believed to be some soluble oligomeric form or forms. The model of the protofibril (lower right, from Petkova *et al.*[79]) is of two β-sheet regions (residues 11–23 and 31–40), separated by a "bend" region in the side chain dimension of the fibril. A single protofibril, the thinnest unit of the β-amyloid fibril observable in scanning transmission electron microscopy, is proposed to consist of two β-amyloid molecules, each of which is a double layer of β-sheets. Note that the hydrogen bonding dimension of the fibrils is oriented along the long fibril axis (roughly perpendicular to the plane of the page in the diagram).

bility), and bradykinesia, of which the last is usually the most troubling to patients. The signs of the triad all reduce to an inability to initiate or stop movements. Pathologically, this triad is attributable to a loss of dopaminergic neurons, especially in the substantia nigra pars compacta. Parkinson's disease, however, is also marked by a second pathological feature, the Lewy body, and Lewy neurites in surviving neurons. Lewy bodies appear as eosinophilic spheres in H&E-stained tissue, and consist of intracellular aggregates of a lipid core (notably sphingomyelin[121]) plus proteins, of which the most abundant is α-synuclein. The pathological definition of Parkinson's disease usually includes the presence of α-synuclein–containing Lewy bodies, in addition to loss of neurons of the substantia nigra and other pigmented neurons. Lewy bodies, however, are also found in other conditions. Among these other conditions are diffuse Lewy body disease (DLBD), where the Lewy bodies also contain α-synuclein, for which reason DLBD is also referred to as a synucleinopathy. In DLBD, however, Lewy bodies are found in many parts of the brain (not only substantia nigra), and the phenotype extends beyond parkinsonism to include dementia and fluctuations of consciousness. Thus, there are conditions with parkinsonism but no defects in α-synuclein, and synucleinopathies with clinical phenotypes different from that of Parkinson's disease. For this reason, neither the clinical symptoms nor the pathologic finding of Lewy bodies is sufficient for the diagnosis of Parkinson's disease.[122] Nevertheless, α-synuclein represents an important case-in-point for understanding the relationship between protein aggregation and disease.

α-Synuclein (FIG. 5) is part of a gene family that also includes β- and γ-synuclein.[123] Synucleins were first isolated from the electric organ of the *Torpedo* fish, which is essentially a massive collection of cholinergic vesicles such as are found in nerve synapses, and at the same time as a cDNA in a rat brain library.[124] α-Synuclein was found to be localized in the synapse and in the nucleus of neurons, but much of the protein is apparently free in cytosol, suggesting an ability to switch between a membrane-bound and a cytosolic form. All of the synucleins contain pseudo-repeats (discussed below) of the sequence –KTKEGV– in the N-terminal domain of the protein. The C-terminus is variable among the synucleins; in α-synuclein, this region is acidic, which tends to inhibit aggregation of the protein. The C-terminal region also contains three phosphorylation sites (two Ser, one Tyr).

Three known point mutations of α-synuclein are associated with autosomal dominantly inherited forms of Parkinson's disease: A53T,[125] A30P (German kindred),[126] and E46K (Spanish kindred).[127] A53T was the first of these mutations to be discovered, and curiously, this position (53 in humans) is T in rodents. In all of these kindreds, there are clinical parkinsonism, Lewy bodies, and neuronal loss in the substantia nigra. These patients also have dementia and widespread Lewy bodies in the cerebral cortices, and the E46K kindred was first reported as diffuse Lewy body disease. In addition to these point mutations, a large kindred from Iowa has Parkinson's disease resulting from triplication of the α-synuclein gene.[128] In this kindred, the Lewy bodies are found not only in neurons, but even in glial cells. These patients, and to some extent, the patients with the above point mutations in the α-synuclein gene, have aggressive Parkinson's disease with an earlier age of onset than sporadic forms of the disease.

α-Synuclein, like the other proteins discussed in this chapter, has a high propensity to aggregate into β-sheet–rich fibrils. It aggregates much more readily than either of the other synucleins, and a comparison of sequences therefore has put the

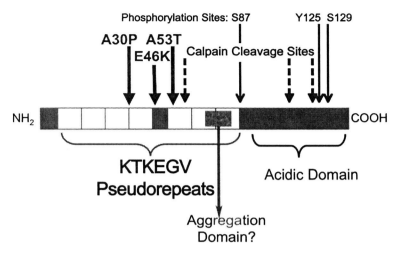

FIGURE 5. Structural motifs in α-synuclein. The protein contains 7 pseudo-repeats, –KTKEGV–, in its N-terminal domain. With surrounding residues, these are believed to form apolipoprotein-like amphiphilic α-helical domains that can bind to lipid. The free peptide in solution is believed to be a "natively unfolded" protein. Upon association with lipid membrane, the peptide undergoes a marked conformational change and becomes highly α-helical. Aggregation is believed to depend strongly on the presence of a hydrophobic region around residues 71–82. The C-terminal domain contains many acidic residues, and inhibits the innate tendency of the peptide to self-association, as can be shown by the increased association when this segment is deleted or excised. The figure shows several point mutations associated with familial Parkinson's disease, as well as phosphorylation sites and calpain I cleavage sites. Both phosphorylation and calpain cleavage favor self-association.

focus on a hydrophobic region of the protein at residues 71–82, encompassing parts of the last two repeat regions as a possible aggregation domain. This is the region of α-synuclein that is most different from corresponding regions of β- and γ-synuclein, both of which are much less hydrophobic in this part of the protein. This region, however, does not include the sites of the point mutations, and there is no direct evidence that this region is an "aggregation domain." In any case, fibrils are widely believed to be at the core of Lewy bodies. For example, immunogold labeling shows that α-synuclein is present at sites along fibrils that can be isolated from Lewy bodies.[129] α-synuclein is also a very sensitive marker for Lewy bodies, and appears to be an invariant component of them. In addition, α-synuclein is sufficient to form fibrils *in vitro*. Lewy bodies, however, are highly complex, and like neuritic plaques, contain variable amounts of many other proteins (e.g., ubiquitin) that are probably co-precipitants or covalent adducts, as well as a high content of lipids.[130] In addition, the α-synuclein in mature Lewy bodies is covalently modified, not only by phosphorylation,[131] but also by tyrosine nitration,[132] methionine oxidation,[133,134] and it is proteolyzed by calpain I.[135] These covalent modifications could enhance aggregation *in vivo*, though it is not entirely clear whether they precede or follow aggregation. In addition, sporadic Parkinson's disease is correlated with a wide variety of environmental exposures such as pesticides and transition metal ions, some of which could cause protein cross-linking through free radical mechanisms.[136]

As might be expected, the three point mutants of α-synuclein mentioned above aggregate more readily than the wild-type protein. The A53T protein forms fibrils, but the A30P protein has been reported not to do so;[a] the latter does form soluble oligomers, however. (No data are available yet on the E46K mutant). The fact that both A53T and A30P mutations lead to disease suggests that it is the oligomeric species rather than fibrils that are neurotoxic, as has also been proposed for β-amyloid and Alzheimer's disease. Fibril formation is probably increased in the Iowa kindred with triplication of the α-synuclein, because of the concentration dependence of fibril formation. Wild-type protein and mutant proteins (A30P and A53T) can be induced to form annular structures in phospholipids bilayers,[137,138] and these could damage membranes or lead to aberrant ion fluxes if they formed in cells. According to this view, then, fibrils and Lewy bodies are not necessarily damaging, but rather, represent markers that appear—only after the cell death, or at least, after cell damage has come and gone—they represent the "ashes" rather than the "fire." The possibility has also been raised that Lewy bodies serve a protective role.[136] One caveat in adopting this view, however, is that the association of the A30P mutation with disease is not as firmly established as is true for the other mutations. The kindred, while clearly afflicted with an inherited form of parkinsonism of high penetrance, is small, and no autopsy results are available; evidence of loss of substantia nigra neurons is limited to PET scans, meaning that the presence of disease has been established on clinical rather than pathologic grounds. Indeed, transgenic animals studies are more supportive of a pathogenic role for both A53T mutant and wild-type α-synuclein than for the A30P mutant protein. Transfection of rodents and primates with wild-type α-synuclein in viruses leads neuronal degeneration in the substantia nigra, but for reasons that are far from clear, results with transgenic species, including *C. elegans*, *Drosophila*, and mice, are strikingly variable and inconsistent. Part of the inconsistency may relate to expression levels. Nevertheless, in models in which expression of α-synuclein in human cell lines is driven by inducible promoters, in all cases tested, the protein has proven to be cytotoxic at the highest expression levels.[139,140] These studies, taken together, suggest that *some* aggregated species of α-synuclein is cytotoxic, though this species is not necessarily fibrils, and may be some small, soluble aggregate. In addition, the point mutants appear to increase aggregation, and since aggregation is concentration-dependent, the Iowa kindred with triplication of the α-synuclein gene also suffers the cytotoxic damage of protein aggregates, even though the protein is of the wild-type sequence.

α-synuclein somewhat resembles the small, exchangeable, amphiphilic α-helical apolipoproteins, such as apolipoprotein A-I.[141,142] It also shows modest homology to the fatty acid binding protein family.[143] The –KTKEGV– pseudo-repeats actually have a periodicity of 11 amino acids, three turns of an α-helix,[141] and can be arrayed as amphiphilic α-helices. This prediction is borne out experimentally. In solution, α-synuclein is essentially unstructured, but it acquires α-helical structure in the presence of lipid surfaces. It binds, however, specifically to small (20–25-nm diameter) but not large (125-nm diameter) unilamellar vesicles, and acidic, but not neutral phospholipids.[141] Upon binding to lipid surfaces, the protein undergoes a major in-

[a]In all of these kindreds—the three point mutations and the gene triplications—there are increased numbers of Lewy bodies in the brain at autopsy. Is this finding consistent with the proposal that the A30P protein does not form fibrils?

crease in α-helicity as estimated from CD spectra (from 3% to 80% in one study, with similar findings in another study).[144] The ability to bind lipid surfaces may account for the presence of this protein at synaptic membranes, and is in accord with its proposed function of regulating the reserve pool of synaptic vesicles in neurons.

From the point of view of aggregation, conditions that favor binding of the protein to lipid surfaces also disfavor fibril formation, suggesting that these two processes are opposed to one another. Zhu and Fink[145] showed that binding of α-synuclein to vesicles depends on phospholipid composition (vesicles with acidic phospholipids bind α-synuclein, whereas vesicles composed entirely of neutral lecithins do not) and the size of vesicles. Furthermore, binding to lipid was associated with induction of α-helical structure and inhibition of fibril formation. On the other hand, either low pH or increased temperature transforms the "natively unfolded" α-synuclein into a β-sheet containing partially compacted (radius of gyration by small angle x-ray scattering) intermediate on the pathway towards fibril formation.[146] Thus, this transformation to a fibrillogenesis-competent intermediate seems to be inhibited by lipid binding. The situation is further complicated by the fact that some lipid (or lipid surfaces) appear to accelerate fibril formation (e.g., vesicles with a low mass ratio of acidic:neutral phospholipids).

Lipids and lipid metabolism also affect aggregation in less direct ways. In one study,[143] treatment of mesencephalic neuronal cells in culture with polyunsaturated fatty acids increased levels of α-synuclein oligomers, as detected through protein crosslinking by a membrane-permeant crosslinker. Saturated fatty acids appeared to have the opposite effect. In addition, the increase in small oligomer formation is not necessarily accompanied by changes in the rate or extent of fibril formation, which was not measured in this study. This effect does not appear to be a direct effect of binding of α-synuclein to the administered fatty acids, and its basis may reside either in effects of administered fatty acids on cellular triglyceride metabolism or a byproduct of fatty acid metabolism (e.g., reactive oxygen species arising during fatty acid oxidation).

PROTEIN AGGREGATION AND PATHOGENICITY

So... Why is Protein Aggregation Harmful to Cells?

Even assuming that the premise of this question is correct, it is far from obvious why this should be so.

In 1901, Alois Alzheimer, a psychiatrist from the town of Markbreit in Bavaria, was working in the psychiatric hospital of Frankfurt, where he met, interrogated, observed and wrote detailed notes on a woman, Auguste D., with memory loss, disorientation, and hallucinations. Her disease resulted in her death at the age of 55 in 1906. In the meantime, Alzheimer had moved to Munich to work as a colleague with his former mentor, Emil Kraepelin, who was then the leading light in what we might now call biological psychiatry. The brain of Auguste D. was sent to Alzheimer in Munich, and he then performed the dissection of his former patient's brain, using the revolutionary new histologic technique of silver impregnation that had been invented by another neuroanatomist and colleague at Munich, Franz Nissl. The autopsy revealed innumerable "miliary foci"[b] and nerve cells filled with "dense bundles of fibrils." The patient had died of what we would now call early-onset Alzheimer's disease—so named by Émil Kraepelin—and was then called presenile dementia. In fact, the diagnosis of Alzheimer's disease is still made with certainty only at autopsy. Alzheimer's bold conclusions went against much of the accepted practice of his day, as he posited a relationship between the palpably physical microscopic findings and the symptoms of a "mental" disease. In a remarkable presentation in 1906, and in several papers in the five years following this presentation, Alzheimer described the microscopic appearance of the two pathognomonic lesions of this disease, senile (amyloid) plaques and neurofibrillary tangles, and recognized quite clearly that there are no fundamental differences between the lesions of presenile dementia that he had described and those of senile dementia.

In microscopic sections of tissues from patients with protein aggregation diseases, including Alzheimer's disease, fibrillar protein deposits can be so blatant and striking that it strains credulity to think that these deposits are innocuous. Nevertheless, as alluded to earlier, experimental evidence suggests that pre-fibrillar aggregates are the most cytotoxic species in the protein aggregation pathway, and some investigators have even gone so far as to propose that fibrils are innocuous. This important question remains hotly debated and unresolved. Although this issue cannot be resolved at present, it might be worthwhile at least to state, in broad strokes, the terms of the debate.

Since the time of Alzheimer, reams of histopathologic data have been produced demonstrating the presence of fibrillar protein deposits in a wide array of neurodegenerative diseases. Added to this, there are also abundant genetic data linking protein aggregation to disease phenotypes. To take but one line of argument from the case of Alzheimer's disease, point mutations within and adjacent to the β-amyloid sequence of β-APP are associated with both a disease phenotype (early-onset Alzheimer's disease and/or cerebral amyloid angiopathy), and increases in histologically demonstrable protein aggregates, which are in most cases fibrillar. This genetic evidence includes transgenic animals in which both fibrillar deposits and disease phenotypes develop *pari passu*. It is remarkable, indeed, that so many different proteins, with so little structural homology at any level, should all lead to both deposits of fibrillar protein aggregates and cell death. The was also a huge body of experimental evidence performed on cells *in vitro*, indicating the neuro- or cytotoxicity of added fibrils, with the initial findings confirmed many times over the years.[147–155] Furthermore, these results were not confined to neurodegenerative diseases: in non-neurological diseases associated with amyloid deposits, cellular dysfunction appeared to be related to crowding of tissues by amyloid at the expense of normal constituents.

Nevertheless, there is also much evidence against the view that fibrils themselves are injurious, and that the most cytotoxic species are some form of pre-fibrillar aggregates. There is little harmony in the literature in defining such aggregates, which is probably a major source of the heat that has been generated in this debate. Most authors, however, try to distinguish between mature fibrils, with a diameter of 10 nm and lengths measured in hundreds of nanometers or more, and smaller, spheroidal aggregates of Aβ and other fibril-forming proteins. These have been identified by electron microscopy, atomic force microscopy, size-exclusion chromatography, and other techniques. They have variously been called "micelles," "protofibrils," and ADDLs

---

[b]The term "miliary," which is most often applied to tuberculosis, refers to the millet seed-sized lesions that develop in end-stage tuberculosis that has spread hematogenously.

(Aβ-derived diffusible ligands), and despite a very wide range of descriptions, refer to soluble aggregates of ≈4–50 molecules. Part of the problem in defining these smaller aggregates is that their properties are highly dependent upon the conditions by which they are formed; size, stability, and presumably also biological activities depend on the peptide concentration, solvents, and other conditions used to form these aggregates. Perhaps the point that most unifies these disparate definitions is that these are intermediates on the fibrillogenesis *pathway*. In other words, these aggregates, if allowed to evolve, would form fibrils sooner or later, and—because these intermediates are temporally unstable—this happened sooner rather than later, making the intermediates difficult to identify, isolate, and study. These difficulties notwithstanding, since 1997, a great deal of experimental data *in vitro* has shown[156,157] that pre-fibrillar aggregates are cytotoxic, while many of these same experiments show a lack of deleterious effect of either the monomeric or fibrillar forms of the protein. It is beyond the scope of this article to describe this sizeable literature, but the reader is referred to other reviews for more information.[158–161]

A few examples will serve to illustrate this type of evidence.[162] Many of the studies in this area focus on the ability of various forms of fibril-forming peptides or proteins to induce cell death in neuronal or other cells. For example, one study[159] examined the cytotoxicity of protein domains that form fibrils, but are not associated with any known disease—the SH3 domain from bovine phosphatidyl-inositol-3′-kinase, and the amino-terminal domain of HypF from *E. coli*. The authors showed that toxicity of these proteins was limited to early stages of aggregation in which oligomers, but not fibrils were present.

Cell death, though a defining aspect of neurodegenerative diseases, is probably only a late event in these diseases. In Alzheimer's disease, for example, cognitive deficits probably precede by many years the overt loss of neurons. For this reason, many recent studies have focused on cellular phenomena. Long-term potentiation (LTP) and long-term depression (LTD) are two electrophysiological measures believed to be at the core of synaptic plasticity, and hence at the core of remembering and forgetting.[163–166] LTP refers to the long-lasting enhancement of synaptic effectiveness that follows after tetanic electrical stimulation of particular types. It has long been investigated in the hippocampus,[167] an area of the brain that is severely affected in Alzheimer's disease. LTD refers to a decrease of synaptic effectiveness after similar types of electrical stimulation. It is perhaps more intuitive or appealing that LTP be at the core of memory, but activity-dependent LTD (e.g., in the cerebellar cortex) is believed to contribute to memory as well, for example in motor learning. It was reported that oligomers of human Aβ obtained from cell media, but not Aβ fibrils or monomers, were able to inhibit hippocampal long-term potentiation (LTP) in rats *in vivo*.[168,169] Several investigators have emphasized the importance of ADDLs in diverse neurotoxic effects, such as the loss of long-term potentiation in hippocampal slices exposed to low doses of ADDLs or in mice stereotactically injected with ADDLs.[157,160,170,171]

It is important to note, also, that the data are not limited to cellular studies. To cite but one point, it has been notoriously difficult to find a consistent correlation between post-mortem cerebral β-amyloid load, and any measure of cognitive defects found ante mortem in patients with Alzheimer's disease.[172,173] There are potential explanations for this lack of correlation. For example, given the extraordinarily complexity of cerebral anatomy, defining the "amyloid load" that is actually relevant to cognition is challenging, to say the least. Indeed, there is also a poor correlation between cerebrovascular lesions and cognitive impairment and other localizing neurological defects. Such explanations aside, the lack of correlation between amyloid load and cognitive defect is certainly disturbing for the amyloid hypothesis, and dovetails with the finding of soluble β-amyloid intermediates in human postmortem material, and with the toxicity that has been observed *in vitro* and *in vivo* for soluble oligomeric species. To address this set of questions, a triple transgenic mouse model harboring presenilin (PS1M146V), β-APP (APPSwe), and tau (tauP301L) mutant transgenes was developed in a uniform genetic background.[174,175] These mice showed age-dependent synaptic dysfunction, including LTP deficits, but significantly, this occurred before they developed either plaques or tangles. Neurological deficits in synaptic plasticity correlated with accumulation of intraneuronal Aβ. These results are in accord with other data that synaptic dysfunction precedes neuronal death. Furthermore, the observed time course (LTP deficits preceding plaques and tangles) suggests, albeit in a correlational way, that pre-fibrillar aggregates might be the most potent neurotoxins in the pathogenesis of Alzheimer's disease.

All of this having been said, controversy remains. Although the pendulum of opinion has swung away from fibrils and towards soluble oligomers as neuro/cytotoxins, the possibility that fibrils could *also* contribute to disease has not been ruled out.

OK... So, How Do Protein Aggregates Harm Cells?

Like the search for the universal structure of all amyloids, the search for the universal effect of protein aggregates is likely to end in the recognition that there is no universal effect. Because proteins aggregate through different types of interactions —the hydrophobic effect in some cases, polar interactions in others—it seems unlikely that "one size fits all," either in structure, function, or dysfunction. Nevertheless, there are some common, recurrent themes in the search for explanations of the cytotoxicity of protein aggregates. One can define the following sets of overlapping, non-exclusive hypotheses for this cytotoxicity:

Disruption of Cell Membranes

A large number of fibril-forming proteins also form membrane-interacting, soluble oligomeric aggregates that could disrupt vital membrane functions, such as the maintenance of ion gradients across membranes. In the case of β-amyloid and α-synuclein among others, these oligomers have been observed to form pores in lipid bilayers, including cell membranes. Lashuel and coworkers[137,138,176] reported the formation of "amyloid pores," that is, annular structures that could be isolated by size-exclusion chromatography and were formed by disease-causing mutant forms of β-amyloid and α-synuclein (the E22G ["Arctic"] mutant form of β-amyloid, and the A53T and A30P mutant forms of α-synuclein). The fraction with the smallest apparent molecular weight by size-exclusion chromatography had a β-sheet-rich secondary structure (circular dichroic spectroscopy), and a relative molecular mass of 320–380 kDa; the structures appeared as rings of 8–12-nm diameter in electron micrographs. In further studies, these authors examined aggregates of $A\beta(1-40)_{ARC}$ (the E22G mutant form of $A\beta(1-40)$) by electron microscopy and observed several types of structures: compact spherical particles of 4–5-nm diameter, annular pore-like "protofibrils," larger spherical particles of 18–25-nm diameter, and short filaments. The authors also observed that Aβ(1–40)$_{ARC}$ could be converted to fibrils more rapidly in mixtures with wild-type Aβ(1–40). Most people with this mutation are heterozygotes, and most mutations within the Aβ sequence lead to autosomal dominant forms of early-onset Alzheimer's disease. Thus, a mixture of wild-type and mutant peptide probably exists in these patients. The authors suggest that this mixture could lead to enhancement of the accumulation of neurotoxic species in the brain. (This begs the question of whether heterozygotes would get more severe and earlier onset of disease than homozygotes with the E22G mutation.)

Soreghan and coworkers[87] observed that β-amyloid is amphiphilic: the protein forms micelle-like aggregates, as can be shown by classic techniques of surface chemistry, such as induction of fluorescence of diphenylhexatriene, or increase of surface pressure in spread or adsorbed monolayers at the air–water interface. That β-amyloid is amphiphilic can be predicted from its sequence: the distribution of lipophilic[c] residues is strikingly non-random, and the C-terminal domain, derived from the presumptive transmembrane domain of the β-amyloid precursor protein, is composed entirely of such residues. Indeed, Soreghan *et al.* showed that the amphiphilicity of β-amyloid peptides, as measured by monolayer collapse pressure, is a function of the length of the C-terminal lipophilic tail in a series of β-amyloid peptides. It has also been shown, by solid-state NMR, that increasing the amphiphilicity of a short peptide derived from β-amyloid (Aβ(16–22)) by octanoylating its N-terminus also causes the β-sheet orientation to flip from antiparallel to parallel, indicating that amphiphilicity favors the parallel orientation of β-sheets in fibrils.[78] The exact structure of micelle-like structures of β-amyloid or other fibril-forming peptides is not known, however, and it is somewhat a matter of speculation as to how such structures could associate with phospholipid bilayers to produce pores. In addition, the relationship of the pore structures described above and these micelle-like structures is not certain: are they identical, interconvertible, or separate?

Amphiphilic peptides such as β-amyloid or α-synuclein, under the "correct" conditions, could form ring-shaped oligomers with exposed lipophilic edges, and these could insert into membrane bilayers.[177–185] The evidence that such ring-shaped structures are functional ion channels is, at present, inconclusive but tantalizing. Furthermore, the exact ("correct") conditions under which such structures might form biologically have not been defined. Many of the studies demonstrating such structures have utilized high peptide concentrations—meaning orders of magnitude higher than the ambient concentrations in the cerebrospinal fluid, though not necessarily higher than those that might be present at select sites in cells—or under non-physiological conditions. As described above, however, it seems clear that many fibril-forming proteins are also potentially amphiphilic and membrane-interacting, and this fact is highly suggestive of a set of mechanisms by which these peptides harm cells. Polyglutamine peptides have also been reported to form ion channels.[182, 186,187] These peptides, though self-associating, are highly polar, and it is not obvious how such peptides would interact with membranes and form an ion channel. Nevertheless, such channel formation has been reported, and is even cation-specific.

---

[c]The term "lipophilic" is more accurate and hence preferable to the more commonly used term "hydrophobic."

Inactivation of Normally Folded Proteins by Protein Aggregates

In the world of protein aggregation diseases, it is possible to overlook the obvious point that the normal function of many proteins is to bind other proteins. Entire functional assemblies of proteins, such as the ribosome, spliceosome, proteasome and so forth, depend on this point. Proteins that bind nucleic acids, such as the large number of transcription factors, DNA repair proteins, histones, and (again) ribosomal proteins, also must interact with other proteins.

The case in point is huntingtin, a protein believed to function normally by interacting with transcription factors that contain their own (non-pathologic) polyglutamine regions. These include the TATA-binding protein and the CREB-binding protein. CREB binding protein (CBP) is a coactivator for CREB-mediated transcription and contains a 15 (mouse) or an 18 (human) glutamine stretch.[188] In a cell culture model of Huntington's disease (expanded polyQ domain of huntingtin transfected into N2A cells), it was shown that CBP, endogenous or overexpressed, is recruited into aggregates and redistributed away from its normal location in the nucleus into huntingtin aggregates.[189] The authors also showed, by immunocytochemistry in transgenic mice expressing the N-terminal domain of huntingtin with an expanded polyglutamine repeat, that CREB-binding protein co-aggregated with huntingtin, and was depleted from its normally diffuse distribution in the nucleus. In addition, similar findings were also seen in post-mortem brain tissue from patients with Huntingon's disease. Finally, these authors also showed that the abnormal distribution of CREB binding protein has functional consequences: in transcription assays using primary cortical neurons, a mutant huntingtin construct inhibited CREB binding protein-dependent transcription, while the huntingtin construct with a normal polyglutamine region had no such effect. Neurodegeneration reminiscent of that occurring in Huntington's disease has recently also been observed in transgenic mice in which two CRE (c-AMP-responsive element) binding proteins (CREB1 and CREM) were disrupted.[190] Finally, gene array studies have suggested that in transgenic mice or tissue culture cells expressing a huntingtin construct with an expanded polyglutamine region, there is downregulation of genes with c-AMP-responsive elements.[191,192] These data all suggest that polyglutamine expansions in huntingtin could be cytotoxic by dint of their ability to sequester or otherwise inhibit transcription factors. This effect need not be limited to huntingtin, and, indeed, similar types of disruption of transcription have been observed in other polyglutamine expansion diseases, such as SCA1, SCA3, and dentatorubral pallidoluysian atrophy.[193,194] In addition, the effect of polyglutamine expansion need not be limited to proteins that interact directly with polyglutamine stretches. Huntingtin, for example, is a huge protein (348 kDa) with many potential protein binding sites distant in the sequence from the N-terminal polyglutamine domain. For example, TAFII130 is a regulator of CREB-mediated transcription, but it does not have its own polyglutamine domain — yet it still binds to SCA3, a protein that has a long polyglutamine region that is expanded in one form of spinocerebellar ataxia.[194] Presumably, the interaction is not simply two polyglutamine domains binding to one another.

A related pathogenic mechanism is interference with nuclear translocation of proteins. The androgen receptor is another protein with a polyglutamine region, and, as with huntingtin and the other proteins of this type, expansion of the polyglutamine region leads to a neurodegenerative disorder, spinobulbar muscular atrophy (SB-

MA). Binding of an androgen to the receptor leads to a conformational change, followed by translocation of the receptor to the nucleus, where the protein acts as a transcription factor.[195] Recent *Drosophila* and murine transgenic models using full-length androgen receptor have shown that androgen binding is necessary for nuclear translocation, and in instances where the polyglutamine domain is expanded, protein aggregation and neurotoxicity.[196] The ligand dependency of all three effects accounts for the fact that only males are affected: in the absence of androgen binding and activation, the protein remains cytoplasmic and non-aggregated.[197] This has also been confirmed by abrogation of the phenotype by castration of male animals, and induction of the phenotype by androgen administration in female animals. Presumably, once in the nucleus, the aggregated protein disrupts transcription regulation of critical genes, but the detailed pathogenic mechanism is not known.

Gumming Up the Works: Binding to Proteins of the Quality Control System of Cells

Anfinsen's classic experiments on the denaturation and refolding of proteins demonstrated, simultaneously, that the information necessary for protein folding was contained in the amino acid sequence of proteins, and that the rate at which this occurs for most proteins is incompatible with life. Consequently, protein folding requires catalysts. This is especially true for large and multidomain proteins, for which spontaneous folding is very inefficient. In addition, proteins are prone to the insults of daily life: oxidation and proteolysis, as well as a host of post-translational modifications, both physiological and non-physiological, that alter proteins covalently and destabilize their folded, functional state. Finally, proteins are dynamic and the polypeptide chain is flexible: proteins can denature or adopt alternate conformations, even without covalent modifications. In some cases—those proteins that readily form fibrils, and the prions, are foremost among them—the energy difference between alternate states may be minute. The ability of a protein to adopt more than one stable conformational state—alternate states—can be termed *conformational plasticity*, and this property seems to be inherent to prions and amyloid-forming proteins and peptides. In any of these "altered states" of proteins (nascent polypeptides, modified proteins, dynamic proteins or protein domains) aggregation-prone segments can become exposed and interact with other such domains, either of the same protein or other proteins.

It is almost a matter of semantics whether such "altered states" ought to be called "misfolded": using this term presumes to know what is the "correctly" folded state of the protein. In some cases (e.g., a small enzyme), this is fairly obvious; in other cases (e.g., β-amyloid or the prion protein, where neither the structure nor function of the "folded state" is known), it is far less obvious. A native state is often defined as the state of the protein with the lowest free energy, but strictly speaking, this is never known—one only ever measures changes in free energy, not absolute quantities of free energy. In addition, in some instances, such as has been proposed for the nucleoporin, Nup2p of *S. cerevisiae*, the native or functional form of the protein is clearly not the one with the lowest free energy, since in these cases, the protein loses its function when it adopts a conformation of lower free energy.[198–200]

Biological systems, however, recognize the existence of altered states of proteins —"misfolded," if you will—in which hydrophobic or other aggregation-prone domains are exposed. This is especially a problem in the endoplasmic reticulum of the cell, where protein concentrations are very high compared to the dilute conditions at which most biochemists work. Nascent proteins exported into the ER often cannot fold until the entire protein is expressed, since in many proteins, perhaps a disproportionate number of cases than one would expect at random, the N- and C-terminal domains of the protein interact with one another in the folded state. For this reason, it is not only a question of accelerating protein folding, but also one of preventing association of nascent proteins with their aggregation-prone sites exposed to solvent. This is where the term "chaperones" got its name: the prevention of *liasons dangereuses* among young protein molecules.

For all of the above reasons—the acceleration of protein folding, the prevention of undesirable protein aggregation, and the riddance of damaged, covalently modified proteins—the cell has a system of quality control that includes, in addition to the chaperones, a number of catalysts of protein folding, and a system for removal and degradation that includes the uniquitination system and the proteasome.[201,202] Not surprisingly, these systems overlap considerably. Chaperones, for example, are involved not only in protein folding, but also in targeting proteins for degradation. For example, the 19S cap of proteasome can acts as a chaperone that promotes protein renaturation,[203] and mammalian Hsp70, best known as a chaperone of the ER where it promotes protein folding when associated with Hsp40, also promotes proteasomal degradation when associated with Bag-1 and CHIP.[204] Chaperones can also participate in signal transduction pathways that lead to transcription regulation of genes in the quality-control system. Thus, these systems represent parts of a coordinated system of response of the cell to a wide range of injuries, including heat and oxidant stress, that tend to damage and/or denature proteins.

Many studies have documented association of chaperones with cellular protein aggregates. Hsp70 chaperones are among the most abundant of the chaperone proteins, and these have been found to be prominently associated with huntingtin and other polyglutamine proteins,[205–207] and with Lewy bodies.[208] The data are not as compelling, but these proteins, and the Hsp16 proteins also interact with intracellular β-amyloid aggregates.[209] This is consistent with the idea that the chaperone proteins are upregulated in response to an increased cellular load of unfolded and/or aggregated protein. Expression of a GFP-polyglutamine fusion protein in *S. cerevisiae* or *C. elegans* leads to increased expression of Hsp104, and high levels of Hsp104 expression attenuate the toxicity of this fusion protein. Hsp70 also suppresses protein aggregation of disease-causing proteins, including GFP-polyglutamine fusion proteins (103,104). Several studies have also shown that high expression levels of some of the heat shock proteins abrogate the aggregation of polyglutamine proteins, and the toxicity associated with this aggregation. Similar findings have been obtained for β-amyloid and α-synuclein. In *Drosophila*, expression of Hsp70 can prevent loss of neurons caused by aggregation of overexpressed α-synuclein.[210]

A major question, then, is how chaperones and related proteins prevent protein aggregation. The answer to this question varies, of course, with the chaperone. In one study using fluorescence resonance energy transfer (FRET), it was observed that an intramolecular conformation change of monomeric huntingtin precedes aggregation.[201] (An intramolecular conformational changes may precede aggregation of β-amyloid as well.[84]) Two heat shock proteins, Hsp70 and Hsp40, are able to inhibit the intramolecular conformational change of huntingtin that precedes aggregation.

Whether an intramolecular conformational change precedes fibril formation in general is an open question, but in any event, a prerequisite for aggregation would appear to be solvent exposure of an aggregation-prone segment of a protein, and from the discussion above, it appears that these are precisely the segments that chaperone proteins are designed to bind.

Finally, the quality control system must also entail the activation of protein degradation machinery, notably the proteasome. Fibrils are highly protease-resistant for at least three reasons: because they are highly structured, and most proteases preferentially cleave unstructured proteins or protein loops; because they are too large to fit into much of the protein-degradation machinery; and because they are insoluble and have significant segments into which water cannot penetrate. Fibrils are somewhat dynamic, and proteins join and leave growth sites of the fibril, but very slowly. Thus, clearance of aggregates must occur before fibril formation occurs. This is a critical issue in the design of therapy against protein aggregation diseases. Initially, therapy was directed against the fibril: the main measure of the effects of immunotherapy in Alzheimer's disease models (i.e., immunotherapy directed against β-amyloid) was the decrease in the plaque burden. Similarly, many peptide-based and nonpeptidic fibrillogenesis inhibitors were developed as potential therapeutic agents. With the emergence of data implicating pre-fibrillar aggregates as neurotoxins, the question arose of whether such therapeutic strategies are well advised or whether, on the contrary, a therapy that prevented formation or led to dissolution of fibrils might increase the concentrations of the more dangerous pre-fibrillar aggregates. An important additional consideration is that aggregated protein, whatever is or are the neurotoxic species, needs to be proteolyzed, and this cannot be accomplished on fibrillar proteins. Thus, it is possible that individuals with a large burden of fibrillar aggregates will inevitably also have a high burden or pre-fibrillar aggregates, if these develop *pari passu*, and that the relief of both of these burdens requires that the aggregates be kept in a form upon which the proteasome and other proteases can act. For this reason, among others, the question of where to direct therapeutic effort remains open. Therapies directed against fibrils could well make matters worse by increasing cellular levels of more toxic soluble oligomers; but on the other hand, solubilizing fibrils could also, under some circumstances, lead to elimination of a cellular burden of fibrils—and hence, also of oligomers—by the proteasome.

Among these circumstances would be a vigorous, indeed activated proteasomal system. One of the mechanisms by which protein aggregates damage cells, however, is by inhibiting the proteasomal system. Protein aggregates, including those of neuritic plaques in Alzheimer's disease, Lewy bodies, and aggregates in Huntington's disease and prion diseases, often contain ubiquitin and 19S and 20S proteasomal components. This might suggest an attempt by the proteasome to clear aggregated proteins. As stated above, however, aggregated proteins are resistant to proteolysis, especially fibrils, but also soluble oligomers. There are several reports of decreased proteasomal activity in diseased tissue, which probably represents inactivation of proteasomes, sometimes even in the face of increased expression of proteasomal components.[211–216] These observations are consonant with observations that defects in the protein degradation system give rise to neurodegenerative disorders, most notably, Parkinson's disease.[217] Two of the genetic loci for Parkinson's disease—PARK2 (associated with early-onset disease[218]) and PARK4 (associated with disease onset at age 50)—are associated with defects in protein degradation: PARK2 encodes an E3 ligase of the ubiquitination system, and PARK4 encodes a ubiquitin ligase/hydrolase.[219] In transgenic mouse models, a defect in E3 ubiquitin ligase resulted in a more severe phenotype of the spinocerebellar ataxia type I model. Thus, defects in the protein degradation system can lead to protein aggregation diseases or worsen disease caused by another defect.

Why the Brain?

A striking feature of protein aggregation diseases is the great frequency with which they involve the central nervous system. Moreover, within the CNS, specific sites are affected while others are spared. α-synuclein and β-amyloid are both ubiquitously expressed, and yet α-synuclein aggregation is manifested predominantly by CNS disease. Furthermore, α-synuclein leads strikingly to death of the pigmented neurons of the substantia nigra pars compacta and locus coeruleus, less so the cholinergic neurons of the basal nucleus of Meynert and a few brainstem nuclei, while leaving most other neuronal groups intact, at least initially. The symptoms of Alzheimer's disease and neuronal death attributed to β-amyloid aggregation are localized especially to the frontal lobes and hippocampus, with lesser atrophy of the parietal lobes, and relative sparing of the occipital lobes and subcortical neurons. In Huntington's disease, there is atrophy of the caudate nucleus and, to a lesser extent, the putamen, while the globus pallidus atrophies only secondarily; the medium-sized, spiny GABAminergic neurons are especially affected. In the spinocerebellar ataxias, there is loss of cerebellar neurons, and to a lesser extent, neurons of the brainstem, spinal cord, and peripheral nerves (as well as other parts of the brain, depending on the particular subtype of SCA); white matter tracts degenerate secondarily. In one form of familial amyotrophic lateral sclerosis, neurodegeneration has been linked to aggregation of a point mutant of the copper-zinc superoxide dismutase gene (*SOD1*) on chromosome 21.[220] Neuronal degeneration and loss are seen in the anterior horn neurons of the entire spinal cord, with similar findings in the hypoglossal, ambiguus, and motor trigeminal cranial nerve nuclei.

Why do these and so many other protein aggregation diseases involve mainly or only the CNS? Some of this susceptibility probably relates to expression levels: although α-synuclein is widely expressed, its expression is highest in brain,[221] and the same is true for β-amyloid. Others have attributed the special susceptibility of the brain to protein aggregation diseases to the high level of oxidant metabolism there. The electron transport chain is, of course, leaky, and this can lead to the generation of reactive (reduced) oxygen and nitrogen species.[132,133] There may be other causes of oxidant stress in the brain as well, and some of the neurotransmitters (e.g., catechols) can themselves generate radical and related species that can crosslink proteins and stabilize oligomeric intermediates of the aggregation pathway.[222] A wide variety of pesticides interact with α-synuclein and are associated with sporadic Parkinson's disease.[223] These factors notwithstanding, the answer to this question is largely unknown.

And why do the protein aggregation diseases occur in these odd distributions? Perhaps the most striking example of differential anatomic distribution of lesions is seen with the prion diseases (spongiform encephalopathies), in which the lesions occurring in different variants of sporadic disease have distinct anatomic distributions, and distinct histopathologic appearances. For example, in patients with either familial or sporadic fatal insomnia, there is severe neuronal fiber loss predominantly in nuclei of the thalamus, but vacuolation is sparse, both findings in marked contrast to the spongiform lesions of Creutzfeldt–Jakob disease. Moreover, these differences can be propagated in studies in which the prions are transmitted to mice. For example, the anatomic distribution of lesions in mice inoculated with prion material from human patients with either familial or sporadic fatal insomnia are quite similar (deposition predominantly in nuclei of the thalamus, marked neuronal fiber loss, little vacuolation), but differ markedly from the pattern seen if the corresponding experiment is performed using brain from patients with Creutzfeldt-Jakob disease.[224]

In the case of the prion diseases, there are point mutations in the prion protein that lead to familial forms of these diseases. In the sporadic and infectious forms of the prion diseases, however, no covalent difference has been found in prion protein from normal subjects and patients with diseases. The prion protein is a non-pathogenic, single-pass transmembrane protein (called PrP^C, for the cellular form of the protein) of unknown function. The prion hypothesis (reviewed in detail elsewhere[225–229]) posits that this protein can be induced to undergo a conformational change to a new form of protein (called PrP^{Sc}, for scrapie, a prion disease occurring widely and naturally in sheep) that can be propagated, or transduced by converting PrP^C molecules into the PrP^{Sc}. The PrP^{Sc} form of the protein aggregates into fibrils: at this point the issue is clouded by controversy, since some of the fibril-forming conformers of PrP have not been shown to be pathogenic (i.e., there is a still undefined relationship between fibril formation and the propagation of disease). Without entering into the controversies of this field, suffice it to say that PrP^C and PrP^{Sc} clearly have conformational differences (for example, the latter is protease-resistant, less soluble, and has a higher β-sheet content than the former). Furthermore, more subtle conformational differences are believed to account for the differences in pathological findings among the prion diseases—a phenomenon that has been called the *strain phenomenon*. "Strains" of infectious prions have been found not only in mammalian prions, but in yeast prions as well.[230–233]

Are similar "strain phenomena" to be found in amyloids as well? Petkova and co-workers[234] recently demonstrated a similar phenomenon for β-amyloid. Small differences in the conditions under which fibrils were formed ("quiescent" versus "agitated") led to discernible differences in morphology both at the level of electron microscopy, and in solid-state NMR spectra at the atomic level. These structural differences, furthermore, could be propagated to daughter and granddaughter fibrils. It seems likely, then, that amyloids more generally can occupy different conformational states.

REFERENCES

1. KYLE, R.A. 2001. Amyloidosis: a convoluted story. Br. J. Haematol.: **114:** 529–538.
2. COHEN, A.S., T. SHIRAHAMA & M. SKINNER. 1982. Electron microscopy of amyloid. *In* Electron Microscopy of Proteins, Vol. 3. J.R. Harris, Ed.: 165–205. Academic Press. New York.
3. GLENNER, G.G., W. TERRY, M. HARADA *et al.* 1971. Amyloid fibrils proteins: proof of homology with immunoglobulin light chains by sequence analysis, Science **172:** 1150–1153.
4. SIPE, J.D. & A.S. COHEN. 2000. Review: history of the amyloid fibril. J. Struct. Biol. **130:** 88–98.

5. EANES, E.D. & G.G. GLENNER. 1968. X-ray diffraction studies on amyloid filaments. J. Histochem. Cytochem. **16:** 673–677.
6. BONAR, L., A.S. COHEN & M.M. SKINNER. 1969. Characterization of the amyloid fibril as a cross-β protein. Proc. Soc. Exp. Biol. Med. **131:** 1373–1375.
7. GLENNER, G.G., E.D. EANES, H.A. BLADEN, et al. 1974. β-pleated sheet fibrils: a comparison of native amyloid with synthetic protein fibrils. J. Histochem. Cytochem. **22:** 1141–1158.
8. SUNDE, M. & C.C.F. BLAKE. 1998. From the globular to the fibrous state: protein structure and structural conversion in amyloid formation, Q. Rev. Biophys. **31:** 1–39, 1998.
9. PRAS, M., M. SCHUBERT, D. ZUCKER-FRANKLIN, et al. 1968. The characterization of soluble amyloid prepared in water. J. Clin. Invest. **47:** 924–933.
10. SELKOE, D.J. & C.R. ABRAHAM. 1986. Isolation of paired helical filaments and amyloid fibers from human brain. Meth. Enzymol. **134:** 388–404.
11. ROHER, A., D. WOLFE, M. PALUTKE & D. KUKURUGA. 1986. Purification, ultrastructure, and chemical analysis of Alzheimer disease amyloid plaque core protein. Proc. Natl. Acad. Sci. **83:** 2662–2666.
12. GEWURZ, H., X.H. ZHANG & T.F. LINT. 1995. Structure and function of the pentraxins. Curr. Op. Immunol. **7:** 54–64.
13. PEPYS, M.B. 1999. The Lumleian lecture: C-reactive protein and amyloidosis—from proteins to drugs? *In* Horizons in Medicine. G. Williams, Ed.: 397–414. Royal College of Physicians of London.
14. GARLANDA, C., B. BOTTAZZI, A. BASTONE & A. MANTOVANI. 2005. Petraxins at the crossroads between innate immunity, inflammation, matrix deposition, and female fertiliy. Annu. Rev. Immunol. **23:** 337–366.
15. HAWKINS, P.N. 2002. Serum amyloid P component scintigraphy for diagnosis and monitoring amyloidosis. Curr. Opin. Nephrol. Hypertens. **11:** 649–655.
16. YAMADA T. 1999. Serum amyloid A (SAA): a concise review of biology, assay methods and clinical usefulness. Clin. Chem. Lab. Med. **37:** 381–388.
17. BENDITT, E.P. & N. ERIKSEN. 1977. Amyloid protein SAA is associated with high density lipoprotein from human serum. Proc. Natl. Acad. Sci. USA **74:** 4025–4028.
18. DWULET, F.E. & M.D. BENSON. 1988. Amino acid structures of multiple forms of amyloid related serum protein SAA from a single individual. Biochemistry **27:** 1677–1682.
19. BETTS, J.C., M.R. EDBROOKE, R.V. THAKKER & P. WOO. 1991. The human acute-phase serum amyloid A gene family: structure, evolution and expression in hepatoma cells. Scand. J. Immunol. **34:** 471–482.
20. WHITEHEAD, A.S., M.C. DE BEER, M. RITS, et al. 1992. Identification of novel members of serum amyloid A protein superfamily as constitutive apolipoproteins of high density lipoprotein. J. Biol. Chem. **267:** 3862–3867.
21. HUSBY, G., G. MARHAUG, B. DOWTON, et al. 1994. Serum amyloid A (SAA): biochemistry, genetics and the pathogenesis of AA amyloidosis. Amyloid: Int. J. Exp. Clin. Invest. **1:** 119–137.
22. ALDO-BENSON, M.A. & M.D. BENSON. 1982. SAA suppression of immune response in vitro: evidence for an effect on T cell-macrophage interaction. J. Immunol. **126:** 2390–2392.
23. ZIMLICHMAN, S., A. DANON, I. NATHAN, et al. 1990. Serum amyloid A, an acute phase protein, inhibits platelet activation. J. Lab. Clin. Invest. **116:** 180–186.
24. LINKE, R.P., V. BOCK, G. VALET & G. ROTHE. 1991. Inhibition of the oxidative burst response of N-formyl peptide-stimulated neutrophils by serum amyloid-A protein. Biochem. Biophys. Res. Commun. **176:** 1100–1105.
25. MITCHELL, T.I., C.I. COON & C.E. BRINCKERHOFF. 1991. Serum amyloid A (SAA3) produced by rabbit synovial fibroblasts treated with phorbol esters or interleukin 1 induces synthesis of collagenase and is neutralized with specific antiserum. J. Clin. Invest. **87:** 1177–1185.
26. BADOLATO, R., J.M. WANG, W.J. MURPHY, et al. 1994. Serum amyloid A is a chemoattractant: induction of migration, adhesion, and tissue infiltration of monocytes and polymorphonuclear leukocytes. J. Exp. Med. **180:** 201–210.
27. MALLE, E. & F.C. DEBEER. 1996. Human serum amyloid A (SAA) protein: a prominent acute-phase reactant for clinical practice. Eur. J. Clin. Invest. **26:** 427–435.

28. BLAKE, C.C., M.J. GEISOW, S.J. OATLEY, et al. 1978. Structure of prealbumin: secondary, tertiary and quaternary interactions determined by Fourier refinement at 1.8Å. J. Mol. Biol. **121:** 339–356.
29. CORNWELL, G.G., 3RD, K. SLETTEN, B. JOHANSSON & P. WESTERMARK. 1988. Evidence that the amyloid fibril protein in senile systemic amyloidosis is derived from normal prealbumin. Biochem. Biophys. Res. Commun. **154:** 648–653.
30. JACOBSON, D.R. & J.N. BUXBAUM. 1991. Genetic aspects of amyloidosis. Adv. Human Genet. **20:** 69–123; 309–311.
31. BLAKE, C.C.F., M.J. GEISOW, I.D.A. SWAN, et al. 1974. Structure of human plasma prealbumin at 2.5 angstrom resolution. J. Mol. Biol. **88:** 1–12.
32. KOO, E.H., P.T. LANSBURY & J.W. KELLY. 1999. Amyloid diseases: abnormal protein aggregation in neurodegeneration. Proc. Natl. Acad. Sci. USA **96:** 9989–9990.
33. KELLY, J.W. 1996. Alternative conformations of amyloidogenic proteins govern their behavior. Curr. Opin. Struct. Biol. **6:** 11–17.
34. KELLY, J.W. 1997. Amyloid fibril formation and protein misassembly: a structural quest for insights into amyloid and prion diseases. Structure **5:** 595–600.
35. WETZEL, R. 1996. 1996. For protein misassembly, its the "I" decade. Cell **86:** 699–702.
36. THOMAS, P.J., B-H QU & P.L. PEDERSON. 1995. Defective protein folding as a cause of human disease. Trends Biochem. Sci. **20:** 456–459.
37. BOOTH, D.R., M. SUNDE, V. BELLOTTI, et al. 1997. Instability, unfolding and aggregation of human lysozyme variants underlying amyloid fibrillogenesis. Nature **385:** 787–793.
38. SEKIJIMA, Y., R.L. WISEMAN, J. MATTESON, et al. 2005. The biological and chemical basis for tissue-selective amyloid disease. Cell **121:** 73–85.
39. FOSS, T.R., M.S. KELKER, R.L. WISEMAN, et al. 2005. Kinetic stabilization of the native state by protein engineering: implications for inhibition of transthyretin amyloidogenesis. J. Mol. Biol. **347:** 841–854.
40. PETERSON, S.A., T. KLABUNDE, H.A. LASHUEL, et al. 1998. Inhibiting transthyretin conformational changes that lead to amyloid fibril formation. Proc. Natl. Acad. Sci. USA **95:** 12956–12960.
41. BAURES, P.W., S.A. PETERSON & J.W. KELLY. 1998. Discovering transthyretin amyloid fibril inhibitors by limited screening. Bioorg. Med. Chem. **6:** 1389–1401.
42. PETRASSI, H.M., T. KLABUNDE, J. SACCHETTINI & J.W. KELLY. 2000. Structure-based design of N-phenyl phenoxazine transthyretin amyloid fibril inhibitors. J. Am. Chem. Soc. **122:** 2178–2192.
43. MCCAMMON, M.G., D.J. SCOTT, C.A. KEETCH, et al. 2002. Screening transthyretin amyloid fibril inhibitors: characterization of novel multiprotein—multiligand complexes by mass spectrometry. Structure **10:** 851–863.
44. SACCHETTINI, J.C. & J.W. KELLY. 2002. Therapeutic strategies for human amyloid diseases. Nature Rev. Drug Disc. **1:** 267–275.
45. KOCH, K.M. 1992. Dialysis-related amyloidosis. Kidney Int. **41:** 1416–1429.
46. BJORKMAN, P.J., M.A. SAPER, B. SAMRAOUI, et al. 1987. Structure of the human class I histocompatibility antigen, HLA–A2. Nature **329:** 506–512.
47. NAIKI, H.S., D. YAMAMOTO, K. HASEGAWA, et al. 2005. Molecular interactions in the formation and deposition of β2-microglobulin-related amyloid fibrils Amyloid **12:** 15–25.
48. YAMAGUCHI, K., S. TAKAHASHI, T. KAWAI, et al. 2005. Seeding-dependent propagation and maturation of amyloid fibril conformation. J. Mol. Biol. **352:** 952–960.
49. GLENNER, G.G. & C.W. WONG. 1984. Alzheimer's disease and Down's syndrome: sharing of a unique cerebrovascular amyloid fibril protein. Biochem. Biophys. Res. Commun. **122:** 1131–1135.
50. MASTERS, C.L., G. SIMMS, N.A. WEINMAN, et al. 1985. Amyloid plaque core protein in Alzheimer disease and Down syndrome. Proc. Natl. Acad. Sci. USA **82:** 4245–4249.
51. VASSAR, R., B.D. BENNETT, S. BABU-KHAN, et al. 1999. β-secretase cleavage of Alzheimer's amyloid precursor protein by the transmembrane aspartic protease BACE. Science **286:** 735–741.
52. VASSAR, R. 2004. BACE1: the β-secretase enzyme in Alzheimer's disease. J. Mol. Neurosci. **23:** 105–114.
53. ALLINSON, T.M., E.T. PARKIN, A.J. TURNER & N.M. HOOPER. 2003. ADAMs family members as amyloid precursor protein α-secretases. J. Neurosci. Res. **74:** 342–352.

54. MURPHY, M.P., L.J. HICKMAN, C.B. ECKMAN, *et al.* 1999. γ-Secretase, evidence for multiple proteolytic activities and influence of membrane positioning of substrate on generation of amyloid beta peptides of varying length. J. Biol. Chem. **274:** 11914–11923.
55. DE STROOPER, B. 2003. Aph-1, Pen-2, and nicastrin with presenilin generate an active γ-secretase complex. Neuron **38:** 9–12.
56. GOLDE, T.E. & C.B. ECKMAN. 2003. Physiologic and pathologic events mediated by intramembranous and juxtamembranous proteolysis. Science: Signal Transduction Knowledge Environment. **172:** RE4.
57. RUSSOA, C., V. VENEZIAA, E. REPETTOA, *et al.* 2005. The amyloid precursor protein and its network of interacting proteins: physiological and pathological implications. Brain Res. Rev. **48:** 257–264.
58. EDBAUER, D., E. WINKLER, J.T. REGULA, *et al.* 2003. Reconstitution of γ-secretase activity. Nat. Cell Biol. **5:** 486–488.
59. FORTINI, M.E. 2002. γ-secretase-mediated proteolysis in cell-surface-receptor signalling. Nat. Rev. Mol. Cell Biol. **3:** 673–684.
60. KOPAN, R. & M.X. ILAGAN. 2004. γ-secretase: proteasome of the membrane? Nat. Rev. Mol. Cell Biol. **5:** 499–504.
61. ESLER, W.P., W.T. KIMBERLY, B.L. OSTASZEWSKI, *et al.* 2000. Transition-state analogue inhibitors of γ-secretase bind directly to presenilin-1. Nat. Cell Biol. **2:** 428–434.
62. LI, Y.M., M. XU, M.T. LAI, *et al.* 2000. Photoactivated γ-secretase inhibitors directed to the active site covalently label presenilin 1. Nature **405:** 689–694.
63. LAVOIE, M.J., P.C. FRAERING, B.L. OSTASZEWSKI, *et al.* 2003. Assembly of the γ-secretase complex involves early formation of an intermediate subcomplex of Aph-1 and nicastrin. J. Biol. Chem. **278:** 37213–37222.
64. MORAIS, V.A., A.S. CRYSTAL, D.S., *et al.* 2003. The transmembrane domain region of nicastrin mediates direct interactions with APH-1 and the γ-secretase complex. J. Biol. Chem. **278:** 43284–43291.
65. HU, Y. & M.E. FORTINI. 2003. Different cofactor activities in γ-secretase assembly: evidence for a nicastrin-Aph-1 subcomplex. J. Cell Biol. **161:** 685–690.
66. TAKASUGI, N., T. TOMITA, I. HAYASHI, *et al.* 2003. The role of presenilin cofactors in the γ-secretase complex. Nature **422:** 438–441.
67. TYCKO, R. 2003. Insights into the amyloid folding problem from solid-state NMR. Biochemistry **42:** 3151–3158.
68. STEJSKAL, E.O. & J.D. MEMORY. 1995. High Resolution NMR in the Solid State: Fundamentals of CP/MAS. Oxford University Press. New York.
69. DUER, M.J. 2004. Introduction to Solid-State NMR Spectroscopy. Blackwell. Malden, MA.
70. BENZINGER, T.L.S., D.M. GREGORY, T.S. BURKOTH, *et al.* 1998. Propagating structure of Alzheimer's β-amyloid (10–35) is parallel β-sheet with residues in exact register. Proc. Natl. Acad. Sci. USA **95:** 13407–13412.
71. BENZINGER, T.L.S., D.M. GREGORY, T.S. BURKOTH, *et al.* 2000. Two-dimentional structure of β-amyloid(10–35) fibrils. Biochemistry **39:** 3491–3499.
72. GREGORY, D.M., T.L.S. BENZINGER, T.S. BURKOTH, *et al.* 1998. Dipolar recoupling NMR of biomolecular self-assemblies: determining inter- and intrastrand distances in fibrilized Alzheimer's β-amyloid peptide. Solid State Nucl. Magn. Reson. **13:** 149–166.
73. ANTZUTKIN, O.N., J.J. BALBACH, R.D. LEAPMAN, *et al.* 2000. Multiple quantum solid-state NMR indicates a parallel, not antiparallel, organization of β-sheets in Alzheimer's β-amyloid fibrils. Proc. Natl. Acad. Sci. USA **97:** 13045–13050.
74. BALBACH, J.J., A.T. PETKOVA, N.A. OYLER, *et al.* 2002. Supramolecular structure in full-length Alzheimer's β-amyloid fibrils: evidence for a parallel β-sheet organization from solid-state nuclear magnetic resonance. Biophys. J. **83:** 1205–1216.
75. TÖRÖK, M., S. MILTON, R. KAYED, *et al.* 2002. Structural and dynamic features of Alzheimer's Aβ peptide in amyloid fibrils studied by site-directed spin labeling. J. Biol. Chem. **277:** 40810–40815.
76. LANSBURY, P.T., JR, P.R. COSTA, J.M. GRIFFITHS, *et al.* 1995. Structural model for the β-amyloid fibril based on interstrand alignment of an antiparallel-sheet comprising a C-terminal peptide. Nat. Struct. Biol. **2:** 990–998.
77. BALBACH, J.J., Y. ISHII, O.N. ANTZUTKIN, *et al.* 2000. Amyloid fibril formation by Aβ16-22, a seven-residue fragment of the Alzheimer's β-amyloid peptide, and structural characterization by solid state NMR. Biochemistry **39:** 13748–13759.

78. GORDON, D.J., J.J. BALBACH, R. TYCKO & S.C. MEREDITH. 2004. Increasing the amphiphilicity of an amyloidogenic peptide changes the β-sheet structure in the fibrils from antiparallel to parallel. Biophys. J. **86:** 428–434.
79. PETKOVA, A.T., Y. ISHII, J.J. BALBACH, et al. 2002. A structural model for Alzheimer's β-amyloid fibrils based on experimental constraints from solid state NMR. Proc. Natl. Acad. Sci. USA **99:** 16742–16747.
80. BUCHETE, N.V., R. TYCKO & G. HUMMER. 2005. Molecular dynamics simulations of Alzheimer's β-amyloid protofilaments. J. Molec. Biol. **353:** 804–821.
81. PETKOVA, E.T., W.-M. YAU & R. TYCKO. 2005. Experimental Constraints on Quaternary Structure in Alzheimer's b-Amyloid Fibrils. Biochemistry **45:** 498–512.
82. KHETERPAL, I., S. ZHOU, K.D. COOK & R. WETZEL. 2000. Aβ amyloid fibrils possess a core structure highly resistant to hydrogen exchange. Proc. Natl. Acad. Sci. USA **97:** 1359713–601.
83. KHETERPAL, I., H.A. LASHUEL, D.M. HARTLEY, et al. 2003. Aβ protofibrils possess a stable core structure resistant to hydrogen exchange. Biochemistry **42:** 14092–14098.
84. SCIARRETTA, K.L., D.J. GORDON, A.T. PETKOVA, et al. 2005. Aβ40-Lactam(D23/K28) models a conformation highly favorable for nucleation of amyloid. Biochemistry **44:** 6003–6014.
85. LEVINE, H., 3RD. 1993. Thioflavine T interaction with synthetic Alzheimer's disease β-amyloid peptides: detection of amyloid aggregation in solution. Protein Sci. **2:** 404–410.
86. KLUNK, W.E., J.W. PETTEGREW & D.J. ABRAHAM. 1989. Quantitative evaluation of congo red binding to amyloid-like proteins with a β-pleated sheet conformation. J. Histochem. Cytochem. **37:** 1273–1281.
87. SOREGHAN, B., J. KOSMOSKI & C. GLABE. 1994. Surfactant properties of Alzheimer's Aβ peptides and the mechanism of amyloid aggregation. J. Biol. Chem. **269:** 28551–28554.
88. CHAUHAN, V.P., A. CHAUHAN & J. WEGIEL. 2001. Fibrillar amyloid β-protein forms a membrane-like hydrophobic domain. Neuroreport **12:** 587–590.
89. YONG, W., A. LOMAKIN, M.D.KIRKITADZE, et al. 2002. Structure determination of micelle-like intermediates in amyloid β-protein fibril assembly by using small angle neutron scattering. Proc. Natl. Acad. Sci. USA **99:** 150–154.
90. PALLITTO, M.M. & R.M. MURPHY. 2001. A mathematical model of the kinetics of β-amyloid fibril growth from the denatured state. Biophys. J. **81:** 1805–1822.
91. MURPHY, R.M. 2002. Peptide aggregation in neurodegenerative disease. Annu. Rev. Biomed. Engineering. **4:** 155–174.
92. KIRKITADZE, M.D., M.M. CONDRON & D.B.TEPLOW. 2001. Identification and characterization of key kinetic intermediates in amyloid β-protein fibrillogenesis. J. Mol. Biol. **312:** 1103–1119.
93. LOMAKIN, A., D.S. CHUNG, G.B. BENEDEK, et al. 1996. On the nucleation and growth of amyloid beta-protein fibrils: detection of nuclei and quantitation of rate constants. Proc. Natl. Acad. Sci. USA **93:** 1125–1129.
94. LOMAKIN, A., D.B. TEPLOW, D.A. KIRSCHNER & G.B. BENEDEK. 1997. Kinetic theory of fibrillogenesis of amyloid β-protein. Proc. Natl. Acad. Sci. USA **94:** 7942–7947.
95. BITAN, G., A. LOMAKIN & D.B. TEPLOW. 2001. Amyloid β-protein oligomerization: prenucleation interactions revealed by photo-induced cross-linking of unmodified proteins. J. Biol. Chem. 276: 35176–35184.
96. BITAN, G. & D.B. TEPLOW. 2004. Rapid photochemical cross-linking: a new tool for studies of metastable, amyloidogenic protein assemblies. Acct. Chem. Res. **37:** 357–364.
97. BITAN, G., M.D. KIRKITADZE, A. LOMAKIN, et al. 2003. Amyloid β-protein (Aβ) assembly: Aβ 40 and Aβ 42 oligomerize through distinct pathways. Proc. Natl. Acad. Sci. USA **100:** 330–335.
98. ZOGHBI, H.Y. & H.T. ORR. 2000. Glutamine repeats and neurodegeneration. Annu. Rev. Neurosci. **23:** 217–247.
99. ROSS, C.A., M.A. POIRIER, E.E. WANKER & M. AMZEL. 2003. Polyglutamine fibrillogenesis: the pathway unfolds. Proc. Natl. Acad. Sci. USA **100:** 1–3.
100. ROSS, C.A. 2002. Polyglutamine pathogenesis: emergence of unifying mechanisms for Huntington's disease and related disorders. Neuron **35:** 819–822.

101. GUSELLA, J.F. & M.E. MACDONALD. 1995. Huntington's disease: CAG genetics expands neurobiology. Curr. Op. Neurobiol. **5:** 656–662.
102. MACDONALD, M.E., S. GINES, J.F. GUSELLA & V.C. WHEELER. 2003. Huntington's disease. NeuroMolec. Med. **4:** 7–20.
103. KROBITSCH, S. & S. LINDQUIST. 2000. Aggregation of huntingtin in yeast varies with the length of the polyglutamine expansion and the expression of chaperone proteins. Proc. Natl. Acad. Sci. USA **97:** 1589–1594.
104. SATYAL, S.H., E. SCHMIDT, K. KITAGAWA, et al. 2000. Polyglutamine aggregates alter protein folding homeostasis in Caenorhabditis elegans. Proc. Natl. Acad. Sci. USA **97:** 5750–5755, 2000.
105. ROIZIN, L., S. STELLAR, N. WILLSON, et al. 1974. Electron microscope and enzyme studies in cerebral biopsies of Huntington's chorea. Trans. Am. Neurol. Assoc. **99:** 240–243.
106. DIFIGLIA, M., E. SAPP, K.O. CHASE, et al. 1997. Aggregation of huntingtin in neuronal intranuclear inclusions and dystrophic neurites in brain. Science **277:** 1990–1993.
107. DAVIES, S.W., M. TURMAINE, B.A. COZENS, et al. 1997. Formation of neuronal intranuclear inclusions underlies the neurological dysfunction in mice transgenic for the HD mutation. Cell **90:** 537–548.
108. ORDWAY, J.M., S. Tallaksen-Greene, C.A. GUTEKUNST, et al. 1997. Ectopically expressed CAG repeats cause intranuclear inclusions and a progressive late onset neurological phenotype in the mouse. Cell **91:** 753–763.
109. JACKSON, G.R., I. SALECKER, X. DONG, et al. 1998. Polyglutamine-expanded human huntingtin transgenes induce degeneration of *Drosophila* photoreceptor neurons. Neuron **21:** 633–642.
110. WARRICK, J.M., H.L. PAULSON, G.L. GRAY-BOARD, et al. 1998. Expanded polyglutamine protein forms nuclear inclusions and causes neural degeneration in *Drosophila*. Cell **93:** 939–949.
111. PERUTZ, M.F., T. JOHNSON, M. SUZUKI & J.T. FINCH. 1994. Glutamine repeats as polar zippers: Their possible role in inherited neurodegenerative disease. Proc. Natl. Acad. Sci. USA **91:** 5355–5358.
112. PERUTZ, M.F., J.T. FINCH, J. BERRIMAN & A. LESK. 2002. Amyloid fibers are water-filled nanotubes. Proc. Natl. Acad. Sci. USA **99:** 5591–5595.
113. SIKORSKI, P. & E. ATKINS. 2005. New model for crystalline polyglutamine assemblies and their connection with amyloid fibrils. Biomacromolecules **6:** 425–432.
114. NELSON, R., M.R. SAWAYA, B. BALBIRNIE, et al. 2005. Structure of the cross-β spine of amyloid-like fibrils. Nature **435:** 773–778.
115. PAULING, L. & R.B. COREY. 1951. The polypeptide-chain configuration in hemoglobin and other globular proteins. Proc. Natl. Acad. Sci. USA **37:** 282–285.
116. ARMEN, R.S., B.M. BERNARD, R. DAY, et al. 2005. Characterization of a possible amyloidogenic precursor in glutamine-repeat neurodegenerative diseases. Proc. Natl. Acad. Sci. USA **102:** 13433–13438.
117. YODER, M.D., N.T. KEEN & F. JURNAK. 1993. New domain motif: the structure of pectate lyase C, a secreted plant virulence factor. Science **260:** 1503–1507.
118. COHEN, F.E. 1993. The parallel β-helix of pectate lyase C: something to sneeze at. Science **260:** 1444–1445.
119. STORK, M., A. GIESE, H.A. KRETZSCHMAR & P. TAVAN. 2005. Molecular dynamics simulations indicate a possible role of parallel β-helices in seeded aggregation of poly-Gln. Biophys. J. **88:** 2442–2451.
120. COOKSON, M.R. 2005. The biochemistry of Parkinson's disease. Annu. Rev. Biochem. **74:** 29–52.
121. DEN HARTOG JAGER, W.A. 1969. Sphingomyelin in Lewy inclusion bodies in Parkinson's disease. Arch. Neurol. **21:** 615–619.
122. MCKEITH, I.G., D. GALASKO, K. KOSAKA, et al. 1996. Consensus guidelines for the clinical and pathologic diagnosis of dementia with Lewy bodies (DLB): report of the consortium on DLB international workshop. Neurology **47:** 1113–1124.
123. CLAYTON, D.F. & J.M. GEORGE. 1998. The synucleins: a family of proteins involved in synaptic function, plasticity, neurodegeneration and disease. Trends Neurosci. **21:** 249–254.

124. MAROTEAUX, L., J.T. CAMPANELLI & R.H. SCHELLER. Synuclein: a neuron-specific protein localized to the nucleus and presynaptic nerve terminal. J. Neurosci. **8:** 2804–2815.
125. POLYMEROPOULOS, M.H., C. LAVEDAN, E. LEROY, et al. 1997. Mutation in the α-synuclein gene identified in families with Parkinson's disease. Science **276:** 2045–2047.
126. KRUGER, R., W. KUHN, T. MULLER, et al. 1998. Ala30Pro mutation in the gene encoding α-synuclein in Parkinson's disease. Nature Genet. **18:** 106–108.
127. ZARRANZ, J.J., J. ALEGRE, J.C. GOMEZ-ESTEBAN, et al. 2004. The new mutation, E46K, of α-synuclein causes Parkinson and Lewy body dementia. Ann. Neurol. **55:** 164–173.
128. SINGLETON, A.B., M. FARRER, J. JOHNSON et al. 2003. α-Synuclein locus triplication causes Parkinson's disease. Science **302:** 841.
129. CROWTHER, R.A., S.E. DANIEL & M. GOEDERT. 2000. Characterisation of isolated α-synuclein filaments from substantia nigra of Parkinson's disease brain. Neurosci. Lett. **292:** 128–130.
130. GAI. W.P., H.X. YUAN, X.Q. LI, et al. 2000. In situ and in vitro study of colocalization and segregation of α-synuclein, ubiquitin, and lipids in Lewy bodies. Exp. Neurol. **166:** 324–333.
131. FUJIWARA, H., M. HASEGAWA, N. DOHMAE et al. 2002. α-Synuclein is phosphorylated in synucleinopathy lesions. Nature Cell Biol. **4:** 160–4.
132. GIASSON., B.I., J.E. DUDA, I.V. MURRAY, et al. 2000. Oxidative damage linked to neurodegeneration by selective α-synuclein nitration in synucleinopathy lesions. Science **290:** 985–989.
133. UVERSKY, V.N., G. YAMIN, P.O. SOUILLAC, et al. 2002. Methionine oxidation inhibits fibrillation of human α-synuclein in vitro. FEBS Lett. **517:** 239–244.
134. HOKENSON, M.J., V.N. UVERSKY, J. GOERS, et al. 2004. Role of individual methionines in the fibrillation of methionine-oxidized α-synuclein. Biochemistry **43:** 4621–4633.
135. MISHIZEN-EBERZ, A.J., R.P. GUTTMANN, B.I. GIASSON, et al. 2003. Distinct cleavage patterns of normal and pathologic forms of α-synuclein by calpain I in vitro. J. Neurochem. **86:** 836–847.
136. HARROWER, T.P., A.W. MICHELL & R.A. BARKER. 2005. Lewy bodies in Parkinson's disease: Protectors or perpetrators? Exp. Neurol. **195:** 1–6.
137. LASHUEL, H.A., D. HARTLEY, B.M. PETRE, et al. 2002. Amyloid pores from pathogenic mutations Nature **418:** 291.
138. LASHUEL, H.A., B.M. PETRE, J. WALL, et al. 2002. α-Synuclein, especially the parkinson's disease-associated mutants, forms pore-like annular and tubular protofibrils. J. Mol. Biol. **322:** 1089–1102.
139. CHIBA-FALEK, O., J.W. TOUCHMAN & R.L. NUSSBAUM. 2003. Functional analysis of intra-allelic variation at NACP-Rep1 in the α-synuclein gene. Hum. Genet. **113:** 426–31.
140. FARRER, M., D.M. MARAGANORE, P. LOCKHART, et al. 2001. α-Synuclein gene haplotypes are associated with Parkinson's disease. Hum. Molec. Genet. **10:** 1847–51.
141. GEORGE, J.M., H. JIN, W.S. WOODS & D.F. CLAYTON. 1995. Characterization of a novel protein regulated during the critical period for song learning in the zebra finch. Neuron **15:** 361–372.
142. DAVIDSON, W.S., A. JONAS, D.F. CLAYTON & J.M. GEORGE. 1998. Stabilization of α-synuclein secondary structure upon binding to synthetic membranes. J. Biol. Chem. **273:** 9443–9449.
143. SHARON. R., M.S. GOLDBERG, I. BAR-JOSEF et al. 2001. α-Synuclein occurs in lipid-rich high molecular weight complexes, binds fatty acids, and shows homology to the fatty acid-binding proteins. Proc. Natl. Acad. Sci. USA **98:** 9110–9115.
144. COLE, N.B., D.D. MURPHY, T. GRIDER, et al. 2002. Lipid droplet binding and oligomerization properties of the Parkinson's disease protein α-synuclein. J. Biol. Chem. **277:** 6344–6352.
145. ZHU, M. & A.L. FINK. 2003. Lipid binding inhibits α-synuclein fibril formation. J. Biol. Chem. **278:** 16873–16877.
146. UVERSKY, V.N., J. LI & A.L. FINK. 2001. Evidence for a partially folded intermediate in α-synuclein fibril formation. J. Biol. Chem. **276:** 10737–10744.

147. PIKE, C.J., A.J. WALENCEWICZ, C.G. GLABE & C.W. COTMAN. 1991. In vitro aging of β-amyloid protein causes peptide aggregation and neurotoxicity. Brain Res **563**: 311–314.
148. PIKE, C.J., D. BURDICK, A.J. WALENCEWICZ, *et al.* 1993. Neurodegeneration induced by β-amyloid peptides in vitro: the role of peptide assembly state. J. Neurosci. **13**: 1676–1687.
149. ROHER, A.E. 1991. β-Amyloid from Alzheimer's disease brains inhibits sprouting and survival of sympathetic neurons. Biochem. Biophys. Res. Commun. **174**: 572–579.
150. KURODA, Y. & M. KAWAHARA. 1994. Aggregation of amyloid β-protein and its neurotoxicity: enhancement by aluminum and other metals. Tohoku J. Exp. Med. **174**: 263–268.
151. LORENZO, A. & B.A. YANKNER. 1994. β-amyloid neurotoxicity requires fibril formation and is inhibited by Congo red. Proc. Natl Acad. Sci. USA **91**: 12243–12247.
152. HOWLETT, D.R., K.H. JENNINGS, D.C. LEE, *et al.* 1995. Aggregation state and neurotoxic properties of Alzheimer's β-amyloid peptide. Neurodegeneration **4**: 23–32.
153. FORLONI, G., O. BUGIANI, F. TAGLIAVINI & M. SALMONA. 1996. Apoptosis-mediated neurotoxicity induced by β-amyloid and PrP fragments. Mol. Chem. Neuropathol. **28**: 163–171.
154. FORLONI, G. 1996. Neurotoxicity of β-amyloid and prion peptides. Curr. Op. Neurol. **9**: 492–500.
155. WELDON, D.T., S.D. ROGERS, J.R. GHILARDI, *et al.* 1998. Fibrillar β-amyloid induces microglial phagocytosis, expression of inducible nitric oxide synthase, and loss of a select population of neurons in the rat CNS in vivo. J. Neurosci. **18**: 2161–2173.
156. HARPER, J.D., S.S. WONG, C.M. LIEBER *et al.* 1997. Observation of metastable Ab amyloid protofibrils by atomic force microscopy. Chem. Biol. **4**: 119–125.
157. WALSH, D.M., I. KLYUBIN & J.V. FADEEVA. 2002. Naturally secreted oligomers of amyloid β-protein potently inhibit hippocampal long-term potentiation *in vivo*. Nature **416**: 535–9.
158. KIRKITADZE, M.D., G.BITAN & D.B. TEPLOW. 2002. Paradigm shifts in Alzheimer's Disease and other neurodegenerative disorders: The emerging role of oligomeric assemblies. J. Neurosci. Res. **69**: 567–577.
159. BUCCIANTINI, M., E. GIANNONI, F. CHITI, *et al.* 2002. Inherent toxicity of aggregates implies a common mechanism for protein misfolding diseases. Nature **416**: 507–511.
160. KLEIN, W.L., W.B. STINE, JR. & D.B. TEPLOW. 2004. Small assemblies of unmodified amyloid β-protein are the proximate neurotoxin in Alzheimer's disease. Neurobiol. Aging **25**: 569–580.
161. BITAN, G., E.A. FRAGINGER, S.M. SPRING & D.B. TEPLOW. 2005. Neurotoxic protein oligomers: what you see is not always what you get. Amyloid **12**: 88–95.
162. STEFANI, M. & C.M. DOBSON. 2003. Protein aggregation and aggregate toxicity: new insights into protein folding, misfolding diseases and biological evolution. J. Mol. Med. **81**: 678–699.
163. BEAR, M.F. & R.C. MALENKA. 1994. Synaptic plasticity: LTP and LTD. Curr. Op. Neurobiol. **4**: 389–399.
164. LISMAN, J.E. & C.C. MCINTYRE. 2001. Synaptic plasticity: A molecular memory switch. Curr. Biol. **11**: R788–R791.
165. MALENKA, R.C. & M.F. BEAR. 2004. LTP and LTD: An Embarrassment of Riches. Neuron **44**: 5–21.
166. ABRAHAM, W.C. & A. ROBINS. 2005. Memory retention – the synaptic stability versus plasticity dilemma. Trends Neurosci. **28**: 73–78.
167. Bliss, T.V.P. & T. Lømo. 1973. Long-lasting potentiation of synaptic transmission in the dentate area of the anaesthetized rabbit following stimulation of the perforant path. J. Physiol. **232**: 331–356.
168. WANG, H-W, J.F. PASTERNAK, H. KUO, *et al.* 2002. Soluble oligomers of β amyloid (1-42) inhibit long-term potentiation but not long-term depression in rat dentate gyrus. Brain Res. **924**: 133–140.
169. KIM, J.H., R. ANWYL, Y.H. SUH, *et al.* 2001. Use-dependent effects of amyloidogenic fragments of β-amyloid precursor protein on synaptic plasticity in rat hippocampus in vivo. J. Neurosci. **21**: 1327–1333.

170. LAMBERT, M.P., A.K. BARLOW, B.A. CHROMY, et al. 1998. Diffusible, nonfibrillar ligands derived from Aβ 1–42 are potent central nervous system neurotoxins. Proc. Natl. Acad. Sci. USA **95:** 6448–6453.
171. KLEIN, W.L. 1998. Aβ toxicity in Alzheimer's disease. *In* Molecular Mechanisms of Neurodegenerative Diseases. M.F. Chesselet, Ed.: 1–49. Humana Press. Totowa, NJ.
172. TERRY, R.D., E. MASLIAH, D.P. SALMON, et al. 1991. Physical basis of cognitive alterations in Alzheimer's disease: synapse loss is the major correlate of cognitive impairment. Ann. Neurol. **30:** 572–580.
173. DICKSON, D.W., H.A. CRYSTAL, C. BEVONA, et al. 1995. Correlations of synaptic and pathological markers with cognition of the elderly. Neurobiol. Aging **16:** 285–298.
174. ODDO, S., A. CACCAMO, J.D. SHEPHERD, et al. 2003. Triple-transgenic model of Alzheimer's disease with plaques and tangles: intracellular Aβ and synaptic dysfunction. Neuron **39:** 409–421.
175. BILLINGS, L.M., S. ODDO, K.N. GREEN, et al. 2005. Intraneuronal Aβ causes the onset of early Alzheimer's disease-related cognitive deficits in transgenic mice. Neuron **45:** 675–688.
176. LASHUEL, H.A., D.M. HARTLEY, B.M. PETRE, et al. 2003. Mixtures of wild-type and a pathogenic (E22G) form of Aβ40 in vitro accumulate protofibrils, including amyloid pores. J. Mol. Biol. **332:** 795–808.
177. ARISPE, N., H.B. POLLARD & E. ROJAS. 1993. Giant multilevel cation channels formed by Alzheimer disease amyloid β-protein [AβP-(1-40)] in bilayer membranes. Proc. Natl. Acad. Sci. USA **90:** 10573–10577.
178. TALAFOUS, J., K.J. MARCINOWSKI, G. KLOPMAN & M.G. ZAGORSKI. 1994. Solution structure of residues 1-28 of the amyloid β-peptide. Biochemistry **33:** 7788–7796.
179. MATTSON, M.P., J.G. BEGLEY, R.J. MARK & K. FURUKAWA. 1997. Aβ25-35 induces rapid lysis of red blood cells: contrast with Aβ1-42 and examination of underlying mechanisms. Brain Res. **771:** 147–153.
180. LIN, H., R. BHATIA & R. LAL. 2001. Amyloid β-protein forms ion channels: implications for Alzheimer's disease pathophysiology. FASEB J. **15:** 2433–2444.
181. VOLLES, M.J., S.J. LEE, J.C. ROCHET, et al. 2001. Vesicle permeabilization by protofibrillar α-synuclein: implications for the pathogenesis and treatment of Parkinson's disease. Biochemistry **40:** 7812–7819.
182. KAGAN, B.L., Y. HIRAKURA, R. AZIMOV, et al. 2002. The channel hypothesis of Alzheimer's disease: current status. Peptides **23:** 1311–1315.
183. LIN, M.C. & B.L. KAGAN. 2002. Electrophysiologic properties of channels induced by Aβ25-35 in planar lipid bilayers. Peptides **23:** 1215–1228.
184. PLEIN, H. 2002. Amyloid β-protein forms ion channels. Trends Neurosci. **25:** 137.
185. HIRAKURA, Y., R. AZIMOV, R. AZIMOVA & B.L. KAGAN. 2000. Polyglutamine induced ion channels: a possible mechanism for the neurotoxicity of Huntington's and other CAG repeat diseases. J. Neurosci Res. **60:** 490–494.
186. MONOI, H., S. FUTAKI, S. KUGIMIYA, et al. 2000. Poly-l-glutamine forms cation channels: relevance to the pathogenesis of the polyglutamine diseases. Biophys. J. **78:** 2892–2899.
187. KAGAN, B.L., Y. HIRAKURA, R. AZIMOV & R. AZIMOVA. 2001. The channel hypothesis of Huntington's disease. Brain Res. Bull. **56:** 281–284.
188. CHRIVIA, J.C., R.P. KWOK, N. LAMB, et al. 1993. Phosphorylated CREB binds specifically to the nuclear protein CBP. Nature **365:** 855–859.
189. NUCIFORA, F.C., JR., M. SASAKI, M.F. PETERS, et al. 2001. Interference by huntingtin and atrophin-1 with CBP-mediated transcription leading to cellular toxicity. Science **291:** 2423–2428.
190. MANTAMADIOTIS, T., T. LEMBERGER, S.C. BLECKMANN, et al. 2002. Disruption of CREB function in brain leads to neurodegeneration. Nature Genet. **31:** 47–54.
191. LUTHI-CARTER, R., A. STRAND, N.L. PETERS, et al. 2000. Decreased expression of striatal signaling genes in a mouse model of Huntington's disease. Hum. Molec. Genet. **9:** 1259–1271.
192. WYTTENBACH, A., J. SWARTZ, H. KITA, et al. 2001. Polyglutamine expansions cause decreased CRE-mediated transcription and early gene expression changes prior to

cell death in an inducible cell model of Huntington's disease. Hum. Molec. Genet. **10:** 1829–1845.
193. LIN, X., B. ANTALFFY, D. KANG, *et al.* 2000. Polyglutamine expansion down-regulates specific neuronal genes before pathologic changes in SCA1.[see comment]. Nature Neurosci. **3:** 157–163.
194. SHIMOHATA, T., T. NAKAJIMA, M. YAMADA, *et al.* 2000. Expanded polyglutamine stretches interact with TAFII130, interfering with CREB-dependent transcription. Nature Genet. **26:** 29–36.
195. TAKEYAMA, K., S. ITO, A. YAMAMOTO, *et al.* 2002. Androgen-dependent neurodegeneration by polyglutamine-expanded human androgen receptor in *Drosophila*. Neuron **35:** 855–864.
196. ADACHI, H., A. KUME, M. LI, *et al.* 2001. Transgenic mice with an expanded CAG repeat controlled by the human AR promoter show polyglutamine nuclear inclusions and neuronal dysfunction without neuronal cell death. Hum. Molec. Genet. **10:** 1039–1048.
197. KATSUNO, M., H. ADACHI, A. KUME, *et al.* 2002. Testosterone reduction prevents phenotypic expression in a transgenic mouse model of spinal and bulbar muscular atrophy. Neuron **35:** 843–854.
198. DENNING, D.P., V. UVERSKY, S.S. PATEL, *et al.* 2002. The *Saccharomyces cerevisiae* nucleoporin Nup2p Is a natively unfolded protein. J. Biol. Chem. **277:** 33447–33455.
199. UVERSKY, V. 2002. Natively unfolded proteins: a point where biology waits for physics. Protein Sci. **11:** 739–756.
200. FINK, A.L. 2005. Natively unfolded proteins. Curr. Op. Struct. Biol. **15:** 35–41.
201. BARRAL, J.M., S.A. BROADLEY, G. SCHAFFAR & F.U. HARTL. 2004. Roles of molecular chaperones in protein misfolding diseases. Sem. Cell Dev. Biol. **15:** 17–29.
202. WELCH, W.J. 2004. Role of quality control pathways in human diseases involving protein misfolding. Sem. Cell Dev. Biol. **15:** 31–38.
203. BRAUN, B.C., M. GLICKMAN, R. KRAFT, *et al.* 1999. The base of the proteasome regulatory particle exhibits chaperone-like activity. Nature Cell Biol. **1:** 221–226.
204. IMAI, J., H. YASHIRODA, M. MARUYA, *et al.* 2003. Proteasomes and molecular chaperones: cellular machinery responsible for folding and destruction of unfolded proteins. Cell Cycle **2:** 585–590.
205. CUMMINGS, C.J., M.A. MANCINI, B. ANTALFFY, *et al.* 1998. Chaperone suppression of aggregation and altered subcellular proteasome localization imply protein misfolding in SCA1. Nature Genet. **19:** 148–154.
206. MUCHOWSKI, P.J., G. SCHAFFAR, A. SITTLER, *et al.* 2000. Hsp70 and hsp40 chaperones can inhibit self-assembly of polyglutamine proteins into amyloid-like fibrils. Proc. Natl. Acad .Sci. USA **97:** 7841–7846.
207. JANA, N.R., M. TANAKA, G. WANG & N. NUKINA. 2000. Polyglutamine length-dependent interaction of Hsp40 and Hsp70 family chaperones with truncated N-terminal huntingtin: their role in suppression of aggregation and cellular toxicity. Hum. Molec. Genet. **9:** 2009–2018.
208. MCNAUGHT, K.S., P. SHASHIDHARAN, D.P. PERL, *et al.* 2002. Aggresome-related biogenesis of Lewy bodies. Eur. J. Neurosci. **16:** 2136–2148.
209. FONTE, V., V. KAPULKIN, A. TAFT, *et al.* 2002. Interaction of intracellular β-amyloid peptide with chaperone proteins. Proc. Natl. Acad. Sci. USA **99:** 9439–9444.
210. AULUCK, P.K., H.Y. CHAN, J.Q. TROJANOWSKI, *et al.* 2002. Chaperone suppression of α-synuclein toxicity in a *Drosophila* model for Parkinson's disease. Science **295:** 865–868.
211. BENNETT, M.C., J.F. BISHOP, Y. LENG, *et al.* 1999. Degradation of α-synuclein by proteasome. J. Biol. Chem. **274:** 33855–33858.
212. KELLER, J.N., K.B. HANNI & W.R. MARKESBERY. 2000. Impaired proteasome function in Alzheimer's disease. J. Neurochem. **75:** 436–439.
213. BENCE, N.F., R.M. SAMPAT & R.R. KOPITO. 2001. Impairment of the ubiquitin-proteasome system by protein aggregation. Science **292:** 1552–1555.
214. MCNAUGHT, K.S. & P. JENNER. 2001. Proteasomal function is impaired in substantia nigra in Parkinson's disease. Neurosci. Lett. **297:** 191–194.
215. VERHOEF, L.G., K. LINDSTEN, M.G. MASUCCI & N.P. DANTUMA. 2002. Aggregate formation inhibits proteasomal degradation of polyglutamine proteins. Hum. Molec. Genet. **11:** 2689–2700.

216. HOLMBERG, C.I., K.E. STANISZEWSKI, K.N. MENSAH, et al. 2004. Inefficient degradation of truncated polyglutamine proteins by the proteasome. EMBO J. **23:** 4307–4318.
217. CUMMINGS, C.J., E. REINSTEIN, Y. SUN, et al. 1999. Mutation of the E6-AP ubiquitin ligase reduces nuclear inclusion frequency while accelerating polyglutamine-induced pathology in SCA1 mice. Neuron 24: 879–92, 1999.
218. KITADA, T., S. ASAKAWA, N. HATTORI, et al. **1999.** Mutations in the parkin gene cause autosomal recessive juvenile parkinsonism. Nature **392:** 605–608.
219. ZHANG, Y., J. GAO, K.K. CHUNG, et al. 2000. Parkin functions as an E2-dependent ubiquitin-protein ligase and promotes the degradation of the synaptic vesicle-associated protein, CDCrel-1. Proc. Natl. Acad. Sci. USA **97:** 13354–13359.
220. BRUIJN, L.I., M.K. HOUSEWEART, S. KATO, et al. **1998.** Aggregation and motor neuron toxicity of an ALS-linked SOD1 mutant independent from wild-type SOD1. Science **281:** 1851–1854.
221. UEDA, K., H. FUKUSHIMA, E. MASLIAH, et al. 1993. Molecular cloning of cDNA encoding an unrecognized component of amyloid in Alzheimer disease. Proc. Natl. Acad. Sci. USA **90:** 11282–11286.
222. CONWAY, K.A., J.C. ROCHET, R.M. BIEGANSKI & P.T. LANSBURY, JR. 2001. Kinetic stabilization of the α-synuclein protofibril by a dopamine-a-synuclein adduct. Science **294:** 1346–1349.
223. DI MONTE, D.A. 2003. The environment and Parkinson's disease: is the nigrostriatal system preferentially targeted by neurotoxins? Lancet Neurol. **2:** 531–538.
224. MASTRIANNI, J.A., R. NIXON, R. LAYZER, et al. 1999. Prion protein conformation in a patient with sporadic fatal insomnia. New Engl. J. Med. **340:** 1630–1638.
225. PRUSINER, S.B., M.R. SCOTT, S.J. DEARMOND & F.E. COHEN. 1998. Prion protein biology. Cell **93:** 337–348.
226. PRUSINER, S.B. 2001. Shattuck lecture: neurodegenerative diseases and prions [comment]. New Engl. J. Med. **344:** 1516–1526.
227. SHORTER, J. & S. LINDQUIST. 2005. Prions as adaptive conduits of memory and inheritance. Nature Rev. Genet. **6:** 435–450.
228. COLLINGE, J. 2005. Molecular neurology of prion disease. J. Neurol. Neurosurg. Psych. **76:** 906–919.
229. WEISSMANN, C. 2005. Birth of a prion: spontaneous generation revisited. Cell **122:** 165–168.
230. KRISHNAN, R. & S.L. LINDQUIST. 2005. Structural insights into a yeast prion illuminate nucleation and strain diversity. Nature **435:** 765–72.
231. TANAKA, M., P. CHIEN, K. YONEKURA et al. 2005. Mechanism of cross-species prion transmission: an infectious conformation compatible with two highly divergent yeast prion proteins. Cell **121:** 49–62.
232. BAGRIANTSEV, S. & S.W. LIEBMAN. 2004. Specificity of prion assembly in vivo. [PSI+] and [PIN+] form separate structures in yeast. J. Biol. Chem. **279:** 51042–51048.
233. MASISON, D.C., M.L. MADDELEIN & R.B. WICKNER. 1997. The prion model for [URE3] of yeast: spontaneous generation and requirements for propagation. Proc. Natl. Acad. Sci. **94:** 12503–8.
234. PETKOVA, A.T., R.D. LEAPMAN, Z. GUO et al. 2005. Self-propagating, molecular-level polymorphism in Alzheimer's β-amyloid fibrils. Science **307:** 262–265.

Thermally Induced Injury and Heat-Shock Protein Expression in Cells and Tissues

MARISSA NICHOLE RYLANDER,[a] YUSHENG FENG,[b] JON BASS,[b] AND KENNETH R. DILLER[a]

Department of Biomedical Engineering[a] and Institute for Computational Engineering and Science,[b] The University of Texas at Austin, Austin, Texas 78712

ABSTRACT: Heat-shock proteins (HSPs) are critical components of a cell's defense mechanism against injury associated with adverse stresses. Initiating insults, such as elevated or depressed temperature, diminished oxygen, and pressure, increase HSP expression and can protect cells against subsequent, otherwise lethal, insults. Although HSPs are very beneficial to the normal cell, cancer cells can also use HSPs in response to stresses associated with various therapies (hyperthermia, chemotherapy, radiation), mitigating injury incurred by these treatments. Hyperthermia is a common treatment option for prostate cancer. HSPs can be induced in regions of the tumor where temperatures are insufficient to cause lethal thermal necrosis. Elevated HSP expression can enhance tumor cell viability and impart increased resistance to subsequent chemotherapy and radiation treatments, thereby promoting tumor recurrence. An understanding of the structure, function, and thermally stimulated HSP kinetics and cell injury for prostate cancer cells is essential to designing effective hyperthermia protocols. Measured thermally induced cellular HSP expression and injury data can be employed to develop a treatment planning model for optimization of the tissue response to therapy based on accurate prediction of the HSP expression and cell damage distribution.

KEYWORDS: thermal injury; damage model; heat-shock protein expression; prostate cancer

INTRODUCTION

Mammalian species have developed numerous mechanisms to cope with stress. Examples at the cellular level include temporary modifications in gene expression to survive changing environments, as well as altering cellular structure and function to deal with more permanent adverse conditions. An important cellular alteration induced by stress involves the synthesis and function of heat-shock proteins (HSPs). These proteins reside in all organisms from bacteria and yeast to humans and exist in various forms. At the cellular level, HSPs exist in the endoplasmic reticulum, mitochondria, cytosol, and nucleus at low levels to respond to everyday stresses within

---

Address for correspondence: Kenneth R. Diller, Chairman, Department of Biomedical Engineering, The University of Texas at Austin, 1 University Station, C0800, Austin, TX 78712-1084. Voice: 512-471-7167; fax: 512-471-0616.
  kdiller@mail.utexas.edu

the cell. HSPs are characterized into families according to their molecular weight. Typically, HSP78, -75, -60, and -10 are found in the organelles and HSP110, -90, -73, -72, and -20 are present in the nucleus and cytosol. Each HSP has many documented functions and can reside in various locations within the cell (TABLE 1).[1]

TABLE 1. HSP locations and specific functions

Family	Chaperone members	Cellular compartments	Functions
HSP100	HSP104	cytoplasm	thermotolerance
HSP90	HSP90	cytoplasm	stabilize inactive forms of certain hormone receptors until hormone is present; interact with certain protein kinases to assist their transit to plasma membranes; prevent aggregation of denatured proteins
	Grp94	endoplasmic reticulum	
HSP70	HSC70	cytoplasm/nucleus	stabilize prefolded/unfolded structures for translocation/folding; assembly of immunoglobins; target aged proteins to lysosomes for degradation; protein secretion; antigen presentation; thermotolerance; interaction with certain immunosuppressants
	HSP70	cytoplasm/nucleus	
	BipGRP78	ER	
	Grp75	mitochondria	
HSP60	HSP60	mitochondria	stabilize prefoled structures for folding/assembly; re-export of precursors to membrane space
HSP40	HSP40	mitochondria	chaperone activity; essential co-chaperone activity with HSP70 to enhance ATPase rate and substrate release
		cytoplasm/nucleus	
Small HSP	HSP27	cytoplasm	prevents polypeptide aggregation; thermotolerance through stabilization of microfilaments; possible roles in cell growth
	αA and αB crystallins	cytoplasm	

There is substantial evidence that HSPs play important physiological roles in normal conditions and in situations involving both systemic and cellular stress.[1] HSPs were first discovered in 1962 as chromosomal puffs in heat-shocked (exposure to elevated temperatures) *Drosophila* salivary gland cells.[2] Since their discovery, it was observed that elevated HSPs can also be triggered by a variety of stressful stimuli including ischemia, hypoxia, pressure overload, heavy metals, free oxygen radicals, protein kinase C, calcium increasing agents, ethanol, amino acid and glucose analogues, inflammation, sodium arsenite, hormones, antibiotics, cytokines, and infection (FIG. 1).[3] Increased HSP expression can be stimulated as a result of normal physiological processes such as development and differentiation. Specifically, HSP70 and -27 have been shown to be induced by elevated temperature, ischemia, oxidative stress, and anticancer drugs.[4]

CELLULAR STRESS RESPONSE

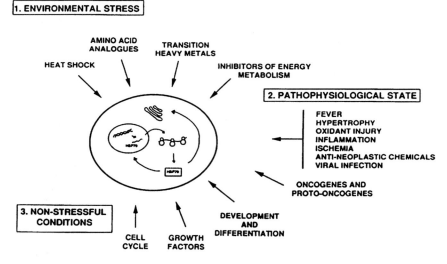

FIGURE 1. Stimuli that induce heat-shock protein expression.[4]

Elevated concentrations of HSPs due to exposure to stress have been shown to provide protection for both cultured cells and animal tissues. One of the first physiological functions associated with the stress-induced accumulation of the inducible HSP was acquired thermotolerance, which is defined as the ability of a cell or organism to become resistant to heat stress after a prior sublethal heat exposure. The phenomenon of acquired thermotolerance is transient in nature and depends chiefly on the severity of the initial heat stress. In general, the greater the initial heat dose, the greater the magnitude and duration of thermotolerance. The expression of thermotolerance following heat will occur within several hours, with maximum expression generally occurring 16–18 h following the initial thermal insult and may last 3–5 days in duration.[5]

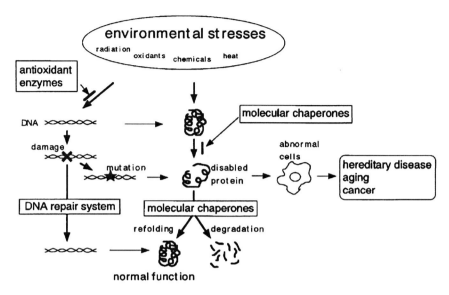

FIGURE 2. Stress-induced chaperone functions of HSP.[1]

The precise mechanisms for enhanced cellular thermotolerance in association with increased HSP levels have not been delineated. However, HSPs have been shown to be involved in preventing protein denaturation and/or processing denaturated proteins for an assisted refolding. Supporting evidence for this scenario came first from a set of *in vitro* experiments by Mizzen and Welch, who demonstrated that heat stress results in translational arrest within a cell. Subsequent resumption of translation resulted in HSP mRNA being translated into HSPs before the synthesis of other proteins took place within the cell.[6] Other researchers have confirmed that the protection HSP provides is based on their ability to act as molecular chaperones to inhibit improper protein aggregation and their capacity to direct newly formed proteins to target organelles for final packaging, degradation, or repair (FIG. 2).[1] HSPs play an important role in self-repair, self-protection mechanisms, and are associated with the refolding of denaturated cellular proteins.[1]

In addition to their chaperone functions, HSPs have numerous other protective roles (FIG. 3). Specifically, HSP70 and -27 have been shown to inhibit apoptosis and thereby increase the survival of cells exposed to a wide range of lethal stimuli.[7] Overexpresion of HSP70 elevates nitric oxide production as a result of cytokine stimulation. Nitric oxide serves to protect cultured cells from TNF-α–induced cell death by inducing HSP70. Overexpression of HSP27 can protect microtubules and actin cytoskeleton in cardiac myocytes and endothelial cells after exposure to ischemia.[7] HSP27, -60, and -70 are important in the progression of cancer both through angiogenesis and their role in apoptosis.

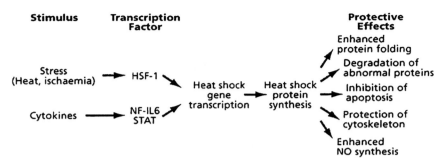

FIGURE 3. Activation of HSP by specific stimuli and their protective effect.[8]

The exact mechanism for stress-induced HSP induction has not been verified. A proposed mechanism, however, for HSP70 induction will be discussed (FIG. 4).[1,9] Transcription of the heat-shock response is controlled by a heat-shock factor (HSF). Under normal conditions, HSFs are bound to HSPs and are inactive. This HSF is responsible for recognizing a target sequence known as the heat-shock element (HSE). The HSE consists of a chain of five repeats of the NGAAN sequences. In the presence of stress, such as heat shock, the HSFs separate from the HSPs. Protein kinase or other serine/threonine kinases phosphorylate the HSFs, which cause them to form trimers in the cytosol. The HSF trimers enter the nucleus and bind to the HSE of the HSP gene. Following binding, the HSFs are further phosphorylated and HSP mRNA is transcribed and exits the nucleus for the destination of the cytoplasm. Upon entering the cytoplasm, new HSPs are synthesized. The HSFs then return to the cytoplasm and bind to the HSPs in the original orientation before exposure to the stress. Multiple HSEs exist on the promoter region of the HSP gene.[1,9] In addition to the HSE, a serum response element resides on the HSP promoter region. The SRE responds to serum stimulation and is responsible for the presence of a basal level of HSP expression in cells.[10] Humans possess multiple copies of the HSP gene. A special feature of HSP DNA coding is that it lacks introns. The HSP mRNA and HSPs can be produced extremely rapidly under stressful stimuli because no RNA splicing is required for HSP mRNA transcription.[9]

Although HSP expression has a myriad of documented protective functions that are beneficial to healthy cells, cancer cells can also use these proteins, preventing effective destruction of tumors with existing therapies. HSPs have been implicated in protective roles in neoplastic tissues including multidrug resistance,[11] regulation of apoptosis,[12] and modulation of p53 functions[13] in a wide range of tumors. This paper focuses on the thermally induced HSP protection pertinent to hyperthermia prostate cancer therapies. Thermally induced HSP27 and -70 have been shown to enhance tumor cell viability by preventing apoptosis and imparting resistance to radiation and chemotherapy following thermal therapy. HSP27 overexpression is a poor prognostic marker for invasive prostatic carcinoma, but the absence of HSP27 is a reliable objective marker in early prostatic neoplasia.[14] Elevated HSP27 levels have been associated with enhanced tumor cell viability by inhibition of apoptosis.

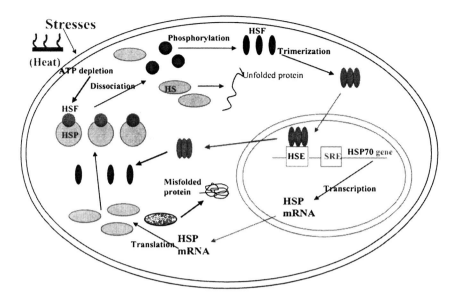

FIGURE 4. Schematic of a proposed mechanism of stress-induced HSP gene transcription.[1,9]

HSP27 modulates reactive oxygen species by means of a glutathione-dependent pathway,[15] providing protection for intracellular proteins and partially explaining the resistance it confers to thermally stressed tumor cells against chemotherapeutic agents.[16,17] Elevated levels of HSP70 have been observed in several types of tumors including breast and cervical cancers[18] and may be involved in cell proliferation, prognosis, and drug resistance.[19] Increased HSP70 expression has been linked to the synergistic effect of hyperthermia on radio and chemotherapies.[20]

Previously characterized thermally induced HSP27 and -70 kinetics data will be employed to design optimized laser therapies through computational predictive models. Prior to discussing the experimental and computational methods involved in optimizing laser therapies, some specific information will be provided on the molecular structure and chaperone activity specific for HSP27 and -70.

HSP27 MOLECULAR STRUCTURE

HSP27 (also known as HSP25 and HSP28) is the smallest molecular mass protein that is induced by heat shock. At present, there is limited knowledge about the structure of HSP27; however, it is known that this protein possesses two homotypic interacting domains located at the N-terminal and C-terminal regions (FIG. 5). The C-terminal domain extends from residue 88–183 where the last 10 residues form a flexible peptide composed of A- and B-crystallins.[21] Through the interaction of two binding domains, HSP27 oligermerization is possible. One of the binding domains occurs between residue 94 and 178 which corresponds to the α-crystallin domain. In several studies this domain has been shown to have a role as the building block of the quaternary structure of other members of this family and has been suggested as the region where major intermolecular interactions occur. It has been suggested that the N-terminal and C-terminal domains are essential in the stable formation of dimers and high-molecular weight multimers.[22] The overall structure of HSP27 consists of a large central cavity where the hydrophobic N terminus is hidden to provide it protection from the environment. The interactions of the N-terminal domain in the central cavity act to stabilize the supramolecular organization of the dimers formed by the α-crystallin domain.[22]

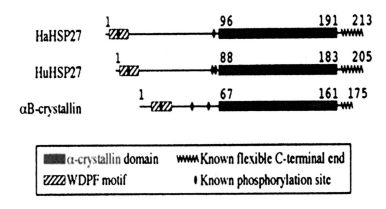

FIGURE 5. Molecular structure of the HSP2 kDa family.[22]

HSP70 MOLECULAR STRUCTURE

The molecular structure of the HSP70 family consists of three domains: 44 kDa, 18 kDa, and 10 kDa (FIG. 6).[23] The 44-kDa domain consists of amino acid residues 1–386 at the N-terminus and contains four domains forming two lobes with a deep cleft between. It is the 44-kDa domain that contains an ATPase domain. The 18-kDa peptide-binding domain consists of amino acid residues 384–543 and is composed of two four-stranded anti-parallel sheets and a single helix. This domain contains the peptide-binding domain responsible for binding folded and unfolded peptides.[23] The 10-kDa domain contains the amino acid residues 542–646 at the C terminus. This segment of the protein is predominately a helix followed by a glycine/proline–rich aperiodic segment next to a highly conserved EEVD terminal sequence.[24]

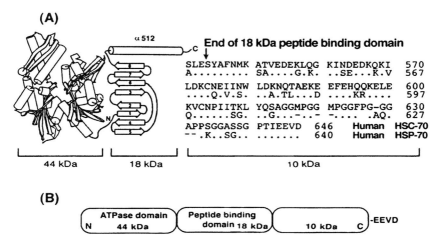

FIGURE 6. Molecular structure of HSP70 kDa family.[23]

HSP27 AND 70 CHAPERONE MECHANISMS

The specific chaperone mechanism is unique for each HSP; therefore this paper will only focus on those associated with HSP27 and -70. HSP27 chaperone function is proposed to behave similarly to highly documented HSP70 protein folding (FIG. 7).[25] HSP70 almost immediately detects improperly folded proteins by recognizing a small stretch of hydrophobic amino acids on a protein's surface. Aided by a set of smaller HSP40 proteins, an HSP70 monomer binds to its target protein and then hydrolyzes a molecule of ATP to ADP, undergoing a conformational change that causes the HSP70 to clamp tightly on the target. After HSP40 dissociates, the dissociation of the HSP70 protein is induced by the rapid rebinding of ATP after ADP release. Repeated cycles of HSP protein binding and release help the target protein to refold.

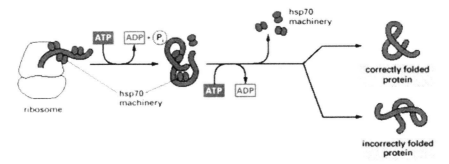

FIGURE 7. Protein-folding mechanism for HSP70.[25]

HSP EXPRESSION AND THERMAL INJURY

Although HSPs permit protection against lethal thermal injury, their induction and related thermotolerance suggests that significant injury has already occurred. The mechanisms contributing to thermal injury vs. thermotolerance are poorly understood. Cells, tissues, and animals show similar kinetics of thermotolerance, which suggests that the morbidity and mortality associated with whole-body heating is due in part to the dysfunction of some critical target tissues.[26–29] It was proposed that the development of thermotolerance results from the improved thermotolerance of the weakest organ and cell systems.[30] For instance, the small intestine is capable of generating thermotolerance[31] and is also reported to be the tissue most sensitive to heat damage.[32] In some instances, HSP expression could be used as a biomarker of cellular injury.[29] In this scenario, cells or tissues most at risk could be detected as the most likely to accumulate HSPs during stress. HSP accumulation could mark a tissue for potential failure.

A better understanding of the thermally induced HSP27 and -70 kinetics and cell injury in prostate tumors would permit prediction and improvement of the therapy outcome by maximizing tumor injury and eliminating HSP expression. Laser therapies are a common treatment option for causing thermally induced prostate tumor destruction. Applied thermal stress tends to elicit offsetting effects of elevated HSP expression and hyperthermia-mediated cell injury. Insufficient thermal damage induces elevated HSP27 and -70 expression that can enhance tumor cell viability, increasing the possibility of tumor recurrence. Thermal therapies are generally employed in conjunction with radiotherapy, chemotherapy, and gene therapy to increase therapeutic efficacy. Elevated HSP expression in surviving tumor cells following hyperthermia imparts increased resistance to subsequent therapies, substantially hindering their effectiveness. We have developed a strategy to predict the thermally induced HSP27 and -70 kinetics and cell injury by using measured prostate cancer cell HSP kinetics data[33] to develop computational treatment planning models.[34]

In this paper, we review basic experimental and computational methods that permit measurement and prediction of thermally induced cell injury and HSP kinetics. We will introduce the mathematical concepts for predicting the temperature profile in the tissue and methods for estimations of the thermally induced damage and HSP expression level corresponding to a certain temperature profile. These mathematical formulations will then be integrated into a finite element model to enable prediction of the tissue response to laser therapy.

TEMPERATURE PROFILE FOR LASER-IRRADIATED TISSUE

A controlled temperature increase in the desired tissue region can be achieved through laser irradiation of tissue. In order to provide an accurate prediction of the tissue response following laser therapy, a computational model was created to predict the temperature, damage, and HSP27 and -70 distributions. The light and thermal distributions during laser heating were modeled using adaptive finite element methods which are capable of minimizing numerical error to specified precision. This model allows specification of the thermal distribution during laser heating and

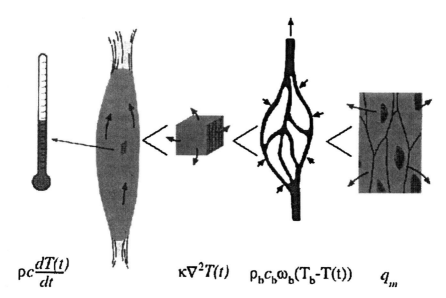

FIGURE 8. Bio-heat transfer equation defines the rate of temperature change in a small volume of tissue.[37]

study of the sensitivity of the thermal behavior to manipulation of individual source parameters.

The mathematical representation of the temperature distribution is based on the Pennes bio-heat transfer equation, which accurately predicts the spatial dependence of the temperature history.[35,36] This expression states that the rate of temperature change in a small volume of tissue is equal to the sum of the heat transfer from the surrounding tissues (conduction) and the heat transfer from blood perfusion as shown in FIGURE 8.[37] Usually, a term (q_m) due to metabolic heat generated by the cells is included in the model. However, such a contribution is negligible for thermal injuries that occur over a short period of time.

Within the present approach we introduce a term to account for the rate of absorbed laser energy per unit volume distributed within the tissue, which is given by $Q(x, y, z) = \mu_a \phi(x, y, z)$ where μ_a (m^{-1}) and (ΦW/m^2) are the irradiation absorption coefficient (wavelength dependent tissue property) and fluence (light distribution generated due to the laser source) respectively.[38,39] External irradiation will be used as the method of inducing a temperature rise and associated thermal injury. The fluence distribution is typically determined by employing Monte Carlo methods where the probability of absorption or scattering of an incident photon on the tumor sphere generated by the laser source can be predicted.[40] Therefore, the temperature profile is obtained by solving the equation

$$c_t \rho_t \frac{\partial T}{\partial t} = \nabla(k(T)\nabla T) + \omega_b(T)c_b(T - T_a) + Q(x, y, z) \quad (1)$$

where ρ_t and c_t are the density and specific heat of the tissue, respectively. The specific heat of blood and the arterial blood temperature are represented as c_b and T_a, respectively. The temperature-dependent thermal conductivity of the tissue and temperature-dependent blood perfusion rate are denoted by k and ω_b, respectively. The nonlinear temperature dependence of thermal conductivity and perfusion were included in the model to increase the accuracy of the temperature prediction following laser heating. The mathematical expressions employed for the nonlinear effects of the temperature-dependent blood perfusion in the tumor are shown in the following equation.[41]

$$\omega_{tumor} = \begin{cases} 0.833 & T < 37.0 \\ 0.833 - (T-37.0)^{4.8}/5.438 \times 10^3, & 37.0 \le T \le 42.0 \\ 0.416 & T > 42.0 \end{cases} \text{ (kg/s/m}^3\text{)} \quad (2)$$

A complete set of data for the temperature-dependent behavior of thermal properties of tissue does not exist. The thermal variation properties of water are well known in the range of 20–100°C and are important because the thermal properties of tissue are dependent on the water content. The temperature-dependent thermal conductivity for water where λ_k is a dimensionless correction factor and w is the water content of prostate tissue, .511, is shown in the following equation.[42]

$$K(T) = 4.19(0.133 + 1.36\lambda_k w)*10^{-1} \quad (W/mK)$$

where $\lambda_k = 1 + 1.78*10^{-3}(T - 20°C)$.
$$(3)$$

The degree of nonlinearity associated with the temperature dependencies of density and specific heat for the tissue and blood was found to be minimal, and therefore was not considered in the model. Sensitivity analysis showed that the temperature dependence of thermal conductivity and perfusion caused a 5% decrease in predicted tissue temperature for the laser irradiation protocols. Parameters employed in the simulation were derived for canine prostate (TABLE 2).

TABLE 2. Optical and thermal properties of canine prostate

Parameter	Symbol	Value
Laser wavelength	λ	810 nm
Absorption coefficient[38]	μ_a	1.5 cm^{-1}
Density of tissue[39]	ρ_t	1045 kg/m^3
Density of blood[39]	ρ_b	1058 kg/m^3
Specific heat of tissue[39]	c_t	3600 J/kgK
Specific heat of blood[39]	c_b	3840 J/kgK
Arterial blood temperature	T_a	310 K

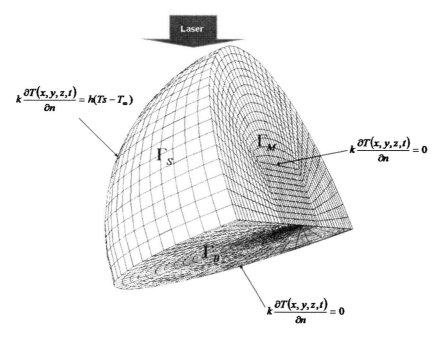

FIGURE 9. Boundary conditions applied to surfaces of spherical tumor model.

BOUNDARY CONDITIONS

The boundary conditions were applied on all tumor surfaces composing the boundary $\Gamma_B \cup \Gamma_M \cup \Gamma_S = \Gamma$, where Γ is the boundary of the tumor (FIG. 9). Surfaces Γ_B and Γ_M were specified insulating boundary conditions since the tumor is symmetric about these boundaries. A convective boundary condition was imposed on the curved outer surface to account for flow of air over the tumor. The mathematical expression employed for calculating the convection coefficient needed for quiescent air surrounding tissue surfaces with convective boundaries is shown below:

$$h = \frac{Nu_{Dav} k}{D} \quad (4)$$

where $Nu_{D,a}$ is the average Nusselt number, k is the thermal conductivity of the tissue and D is the diameter of the tumor. The radius of the simulated tumor was 6 mm since typicalprostrate tumors (derivied from PC3prostrate cancer cells) grown on mice attain a maximum spherical volume of 1 cm^3 prior to exhibiting necrotic cores. The Nu_{Da} can be calculated according to the following expression for flow over a sphere:

$$Nu_{Dav} = \left\{ 0.60 + \frac{0.387 Ra_D^{1/6}}{[1 + (0.59/Pr)^{9/16}]^{8/27}} \right\}^2 \quad (5)$$

where Pr is the Prandtl number and Ra_D is the Rayleigh number as defined below:

$$Ra_D = \frac{g\beta(T_s - T_\infty)D^3}{\nu\alpha} \tag{6}$$

where g is the local acceleration due to gravity (m/s), β is the volume coefficient of air expansion (1/K) evaluated at the mean value of the surface and air temperatures, ν is the kinematic viscosity (m^2/s), and α is thermal diffusivity (m^2/s). The values for T_s and T_∞ were stipulated as 310 K and 295 K, respectively. The parameters Pr, ν, and α were evaluated for air at 295 K with values of .849, 1.55×10^{-5} m/s^2, and 2.18×10^{-5} m^2/s respectively.[43]

INJURY AND HSP EXPRESSION MODELS

The computational models were based on measured thermally induced cell injury and HSP expression for normal (RWPE-1) and cancerous prostate (PC3) cells induced by water-bath heating. In order to acquire this data, PC3 and RWPE-1 cells were cultured in phenolic culture flasks. Upon reaching confluence, the flasks were submerged in a constant temperature water bath at a predetermined temperature and duration in the range of 44 to 60°C and 1–30 min.[33] The maximum experimental temperature caused complete cell death for the shortest heating duration. Following heating, the flasks were returned to an incubator heated at 37°C for subsequent manifestation of damage and HSP elevation.

HSP 27 AND HSP70 EXPRESSION MEASUREMENT AND MODEL

After an incubation time of 16–18 h post-heating (shown to be an effective evaluation period for measuring maximum HSP70 expression[33]), cells were lysed (cell membranes are broken with enzymes to extract protein) in buffer solution containing protease inhibitors and 10% SDS. The HSP 27 and -70 expression levels in each supernatant solutions were analyzed via gel electrophoresis and Western blotting.[33] The measured HSP27 and -70 kinetics data enabled formulation of an empirical model for prediction of HSP expression as described in previous work.[33] The proposed model describes HSP expression as a function of only temperature and heating duration which is adequately supported by our experimental data.

Mathematical Model

Employing Maple® permitted a wide array of functions to be explored for determination of the most appropriate mathematical formulation to accurately fit the entire data set for all measured temperatures and heating durations. HSP expression induced by a transient temperature field was found to obey the following relationship previously described:[33]

$$\frac{\partial H(t, T)}{\partial t} = f(t, T) \cdot H(t, T) \tag{7}$$

where $f(t, T)$ is a general rate function that can take various forms. In our case, we select $f(t, T) = (\alpha - \beta_1 t^{r-1})$, which captures the characteristics of HSP expression denoted as $H = H(T, t)$. The parameters $\alpha, \beta_1 (= \beta \cdot \gamma)$, and γ are are independent of time, but may be dependent on temperature, with $\gamma > 1$. $H(t, T)$ was found to be represented by the following function:

$$H(t, T) = A e^{\alpha t - \beta t^r} \tag{8}$$

where A is a temperature-dependent constant. Since the basal value of $H(t, T) = 1$ at $t = 0$ due to normalization, $A = 1$ for the measured data set. A least squares approach was employed for parameter estimation. The model is valid for predicting both HSP27 and HSP70 expression, but unique expression parameters, $\alpha, \beta_1 (= \beta \cdot \gamma)$, and γ describe HSP27 and HSP70 expression. The HSP expression model was integrated into the finite element model to enable prediction of the thermally-induced HSP response due to laser heating.

CELL INJURY MEASUREMENT AND CELL DAMAGE MODEL

Cell viability was assessed following 72 h post-heating (shown to be an effective evaluation period for measuring the extent of cell death[44]). Cells underwent propidium-iodide staining. Propidium iodide is permeable to only dead or dying cells, enabling injured cell populations to appear fluorescent. A flow cytometer was utilized to detect the fluorescence levels for all thermally stressed cell samples to permit quantification of cell viability.[33]

The availability of both measured cell injury and thermal history data enabled determination of the constitutive parameter values for an Arrhenius damage model:[45]

$$\Omega(\tau) = \ln(C_0/C_\tau) = A \int_0^\tau e^{-(E_a/\Re T(t))} dt \tag{9}$$

where Ω is damage, defined as the logarithm of the ratio of the initial concentration of healthy cells, C_0, to the concentration of healthy cells remaining after thermal stimulation, C_τ, for a stimulation duration of τ (s). $A(1/s)$ is a scaling factor, E_a (Jmol^{-1}) is injury process activation energy, $\Re$(Jmol^{-1}K^{-1}) is the universal gas constant, and $T(K)$ is instantaneous absolute temperature of the cells during stress, which is a function of time, t(s). The cell viability values for PC3 cells 72 h following hyperthermia were used to determine the damage parameters employed in the Arrhenius damage model.[33] The Arrhenius damage integral was fit to the cell injury data to characterize the response to thermal stress temperature and duration. At each temperature the threshold time (τ) was determined for $\Omega = 1$ for which $C_\tau = 1/e$ of C_0. For isothermal stress conditions and when $\Omega = 1$, the damage equation simplifies to the logarithmic form,

$$\ln(\tau) = \left(\frac{E_a}{R}\right)\left(\frac{1}{T}\right) - \ln(A). \tag{10}$$

The thermal damage kinetic coefficients of A and E_a were determined from the intercept and slope respectively of the best-fit linear function to fit the experimental data.[33]

According to the Arrhenius formulation, increasing temperature and heating duration will cause the damage value to increase indefinitely with complete cell damage represented by infinity. In order to more meaningfully represent the damage, another parameter was employed for determining tissue injury in the finite element simulations. The damage fraction, F_D of damaged to undamaged tissue is given by

$$F_D = \frac{C_o - C_f}{C_o} = 1 - \exp^{-\Omega}. \qquad (11)$$

Native tissue is represented by $F_D = 0$ ($\Omega = 0$) and fully denatured tissue is given by $F_D = 1$ ($\Omega =$ infinity).[42]

THERMAL, HSP, AND DAMAGE PREDICTIVE FINITE ELEMENT MODELS

Although HSP expression and damage were induced experimentally through water-bath heating, the ultimate goal is development of laser protocols to minimize HSP expression and maximize damage in the tumor region. Ultimately, HSP expression and damage data from laser-irradiated tissue will be employed to refine the accuracy of the model. Current cellular data will demonstrate the power of the model for predicting and optimizing HSP expression. Optimal heating protocols were identified by determining the most advantageous energy deposition pattern of a laser source necessary to produce a temperature distribution resulting in the desired HSP expression and tissue damage pattern. This involved determination of the appropriate laser source parameters consisting of wavelength, number of probes, optical fiber orientation in the tissue, power, and thermal exposure time. In the following simulations, only the laser power and thermal exposure time will be varied when investigating protocols. Determination of the desired temperature distribution was based on optimal heating protocols (time/temperature) determined through cellular HSP expression and damage kinetic models. The defined energy dissipation parameters will ultimately be applied to the laser source to produce destruction of the tumor while minimizing HSP expression in the targeted regions.

SIMULATION RESULTS

All simulations model external irradiation incident on the tumor surface and employ a laser wavelength of 810 nm. The parameters employed for the first simulation are laser power of 2 W and pulse duration of 1 min. The laser source was applied externally because our current laser heating experiments *in vivo* initially employ external radiation in order to prevent HSP expression induced by insertion of the laser probe. The model can accurately predict the temperature, damage fraction, and HSP expression in the tumor for the given specified laser parameters (FIG. 10).

Significant elevations in temperature occurred at the tumor boundary in close proximity with the laser source, but a substantial portion of the tumor experienced minimal or no elevation in temperature. Nominal damage was induced throughout

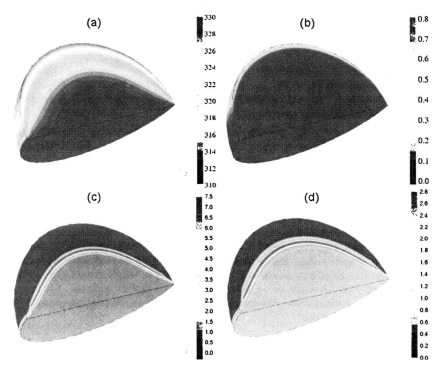

FIGURE 10. Finite element model prediction of (a) temperature (K), (b) damage, (c) HSP27 (mg/mL), and (d) HSP70 (mg/mL) distribution for externally irradiated prostate tumor with a laser power of 2W and pulse duration of 1 min.

the majority of the tumor volume, with the greatest injury occurring near the laser probe and diminishing with increasing distance from the laser source. All HSP expression was normalized with respect to the basal level of HSP expression for unstressed conditions and is represented as 1 mg/mL. As a result, HSP expression higher than 1 mg/mL represents an elevation in expression. There are three observable zones of interest with regard to HSP expression in the tumor. The topmost blue region represents a zone where significant damage was induced causing denaturation of all proteins and rendering the cell machinery incapable of HSP expression. The middle region with HSP expression greater than one represents the thermal regime where temperatures caused considerable induction of HSP expression. This is the region of concern where HSP expression must be minimized to prevent tumor cell viability following treatment. The final bottommost region represents a zone where temperatures were insufficient to elicit significant HSP expression with expression levels at or near the basal level.

In order to evaluate the success of the therapy, it is essential to clearly define characteristics of an optimal thermal treatment. The objective for designing a laser therapy should be complete tumor destruction and preservation of healthy surrounding tissue formulated according to the following mathematical criteria:

(1) Complete tumor destruction: Tumor must experience maximum cell injury (e.g., $F_D \geq .99$), and HSP expression must be diminished below its basal level of 1 ($HSP_{27,70} \leq 1$ mg/mL).

(2) Preservation of healthy tissue: Healthy tissue must receive minimal thermal damage (e.g., $F_D \leq .01$) and induction of increased levels of HSP expression ($HSP_{27,70} > 1$ mg/mL) to mitigate injury to subsequent thermal or chemotheraphy and radiation treatments.

The simulations in this paper consider only a spherical solid tumor volume so only criterion 1 is applicable. Evaluating this therapy outcome according to criterion 1 would permit us to realize that this is an unsatisfactory treatment due to insufficient thermal damage incurred throughout the entire tumor and the presence of a large region of elevated HSP27 and -70 expression. The elevated level of HSP expression would enhance the tumor cell viability, increasing the possibility of tumor recurrence and imparting tumor resistance to chemotherapy and radiation.

In order to achieve a more desirable treatment outcome, another therapy was considered in which the laser power was increased to 4 W and the pulse duration was lengthened to 2 min (FIG. 11). This therapy incurred more substantial temperature elevations and extensive thermal injury. Dramatic tissue injury extended to a greater tumor depth corresponding to a larger zone of protein denaturation. The topmost dark blue zone representing a region experiencing extreme thermal damage without

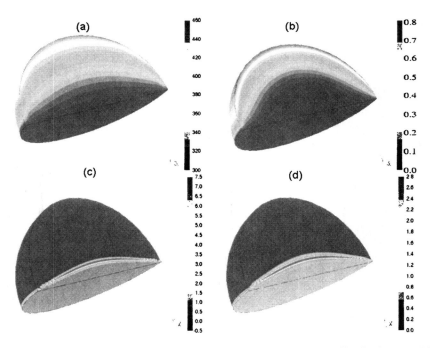

FIGURE 11. Finite element model prediction of (a) temperature (K), (b) damage, (c) HSP27 (mg/mL), and (d) HSP70 (mg/mL) distribution for externally irradiated prostate tumor with a laser power of 4W and pulse duration of 2 min.

elevation in HSP expression still exists, but has enlarged in size. Below this denatured zone is a region of high HSP expression. This region has been shifted in depth and narrowed in size, but still represents a zone of concern for tumor regrowth. With increasing depth, a zone with minimal HSP expression is observed due to insufficient temperature elevation.This therapy is still considered unacceptable due to the region of HSP expression since it does not adequately satisfy criterion 1.

The simulated cases have not yet identified an ideal therapy where HSP expression has been eliminated. The laser power is increased further to 5 W and the pulse duration is lengthened to 3 min. If we achieve HSP expression reduction below its unstressed basal level, the damage will have been so extensive causing complete tumor destruction. As a result, HSP27 and -70 elimination is the strictest criterion, and they will be the only quantities simulated. The HSP27 and -70 expression distributions were identical and therefore were represented by a single figure (FIG. 12). The outcome of this therapy yields a tumor with not only a lack of HSP expression increase, but also extreme denaturation of all proteins such that HSP expression is nearly nonexistent. The complete elimination of HSP expression yields a tumor with complete cell death so the cell damage plot is not shown. The tumor will not experience re-growth or thermally induced resistance to subsequent chemotherapy or radiation. This therapy would be considered a success theoretically; however, we have neglected the existence of surrounding healthy tissue that would have been damaged due to elevated temperatures utilized to achieve this complete kill.

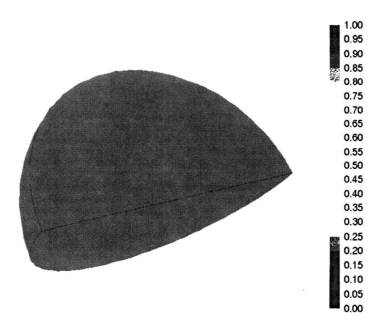

FIGURE 12. Finite element model prediction of HSP27 (mg/mL) and HSP70 (mg/mL) distribution for externally irradiated prostate tumor with a laser power of 5W and pulse duration of 3min.

CONCLUSION

An accurate HSP expression model was developed based on measured cellular data. This is the first study utilizing HSP expression computational models for prediction of HSP expression for use in thermal therapy planning. The Arrhenius injury model enabled the damage in the tumor to be predicted with precision based on a given temperature profile in the tumor. Previous studies have demonstrated a much lower thermal threshold for destruction of AT-1 canine prostate tumors *in vivo* compared to their *in vitro* counterparts under similar conditions due to the presence of the vascular network *in vivo*.[46,47] Thus, it will be essential to investigate the hyperthermia-induced HSP expression kinetics and cell viability modifications for PC3 tumors *in vivo* before final dosimetry guidelines are developed for prostate cancer therapy.

The integration of thermal, damage, and HSP expression models into a single finite element model enabled prediction of the prostate tumor response to a given laser therapy. After considering a wide range of laser parameters commonly employed in surgical procedures in the finite element modeling simulations, definite zones of HSP expression were identified in the irradiated prostate tumors after therapy. The existence of these zones is of great concern, and minimization of HSP expression should be considered in patient treatment design. In designing future therapies there will be a tradeoff between complete tumor eradication and damage minimization to healthy tissue. Further exploration into the role HSP expression in tumor survival is essential in designing successful therapies. The model developed in this study will enable the damage and HSP expression to be predicted for a wide range of laser parameters of the physician's choice prior to surgery to improve understanding of the tissue response and to enable optimized design of the patient therapy.

REFERENCES

1. KIANG, J.G. & G.C. TSOKOS. 1998. Heat shock protein 70kDa: molecular biology, biochemistry, and physiology. Pharmacol. Ther. **80:** 183–201.
2. RITOSSA, F. 1962. A new puffing pattern induced by temperature shock and DNP in Drosophila. Experientia **18:** 571–573.
3. WEICH, H. *et al.* 1992. HSP90 chaperones protein folding in vitro. Nature **358:** 169–170.
4. SCHLESINGER, M. *et al.* 1990. Stress Proteins: Induction and Function. Springer-Verlag. Berlin.
5. WANG, S., K.R. DILLER & S. AGGARWAL. 2003. Heat shock protein 70 expression kinetics. J. Biomech. Eng. **125:** 794–797.
6. MIZZEN, L.A. & W.J. WELCH. 1988. Characterization of the thermotolerant cell. I. Effects on protein synthesis activity and the regulation of heat-shock protein 70 expression. J. Cell Biol. **106:** 1105–1116.
7. GARRIDO, C. 2001. Heat shock proteins: endogenous modulators of apoptotic cell death. Biochem. Biophys. Res. Commun. **286:** 433–442.
8. LATCHMAN, D. 2001. Stress proteins. Cardiovasc. Res. **51:** 637–646.
9. MORIMOTO, R.I. & M.G. SANTORO. 1998. Stress-inducible responses and heat shock proteins: new pharmacologic targets for cytoprotection. Nature Biotechnol. **16:** 833–838.
10. WU, B.J. & R.I. MORIMOTO. 1985. Transcription of the human HsP70 gene is induced by serum stimulation. Proc. Natl. Acad. Sci. USA **82:** 6070–6074.
11. CIOCCA, D.R. *et al.* 1993. Heat shock protein hsp70 in patients with axillary lymph node-negative breast cancer: prognostic implications. J. Natl. Cancer Inst. **85:** 570–574.

12. TOMEI, L.D. & F.O. COPE. 1991. Apoptosis: The Molecular Basis of Cell Death. Cold Spring Harbor Laboratory Press. Cold Spring Harbor, NY.
13. LEVINE, A.J. et al. 1991. The p53 tumor suppressor gene. Nature 351: 453–456.
14. CORNFORD, P.A. et al. 2000. Heat shock protein expression independently predicts clinical outcome in prostate cancer. Cancer Res. 60: 7099–7105.
15. MEHLEN, P. et al. 1996. Human hsp27, Drosophila hsp27, and human αβ-crystallin expression-mediated increase in gluthathione is essential for the protective activity of these proteins against TNF-α-induced cell death. EMBO J. 15: 2695–2706.
16. OESTERREICH, S. et al. 1993. The small heat shock protein hsp27 is correlated with growth and drug resistance in human breast cancer cell lines. Cancer Res. 53: 4442–4448.
17. RICHARDS, E.H. et al. 1996. Effects of overexpression of the small heat shock protein HSP27 on the heat and drug sensitivities of human testis tumor cells. Cancer Res. 56: 2446–2451.
18. GEORGOPOULOUS, C. & W.J. WELCH. 1993. Role of the major heat shock proteins as molecular chaperones. Annu. Rev. Cell Biol. 9: 601–634.
19. CRAIG, E.A. et al. 1994. Heat shock proteins and molecular chaperones: mediators of protein conformation and turnover in the cell. Cell 78: 365–372.
20. VARGUS-ROIG, L.M. et al. 1997. Heat shock proteins and cell proliferation in human breast cancer biopsy samples. Cancer Detect. Prev. 21: 441–451.
21. HALEY, D. et al. 1998. The small heat shock protein, αβ-crystallin, has a variable quaternary structure. J. Mol. Biol. 277: 27–35.
22. LAMBERT, H. et al. 1999. HSP27 multimerization mediated by phosphorylation-sensitive intermolecular interactions at the amino terminus. J. Biol. Chem. 274: 9378–9385.
23. MORSHAUSER, R.C. et al. 1995 The peptide-binding domain of the chaperone protein HSC70 has an unusual secondary structure topology. Biochemistry 34: 6261–6266.
24. HIGHTOWER, L.E. et al. 1994. Interaction of vertebrate HSC70 and HSP70 with unfolded proteins and peptides in: the biology of heat shock proteins and molecular chaperones. Cold Spring Harbor Laboratory Press. Cold Spring Harbor, NY.
25. ALBERTS, B. et al. 1994. Molecular Biology of the Cell (3rd ed.). Garland Publishing. New York.
26. URANO, M. 1986. Kinetics of thermotolerance in normal and tumor tissues: a review. Cancer Res. 46: 474–482.
27. WESHLER, Z. et al. 1984. Development and decay of systemic thermotolerance in rats. Cancer Res. 44: 1347–1351.
28. MOSELEY, P.L. 1997. Heat shock proteins and heat adaptation of the whole organism. J. Appl. Physiol. 83: 1413–1417.
29. HALL, D.M. 2000. Caloric restriction improves thermotolerance and reduces hyperthermia-induced cellular damage in old rats. FASEB J. 14: 78–86.
30. KREGEL, K.C. 2002. Heat shock proteins: modifying factors in physiological stress responses and acquired thermotolerance. J. Appl. Physiol. 92: 2177–2186.
31. HUME, S.P. & J.C.L. MARIGOLD. 1980. Transient, heat-induced thermal resistance in the small intestine of mouse. Radiat. Res. 82: 526–535.
32. HENSCHEL, A. et al. 1969. An analysis of the deaths in St. Louis during July 1966. Am. J. Public Health 59: 2232–2240.
33. RYLANDER, M.N. et al. HSP27, 60, and 70 expression kinetics and cell viability in normal and cancerous prostate cells. Under review.
34. RYLANDER, M.N. et al. 2005. Optimizing HSP expression in prostate cancer laser therapy through predictive computational models. J. Biomed. Optics. Under review.
35. PENNES, H.H. 1948. Analysis of tissue and arterial blood temperatures in the resting forearm. J. Appl. Physiol. 1: 93–122.
36. WISSLER, E.H. 1998. Pennes 1948. Paper revisited. J. Appl. Physiol. 85: 35–41.
37. LEE, R. & W. DOUGHERTY. 2003. Electrical injury: mechanisms, manifestations, and therapy. IEEE 10: 810–819.
38. STAR, W. 1995. Diffusion theory of light transport. In Optical-Thermal Response of Laser-irradiated Tissue. A.J. Welch & M. Gemert, Eds.: 166–169. Plenum Press. New York.

39. KIM, B. et al. 1996. Nonlinear finite-element analysis of the role of dynamic changes in blood perfusion and optical properties in laser coagulation of tissue. IEEE J. Selected Topics Quantum Electron. **2:** 922–933.
40. JACQUES, S. & L. WANG. 1995. Monte Carlo modeling of light transport in tissues. *In* Optical-Thermal Response of Laser-irradiated Tissue. Plenum Press. New York.
41. LANG, B. et al. 1999. Impact of nonlinear heat transfer on temperature control in regional hyperthermia. IEEE Trans. Biomed. Eng. **46:** 1129–1138.
42. ZHU, D. et al. 2002. Kinetic thermal response and damage in laser coagulation of tissue. Lasers Surg. Med. **31:** 313–321.
43. INCROPERA, F. & D. DEWITT. 1996. Fundamentals of Heat and Mass Transfer. John Wiley and Sons. New York.
44. RYLANDER, M.N. et al. 2005. Correlation of HSP 70 expression and cell viability following thermal stimulation of bovine aortic endothelial cells. J. Biomech. Eng. **127:** 751–757.
45. HENRIQUES, F.C. 1947. Studies of thermal injury V. The predictability and the significance of thermally induced rate processes leading to irreversible epidermal injury. Arch. Pathol. **43:** 489–502.
46. BHOWMICK, S. & J.C. BISCHOF. 1998. Supraphysiological thermal injury in Dunning AT-1 prostate tumor cells. J. Biomech. Eng. **122:** 51–59.
47. BHOWMICK, S. et al. 2004. *In vitro* thermal therapy of AT-1 Dunning prostate tumors. Int. J. Hyperthermia **20:** 73–92.

Cellular Response to DNA Damage

JOHNNY KAO,[a] BARRY S. ROSENSTEIN,[a,b] SHEILA PETERS,[a] MICHAEL T. MILANO,[c] AND STEPHEN J. KRON[d]

[a]*Department of Radiation Oncology, Mount Sinai School of Medicine, New York, New York 10029, USA*

[b]*Department of Radiation Oncology, New York University School of Medicine, New York, New York 10016, USA*

[c]*Department of Radiation Oncology, University of Rochester, Rochester, New York, USA*

[d]*Center for Molecular Oncology, University of Chicago, Chicago, Illinois, USA*

ABSTRACT: Eukaryotic cells, from yeast to man, possess evolutionarily conserved mechanisms to accurately and efficiently repair the overwhelming majority of DNA damage, thereby ensuring genomic integrity. Important repair pathways include base excision repair, nucleotide excision repair, mismatch repair, non-homologous end-joining, and homologous recombination. Defects in DNA repair processes generally result in susceptibility to cancer and, often, abnormalities in multiple organ systems. While signal transduction pathways have been intensely studied, epigenetic changes occurring in response to DNA damage are rapidly increasing in importance. Effective radiation and chemotherapy sensitization could result from selective inhibition of DNA repair in tumor cells. DNA damage repair is a dynamic field of research where the fruits of basic research often have important clinical implications.

KEYWORDS: DNA damage; DNA repair; radiation; non-homologous end-joining; homologous recombination

DNA: STRUCTURE AND FUNCTION

The structure of the deoxyribonucleic acid (DNA) consisting of two long nucleotide chains organized in a double-stranded helix with hydrogen bonding between complimentary bases was initially described by Watson and Crick in 1953 (FIG. 1a).[1] Subsequently, the genetic code was elucidated to reveal that DNA bases (adenine, cytosine, guanine, thymine) encode genes that are transcribed to messenger ribonucleic acid (RNA), which are subsequently translated into proteins. Therefore, the DNA sequence contains instructions that determine the timing and amount of protein expression in a given cell type.[2] The overwhelming majority of human DNA consists of introns that do not code for protein. The remainder of DNA consists of exons, which code for expressed proteins and regulatory DNA. DNA forms a complex and

Address for correspondence: Stephen J. Kron, M.D., Ph.D., Associate Professor, University of Chicago, Center for Molecular Oncology, 924 East 57th Street, Knapp Center Room R320, Chicago, IL. Voice: 773-834-0256; fax: 773-702-4394.
  skron@uchicago.edu

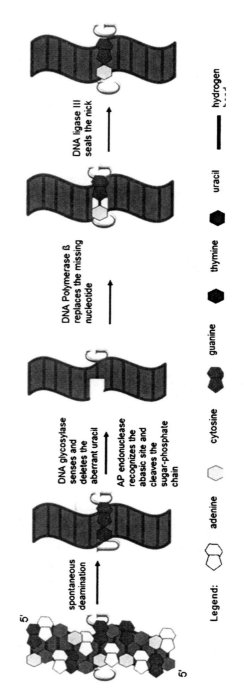

FIGURE 1. Base excision repair.

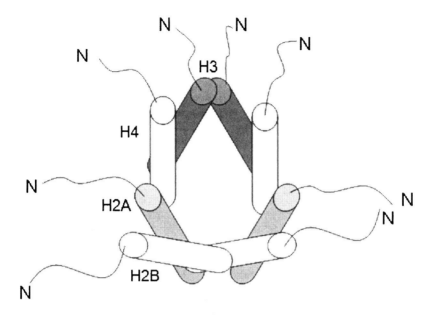

FIGURE 2. The histone octomer consists of H2A, H2B, H3 and H4 dimers that form a globular core, whereas the N-terminal tails protrude and are accessible for post-translational modification.

highly regulated protein–DNA structure called chromatin. The most basic unit of DNA packaging is the nucleosome, consisting of 146 nucleotide length double-stranded DNA wrapped around an octamer of four pairs of histones H2A, H2B, H3 and H4 (FIG. 2) and connected DNA linkers to form a structure described as "beads on a string." The human genome consists of 2.9×10^9 nucelotide pairs that are tightly packaged in 22 paired autosomal chromosomes and 2 sex chromosomes.[3]

DNA DAMAGE REPAIR IS A HIGHLY EFFICIENT PROCESS

DNA plays a vital role in maintaining genetic stability. As a result, the processes of DNA replication and the repair of DNA damage are highly regulated to maintain an acceptable mutation rate to ensure that genetic stability will not be disrupted. Some common threats to this stability include heat, spontaneous metabolic events (oxidative damage, hydrolytic attack, and uncontrolled methylation), spontaneous base loss, ultraviolet radiation, ionizing radiation, and reactive oxidative species and environmental chemicals (e.g., cigarette smoke), all of which may result in DNA damage.[4,5] For example, DNA damage such as base loss resulting from spontaneous hydrolysis of DNA glycosyl bonds occurs at a daily rate of 10^4 per cell.[6] However, more than 99.9% of accidental base changes will be successfully repaired by DNA repair mechanisms.[7] Still, the importance of DNA repair is highlighted by a number of diseases caused by an underlying deficit in specific repair processes.

Although there are many mechanisms by which DNA is damaged and repaired, the remainder of this article will focus primarily on cell response to DNA damage from ionizing radiation, the prototypical DNA damaging agent. Ionizing radiation is widely utilized in medicine for imaging (via X-rays, computed tomography scans, nuclear medicine, and fluoroscopy) and cancer therapy.[8] Ionizing radiation produces a wide array of DNA lesions, including DNA base damage (abasic sites or base modification), single-strand breaks, double-strand breaks, sugar damage, DNA–DNA cross-links and DNA–protein cross-links.[9] Importantly, radiation damage occurs in clusters rather than as single events more typical of naturally occurring DNA damage.[10] All organisms are exposed to low levels of naturally occurring ionizing radiation damage, predominantly resulting from radon gas.[11] Model organisms such as yeast are useful for exploring evolutionarily conserved DNA repair processes. As a result of this basic research, DNA damage repair is currently under intense investigation as a potential target for improving the efficacy of cancer therapy.[12]

The standard international unit of ionizing radiation dose is the Gray (Gy). Every cell exposed to 1 Gy of ionizing radiation will sustain 1000–2000 damaged bases, 800–1000 sugar damages, 1000 single-strand breaks, approximately 40 double-strand breaks, 30 DNA–DNA cross-links and 150 DNA–protein cross-links.[9] The cell may have three general responses to this damage: Most frequently, the cell will successfully repair the damage and survive without any consequences. Alternatively, the damage is not repaired, and the cell will die by mitotic death, apoptosis, or permanent growth arrest. The final pathway is misrepair, in which the cell survives with genetic changes.[13]

DNA repair responses are highly complex and remain incompletely understood. More than 130 known DNA repair genes have been identified from the human genome.[14] A review of the function of each of these genes is beyond the scope of this paper. Instead, key genes will be highlighted to illustrate integral concepts of DNA damage response. When DNA damage occurs, a sensor protein must first detect the damaged site. This is followed by recruitment of a signaling protein, generally an upstream kinase. Finally, mediator/regulator proteins transmit the signal to effector proteins, which determine the cell's fate and/or the outcome of DNA damage repair.[5] Some known examples of DNA damage repair processes are illustrated in the following sections of this paper.

DIRECT REVERSAL OF BASE DAMAGE

A common example of spontaneous base damage is alkylation (most commonly methylation) at specific base sites. Additionally, some chemotherapy agents, including nitrogen mustard and temozolomide add alkyl groups at the O6 position of guanine.[15] If unrepaired, DNA alkylation damage may result in direct cellular toxicity, mutations, or gene silencing. Therefore, repair proteins, such as the MGMT gene product, O6-methylguanine-DNA methyltransferase, can repair alkylation damage without having to break the sugar phosphate chain. In patients with impaired MGMT, temozolomide and nitrogen mustards in combination with radiation therapy are significantly more effective than in patients with functioning MGMT.[16]

BASE EXCISION REPAIR

When a mutated base is encountered (most commonly, cytosine spontaneously deaminated to uracil), it is recognized and removed by the enzyme uracil DNA glycosylase, resulting in an abasic site (FIG. 1).[7] The enzyme AP (apurinic/apyrimidinic) endonuclease recognizes the abasic site and in conjunction with phosphodiesterase cleaves the sugar phosphate chain. DNA polymerase ß fills in cytosine, using the complimentary strand as a template, and the nick is sealed by DNA ligase III. The ends of DNA are capped by telomeres, which have a unique structure, preventing them from being recognized as single-strand breaks.[17] Variations of this general mechanism are used to repair up to 10 damaged bases, sugar backbone, and single-strand DNA breaks.

NUCLEOTIDE EXCISION REPAIR

More complex types of DNA damage, such as pyrimidine dimers, are corrected by a separate repair pathway. Thymine dimers caused by ultraviolet (UV) irradiation are detected by the xeroderma pigmentosum (XP)C-associated enzyme complex, which recognizes the structural distortion to the double helix (FIG. 3).[18] A nuclease cleaves the phosphodiester backbone on both sides of the thymine dimer and the oligonucleotide (roughly 30 nucleotides in length) is removed by a protein called DNA helicase. The gap is repaired by DNA polymerase and DNA ligase, using the complimentary strand as a template. Interestingly, separate pathways are used to repair damage to genes undergoing active gene expression and transcriptionally silent genes.

Xeroderma pigmentosum is a condition with defective nucleotide excision repair. The xeroderma pigmentosum genes encode at least seven proteins (XPA through XPG) that play important roles in nucleotide excision repair.[19] Patients with this condition are extremely sensitive to ultraviolet radiation and have a ~10,000-fold increased risk of UV-induced skin cancer and progressive neurologic complications.[9] Cockayne's syndrome (mutation in genes CSA or CSB), a condition with defective transcription-coupled repair, results in an increased risk of UV-induced skin cancer and other clinical symptoms, including mental retardation, growth retardation, and retinal degeneration.[20]

DOUBLE-STRAND BREAK REPAIR: RECOGNITION AND SIGNALING PATHWAYS

Double-strand breaks (DSBs) may occur naturally through background ionizing radiation, reactive oxygen species, and replication errors (such as stalled replication forks at sites of DNA damage).[21,22] DSBs also occur in the context of normal physiology, including meiosis, where they initiate the process of meiotic recombination and in lymphocytes where VD(J) recombination and class switching are necessary for proper immune function.[23,24] DSBs pose a greater threat to genomic stability because the complementary strand cannot be used as a template for high-fidelity repair. In yeast, a single unrepaired double-strand break (of a genome of 1.5×10^7 base

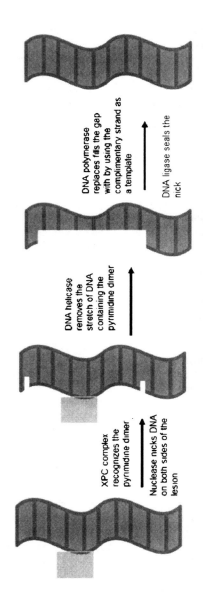

FIGURE 3. Nucleotide excision repair.

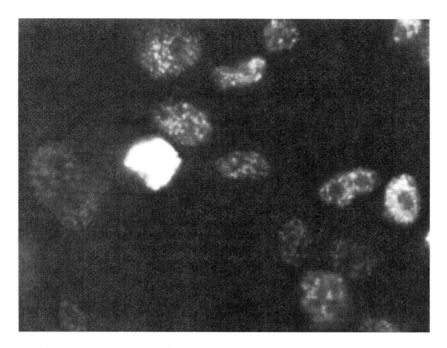

FIGURE 4. Gamma-H2AX foci visualized 20 minutes after 8 Gy in PC-3 prostate cancer cells.

pairs) is readily detected by DNA damage-response mechanisms and may result in global cell response and cell death.[25,26] DSBs also are widely acknowledged as the critical lesions determining cytotoxicity after ionizing radiation (IR).[9] There is a complex repair machinery to recognize and attempt to respond to double-strand breaks by repair, modified gene expression, or apoptosis.

After induction of a double-strand break by IR, the Mre11/Rad50/NBS1 (MRN) complex is recruited by free DNA ends and triggers autophosphorylation of ATM (ataxia telangiectasia mutated protein) at Ser 1981, resulting in ATM dimer dissociation and activation (FIG. 1).[27,28] Methylation of lysine 79 of histone H3 by the enzyme DOT1 is a sentinel event,[28] and the methylated H3 recruits the signaling protein 53BP1.[28] ATM, and related proteins, ATR (ataxia telangiectasia– and Rad-3–related) and DNA-PKcs (DNA-dependent protein kinase catalytic subunit), are kinases which may have important functions in DNA damage repair and checkpoint response by phosphorylating proteins on SerGlu and ThrGlu sites.[13] ATM plays a central role in DNA damage response by amplifying the damage signal. One of many important targets of ATM is histone H2AX Serine 139. Minutes after IR, γ-H2AX foci may be detected at DNA DSB sites that can be detected by a phospho-specific antibody (FIG. 4).[29,30] After ATM activation, related signal transduction pathways may result in DNA damage repair (allowing for cell survival if the damage is successfully repaired), cell cycle arrest (preventing cells with damaged DNA from replicating), or programmed cell death (removing cells by apoptosis) (FIG. 5).

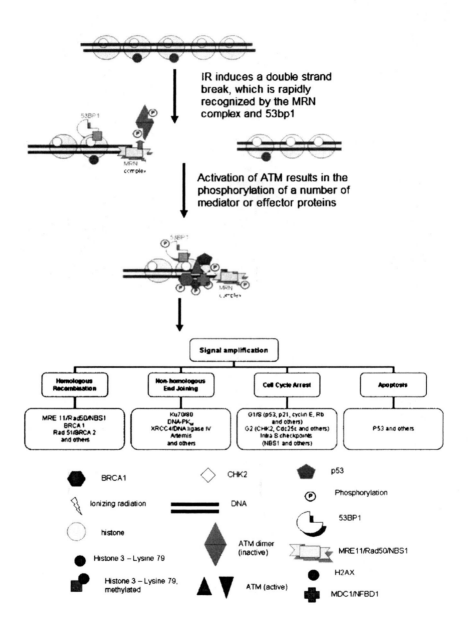

FIGURE 5. DNA damage-response signal transduction.

The important and diverse roles of ATM in DNA damage signaling, cell cycle control, and maintaining genomic stability are illustrated by studying patients with ataxia telangiectasia (homozygous ATM mutations). Ataxia telangiectasia is associated with cerebellar ataxia, immune deficiency, severe radiation sensitivity, oculocutaneous telangiectasia, neurologic deficits, and increased risk of cancer, particularly lymphoma.[31] Additionally, there is evidence of increased cancer risk and radiation sensitivity among ATM heterozygotes.[32] The MRN complex is another upstream signaling mechanism of DNA damage (FIG. 5).[33] Similar to mutations to ATM, NBS1 mutations result in the Nijmegen breakage syndrome associated with G1/S checkpoint defects, impaired upregulation of p53 after radiation, neurologic deterioration and radiosensitivity.[34] A mutation in Mre11 results in the ataxia telangiectasia–like disorder.[35]

NON-HOMOLOGOUS END-JOINING

If the cell attempts to repair the double-strand break, the primary mechanisms are non-homologous end-joining and homologous recombination. Non-homologous end-joining is the simpler, but more error-prone mechanism. The Ku70/80 complex senses and binds to DNA ends and recruits DNA-PKcs (FIG. 6).[36] The DNA-PKcs stabilizes DNA ends by holding them in close proximity in a process called synapsis.[37, 38] The endonuclease Artemis is activated by ATM and processes the DNA double-strand ends, allowing them to be rejoined by non-homologous end-joining.[39] In the majority of cases, one or more nucleotides, particularly 5' or 3' overhangs, are trimmed from the ends by the Artemis/DNA–PKcs complex and/or the MRE11 exonuclease. The XRCC4 gene product binds to DNA ligase IV, which ligates the broken ends.[40] This process is highly error-prone, and the deleted DNA will result in mutation unless the sequence is non-coding and non-essential. In species with relatively small genomes, including yeast, bacteria or *Drosophila*, non-homologous end-joining is an uncommon mode of double-strand break repair.[7] However, in mammalian cells, where more than 90% of DNA is non-coding, non-homologous

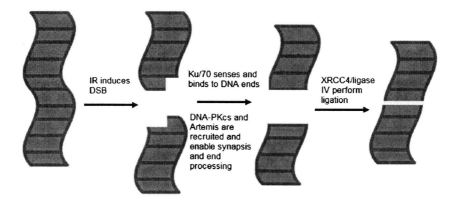

FIGURE 6. Non-homologous end-joining.

end-joining is a common approach to repairing DNA damage and is better tolerated. This process is well characterized, particularly in the context of VD(J) recombination, where the high mutation rate results in the evolutionary advantage of immunologic diversity capable of responding to a wide array of antigens.[41] Since a sister chromatid is not needed for this repair pathway, non-homologous end-joining predominates in G1.

Mutations in any of the key genes regulating non-homologous end-joining result in significant sensitivity to ionizing radiation.[42] The phenotype of mice with a DNA–PKcs deficiency is that of severe combined immunodeficiency (SCID) syndrome due to impaired VD(J) recombination and radiation sensitivity.[37] Deficiencies in Artemis also yield a SCID-like radiosensitive phenotype,[43] and mutations in Ku70/80, XRCC4, and ligase IV demonstrate significant radiation hypersensitivity.[9,44] However, deficits in non-homologous end-joining do not significantly increase the risk of carcinogenesis.

HOMOLOGOUS RECOMBINATION

Homologous recombination is much more accurate and complex than non-homologous end-joining. Here, we present a highly simplified description of the process. The double-strand breaks are initially sensed by the MRN complex and the ends are processed by Mre11 and RPA, among other proteins (FIG. 7).[45] The mediator protein Rad52 plays a role in the search for a homologous DNA sequence. After end processing, the 3' single strands and the homologous DNA are loaded to the recombinase machinery. The primary recombinase in eukaryotic cells is Rad51. Emerging data suggest that BRCA1 and BRCA2 also play important roles in homologous recombination.[46,47] Crossed DNA strands or "Holliday junctions" further stabilize the joint molecule,[46] and high-fidelity DNA synthesis occurs using the intact homologous DNA as a template (FIG. 7). After completion of synthesis, Rad54 allows for separation of the DNA strands and certain nucleases free Holliday junctions.[46]

Homologous recombination is favored in S and G2 phases because an intact sister chromatid is readily available to serve as a template. Interestingly, the S phase corresponds to the most radio-resistant phase of the cell cycle suggesting that double-strand breaks are accurately and rapidly repaired by intact homologous recombination. In G1, homologous recombination is possible, but much less common, because the broken ends must be juxtaposed with homologous non-sister chromatid.

While homologous recombination is considered more accurate than non-homologous end-joining in repairing homologous recombination, unregulated homologous recombination can play a role in carcinogenesis by means of a heterozygous mutated allele as a template for gene conversion. This process, described as loss of heterozygosity, is seen in the natural history of cancer from dysplasia to invasive cancer. For example, in patients with Bloom syndrome, the BLM helicase is mutated, resulting in excessive homologous recombination.[48] These patients are growth-retarded and immune-deficient; they demonstrate impaired spermatogenesis and are susceptible to developing cancer.[9] A related condition, Werner's syndrome, also involves a mutated helicase and is associated with premature aging and cancer susceptibility.[49]

In general, cells with defective homologous recombination, but intact non-homologous end-joining, are moderately radiosensitive, but demonstrate significant

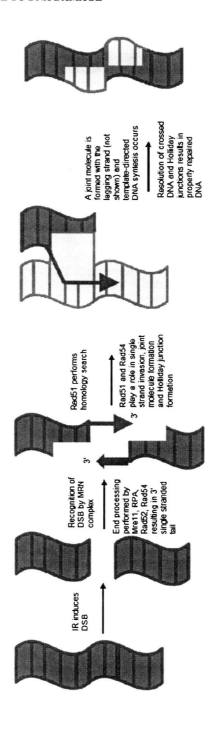

FIGURE 7. Homologous recombination.

cancer susceptibility. BRCA1 and BRCA2 are associated with an approximately 85% lifetime risk of breast cancer and a 10–40% lifetime risk of ovarian cancer.[50] BRCA1 and 2 have additional functions unrelated to homologous recombination.[51] The cancer susceptibility is partly related to defects in homologous recombination and partly due to using more error-prone pathways, such as single-strand annealing (a double-strand break repair mechanism that utilizes local homology of the 3′ single strand) to repair double-strand breaks.[52] While BRCA1 and 2 cells are only moderately radiosensitive, they are extremely sensitive to drugs, such as cisplatin and mitomycin C. that work by DNA intra-strand cross-linkers.[53] An unresolved issue is why BRCA1 and 2, genes expressed in most cell types, are associated primarily with an increase in breast and ovarian cancers.

All mutants that interfere with non-homologous end-joining (either upstream signaling mutations such as ATM and NBS1 or downstream effector mutations such as XRCC4 or ligase IV) render cells markedly radiosensitive, whereas the effect of mutations affecting homologous recombination on radiosensitivity is less dramatic. It is possible that the clinically significant IR-induced double-strand breaks are too complex for effective homologous recombination. Armed with these data, it is tempting to speculate that non-homologous end-joining is the biologically most significant process for handing IR-induced double-strand breaks. However, this hypothesis remains to be verified by experimental data.

MISMATCH REPAIR

During the process of DNA replication and homologous recombination, base–base mismatches, base insertion and deletions, may be introduced.[54] For instance, single base-pair mismatches occurring during replication are occasionally not corrected by the "proofreading" function of DNA polymerase. Misrepairs may result from homologous recombination, while insertions and deletions may result from slippage of DNA polymerase during replication. Clinically, these small insertions and deletions are called microsatellite instability, because microsatellite regions consist of simple repetitive sequences that tend to induce DNA polymerase slippage during replication. Failure to remove these errors results in the patient's susceptibility to accumulating spontaneous mutations. The MutS protein (consisting of MSH2 and MSH6) recognizes and binds to mismatched bases.[55] Another protein, MLH1, contributes to excising the mismatched base. In addition to these specialized genes, mismatch repair overlaps significantly with enzymes involved with base excision repair and nucleotide excision repair to accomplish synthesis of correct bases and religation of DNA ends. Mutations to MSH2 and MLH1 are associated with hereditary non-polyposis colon cancer syndromes (Lynch I and II) that increase the risk of colorectal cancer, endometrial cancer, and other malignancies.[56]

EXPERIMENTAL METHODS FOR STUDYING DNA BREAKS

Classical methods for studying DNA breaks include pulsed-field gel electrophoresis and the Comet assay.[11] In pulsed-field gel electrophoresis, alternating direction of current while varying pulse time can separate large pieces of DNA, thereby identifying DNA breaks. The Comet assay or single-cell gel electrophoresis quantifies the amount of damaged DNA, which travels more rapidly in the "tail," whereas the intact DNA forms the head. Running these assays at alkaline conditions will focus on single-strand breaks, whereas neutral conditions are used to study double-strand breaks. While these assays represented a significant advance over the sucrose gradient sedimentation, neutral filter elution and nucleoid sedimentation technique, these assays are all rather cumbersome.

Quantifying gamma-H2AX after radiation represents a relatively convenient method for quantifying IR-induced DNA double-strand breaks (FIG. 4).[57,58] By using the phosphospecific antibody to Serine 139 on H2AX, a linear correlation between dose, measured double-strand breaks by pulsed-field gel electophoresis, and H2AX foci have been demonstrated.[59] Additionally, the H2AX foci are cleared over time, suggesting that DNA double-strand break repair can be followed *in vitro*.[30] However, the identification of gamma-H2AX after forms of cell stress not associated with double-strand breaks (e.g., hypoxia and hydrogen peroxide) calls into question the specificity of gamma-H2AX as a reporter of double-strand breaks in certain circumstances.[60] Other molecular methods for studying DNA damage repair have employed fluorescent antibody– and green fluorescence protein–based approaches for visualizing subnuclear repair foci at sites of DNA damage.[61]

CLUSTERED LESIONS: A CRITICAL ENDPOINT FOR CANCER THERAPY?

It is hypothesized that multiply damaged sites, which are most commonly produced by ionizing radiation, may be particularly difficult to repair.[10] Indeed, attempted repair of clustered base damage or abasic sites has been shown to result in double-strand breaks.[62] Additionally, certain types of ionizing radiation, such as high-energy neutrons, produce a particularly high incidence of multiply damaged sites. Neutron radiation is associated with a particularly high rate of treatment-related complications, demonstrating the challenge of repairing these lesions.[9] Double-strand breaks are considered a subset of clustered lesions, but the mechanisms of repair and the biologic significance of these lesions remains an active area of investigation.[63] These data raise the possibility that mechanisms for repairing physiologic double-strand breaks may not be those employed in the repair of IR-induced double strand breaks and multiply damaged sites. It may be interesting to characterize the types of DNA lesions found at sites of repair foci that persist several hours after double-strand break induction (gamma-H2AX or type III MRN foci), which may correspond to unrepaired or difficult-to-repair lesions.[64] Emerging experimental data demonstrate that easily repaired lesions (e.g., abasic sites, base damage, endogenous double-strand breaks) are handled without activating ATM or ATR.[5] In contrast, ATM is routinely activated by exogenous double-strand breaks, although a global DNA damage response (i.e., activation of checkpoint control or apoptosis) does not necessarily occur. However, full activation of ATR or ATM is observed when DNA repair is blocked experimentally.[5]

CONCLUSION

Ionizing radiation and many cytotoxic chemotherapy agents are effective in the treatment of cancer primarily due to the creation of DNA damage and the resulting lethality in tumor cells. However, aberrant repair of DNA damage leads to genomic instability and susceptibility to cancer. Genetic syndromes involving DNA repair genes have clinical phenotypes that involve abnormalities in multiple organ systems in addition to increased cancer risk and, in some cases, radiosensitivity. The marked increase in knowledge of DNA damage repair accrued over the past decade may ultimately allow for targeting these pathways to improve the efficacy of radiation and chemotherapy in treating cancer. Gaps in the current understanding of DNA damage repair remain, but inspire further research. Finally, elucidation of the histone code regulating DNA damage repair is an emerging area of research that will undoubtedly provide novel insights.

REFERENCES

1. WATSON, J.D. & F.H. CRICK. 1953. Molecular structure of nucleic acids: a structure for deoxyribose nucleic acid. Nature **171:** 737–738.
2. LANDER, E.S. *et al.* 2001. Initial sequencing and analysis of the human genome. Nature **409:** 860–921.
3. VENTER, J.C. *et al.* 2001. The sequence of the human genome. Science **291:** 1304–1351.
4. SCHAR, P. 2001. Spontaneous DNA damage, genome instability, and cancer: when DNA replication escapes control. Cell **104:** 329–332.
5. ROUSE, J. & S.P. JACKSON. 2002. Interfaces between the detection, signaling, and repair of DNA damage. Science **297:** 547–551.
6. DINNER, A. R., G. M. BLACKBURN & M. KARPLUS. 2001. Uracil-DNA glycosylase acts by substrate autocatalysis. Nature **413:** 752–755.
7. ALBERTS, B. 2002. Molecular Biology of the Cell. Garland Science. New York.
8. VIJAYAKUMAR, S. & S. HELLMAN. 1997. Advances in radiation oncology. Lancet **349** Suppl. **2:** SII1–3.
9. STEEL, G.G. 2002. Basic Clinical Radiobiology. Oxford University Press. London–New York.
10. WARD, J.F. 1994. The complexity of DNA damage: relevance to biological consequences. Int. J. Radiat. Biol. **66:** 427–432.
11. HALL, E.J. 2000. Radiobiology for the Radiologist. Lippincott Williams & Wilkins. Philadelphia.
12. KAO, J. *et al.* 2006. Gamma-H2AX as a therapeutic target for improving the efficacy of radiation therapy. Current Cancer Drug Targets. In press.
13. SHILOH, Y. 2003. ATM and related protein kinases: safeguarding genome integrity. Nat. Rev. Cancer **3:** 155–168.
14. WOOD, R.D., *et al.* 2001. Human DNA repair genes. Science **291:** 1284–1289.
15. DRABLOS, F. *et al.* 2004. Alkylation damage in DNA and RNA: repair mechanisms and medical significance. DNA Repair (Amst.) **3:** 1389–1407.
16. HEGI, M.E., *et al.* 2005. MGMT gene silencing and benefit from temozolomide in glioblastoma. N. Engl. J. Med. **352:** 997–1003.
17. LEI, M. *et al.* 2003. DNA self-recognition in the structure of Pot1 bound to telomeric single-stranded DNA. Nature **426:** 198–203.
18. DIP, R., U. CAMENISCH & H. NAEGELI. 2004. Mechanisms of DNA damage recognition and strand discrimination in human nucleotide excision repair. DNA Repair (Amst.) **3:** 1409–1423.
19. BERNEBURG, M. & A.R. LEHMANN. 2001. Xeroderma pigmentosum and related disorders: defects in DNA repair and transcription. Adv. Genet. **43:** 71–102.

20. DE WAARD, H. et al. 2004. Different effects of CSA and CSB deficiency on sensitivity to oxidative DNA damage. Mol. Cell. Biol. **24:** 7941–7948.
21. FRIEDBERG, E.C. 2003. DNA damage and repair. Nature **421:** 436–440.
22. VILENCHIK, M.M. & A.G. KNUDSON. 2003. Endogenous DNA double-strand breaks: production, fidelity of repair, and induction of cancer. Proc. Natl. Acad. Sci. USA **100:** 12871–12876.
23. RICHARDSON, C., N. HORIKOSHI & T.K. PANDITA. 2004. The role of the DNA double-strand break response network in meiosis. DNA Repair (Amst.) **3:** 1149–1164.
24. BASSING, C.H. & F.W. ALT. 2004. The cellular response to general and programmed DNA double strand breaks. DNA Repair (Amst.) **3:** 781–796.
25. LEE, S.E. et al. 1998. Saccharomyces Ku70, mre11/rad50 and RPA proteins regulate adaptation to G2/M arrest after DNA damage. Cell **94:** 399–409.
26. BENNETT, C.B. et al. 1993. Lethality induced by a single site-specific double-strand break in a dispensable yeast plasmid. Proc. Natl. Acad. Sci. USA **90:** 5613–5617.
27. BAKKENIST, C.J. & M.B. KASTAN. 2003. DNA damage activates ATM through intermolecular autophosphorylation and dimer dissociation. Nature **421:** 499–506.
28. STUCKI, M. & S.P. JACKSON. 2004. Tudor domains track down DNA breaks. Nat. Cell Biol. **6:** 1150–1152.
29. ROGAKOU, E.P. et al. 1998. DNA double-stranded breaks induce histone H2AX phosphorylation on serine 139. J. Biol. Chem. **273:** 5858–6868.
30. TANEJA, N. et al. 2004. Histone H2AX phosphorylation as a predictor of radiosensitivity and target for radiotherapy. J. Biol. Chem. **279:** 2273–2280.
31. CHUN, H.H. & R.A. GATTI. 2004. Ataxia-telangiectasia, an evolving phenotype. DNA Repair (Amst.) **3:** 1187–1196.
32. CESARETTI, J.A. et al. 2005. ATM sequence variants are predictive of adverse radiotherapy response among patients treated for prostate cancer. Int. J. Radiat. Oncol. Biol. Phys. **61:** 196–202.
33. LEE, J.H. & T.T. PAULL. 2004. Direct activation of the ATM protein kinase by the Mre11/Rad50/Nbs1 complex. Science **304:** 93–96.
34. DIGWEED, M. & K. SPERLING. 2004. Nijmegen breakage syndrome: clinical manifestation of defective response to DNA double-strand breaks. DNA Repair (Amst.) **3:** 1207–1217.
35. TAYLOR, A.M., A. GROOM & P.J. BYRD. 2004. Ataxia-telangiectasia-like disorder (ATLD): its clinical presentation and molecular basis. DNA Repair (Amst.) **3:** 1219–1225.
36. WALKER, J.R., R.A. CORPINA & J. GOLDBERG. 2001. Structure of the Ku heterodimer bound to DNA and its implications for double-strand break repair. Nature **412:** 607–614.
37. COLLIS, S.J. et al. 2005. The life and death of DNA-PK. Oncogene **24:** 949–961.
38. LIEBER, M.R. et al. 2003. Mechanism and regulation of human non-homologous DNA end-joining. Nat. Rev. Mol. Cell Biol. **4:** 712–720.
39. RIBALLO, E. et al. 2004. A pathway of double-strand break rejoining dependent upon ATM, Artemis, and proteins locating to gamma-H2AX foci. Mol. Cell. **16:** 715–724.
40. GRAWUNDER, U. et al. 1997. Activity of DNA ligase IV stimulated by complex formation with XRCC4 protein in mammalian cells. Nature **388:** 492–495.
41. LIEBER, M.R. et al. 2004. The mechanism of vertebrate nonhomologous DNA end joining and its role in V(D)J recombination. DNA Repair (Amst.) **3:** 817–826.
42. WILLERS, H., J. DAHM-DAPHI & S.N. POWELL. 2004. Repair of radiation damage to DNA. Br. J. Cancer **90:** 1297–1301.
43. MOSHOUS, D. et al. 2001. Artemis, a novel DNA double-strand break repair/V(D)J recombination protein, is mutated in human severe combined immune deficiency. Cell **105:** 177–186.
44. RIBALLO, E. et al. 2001. Cellular and biochemical impact of a mutation in DNA ligase IV conferring clinical radiosensitivity. J. Biol. Chem. **276:** 31124–3132.
45. VAN DYCK, E. et al. 1999. Binding of double-strand breaks in DNA by human Rad52 protein. Nature **398:** 728–731.
46. WYMAN, C., D. RISTIC & R. KANAAR. 2004. Homologous recombination-mediated double-strand break repair. DNA Repair (Amst.) **3:** 827–833.

47. BHATTACHARYYA, A. *et al.* 2000. The breast cancer susceptibility gene BRCA1 is required for subnuclear assembly of Rad51 and survival following treatment with the DNA cross-linking agent cisplatin. J. Biol. Chem. **275:** 23899–23903.
48. WU, L. & I.D. HICKSON. 2003. The Bloom's syndrome helicase suppresses crossing over during homologous recombination. Nature **426:** 870–874.
49. GRAY, M.D. *et al.* 1997. The Werner syndrome protein is a DNA helicase. Nat. Genet. **17:** 100–103.
50. HABER, D. 2000. Roads leading to breast cancer. N. Engl. J. Med. **343:** 1566–1568.
51. VENKITARAMAN, A.R. 2002. Cancer susceptibility and the functions of BRCA1 and BRCA2. Cell **108:** 171–182.
52. STARK, J.M. *et al.* 2004. Genetic steps of mammalian homologous repair with distinct mutagenic consequences. Mol. Cell. Biol.. **24:** 9305–9316.
53. CONNELL, P.P. *et al.* 2004. A hot spot for RAD51C interactions revealed by a peptide that sensitizes cells to cisplatin. Cancer Res. **64:** 3002–3005.
54. STOJIC, L., R. BRUN & J. JIRICNY. 2004. Mismatch repair and DNA damage signalling. DNA Repair (Amst.) **3:** 1091–1101.
55. FLORES-ROZAS, H., D. CLARK & R.D. KOLODNER. 2000. Proliferating cell nuclear antigen and Msh2p-Msh6p interact to form an active mispair recognition complex. Nat. Genet. **26:** 375–378.
56. CHUNG, D.C. & A.K. RUSTGI. 2003. The hereditary nonpolyposis colorectal cancer syndrome: genetics and clinical implications. Ann. Intern. Med. **138:** 560–570.
57. SEDELNIKOVA, O.A. *et al.* 2002. Quantitative detection of (125)IdU-induced DNA double-strand breaks with gamma-H2AX antibody. Radiat. Res. **158:** 486–492.
58. MACPHAIL, S.H. *et al.* 2003. Expression of phosphorylated histone H2AX in cultured cell lines following exposure to X-rays. Int. J. Radiat. Biol. **79:** 351–358.
59. ROTHKAMM, K. & M. LOBRICH. 2003. Evidence for a lack of DNA double-strand break repair in human cells exposed to very low x-ray doses. Proc. Natl. Acad. Sci. USA **100:** 5057–5062.
60. BANATH, J.P. & P.L. OLIVE. 2003. Expression of phosphorylated histone H2AX as a surrogate of cell killing by drugs that create DNA double-strand breaks. Cancer Res. **63:** 4347–4350.
61. LISBY, M., U.H. MORTENSEN & R. ROTHSTEIN. 2003. Colocalization of multiple DNA double-strand breaks at a single Rad52 repair centre. Nat. Cell Biol. **5:** 572–527.
62. BLAISDELL, J.O. & S.S. WALLACE. 2001. Abortive base-excision repair of radiation-induced clustered DNA lesions in *Escherichia coli*. Proc. Natl. Acad. Sci. USA **98:** 7426–7430.
63. SUTHERLAND, B.M. *et al.* 2002. Clustered DNA damages induced by x rays in human cells. Radiat. Res. **157:** 611–616.
64. MIRZOEVA, O.K. & J.H. PETRINI. 2001. DNA damage-dependent nuclear dynamics of the Mre11 complex. Mol. Cell. Biol. **21:** 281–288.

Autophagy

AMEETA KELEKAR

Department of Laboratory Medicine and Pathology, University of Minnesota, Minneapolis, Minnesota 55455, USA

ABSTRACT: Autophagy is a major intracellular pathway for the degradation and recycling of long-lived proteins and cytoplasmic organelles. Like apoptotic programmed cell death, autophagy is an essential part of growth regulation and maintenance of homeostasis in multicellular organisms. Autophagic vacuole formation is also activated as an adaptive response to a variety of extracellular and intracellular stimuli, including nutrient deprivation, hormonal or therapeutic treatment, bacterial infection, aggregated and misfolded proteins and damaged organelles. Mediators of class I and class III PI3 kinase signaling pathways and trimeric G proteins play major roles in regulating autophagosome formation during the stress response. Defective autophagy is the underlying cause of a number of pathological conditions, including vacuolar myopathies, neurodegenerative diseases, liver disease, and some forms of cancer. This chapter provides an overview of the morphology and molecular basis of autophagosome formation and offers a glimpse into the role of autophagy in normal growth and development, while discussing the pathological implications of its deregulation.

KEYWORDS: autophagy; growth regulation; homeostasis; stress

INTRODUCTION

Autophagy (Greek for "the eating of oneself") refers to the evolutionarily conserved, regulated turnover of cellular constituents that occurs during development, and as a response to stress. While short-lived proteins are degraded via the ubiquitin/proteasome pathway in higher eukaryotes, most long-lived proteins and even whole organelles are recycled through an autophagic process. The morphology of autophagy was first characterized in mammalian cells, although most of the molecular analyses were carried out using yeast genetics. The three major forms of autophagy described in mammalian cells are microautophagy, macroautophagy, and chaperone-mediated autophagy.[1] Micro- and macroautophagy are both highly conserved from yeast to mammals while chaperone-mediated autophagy appears largely confined to mammalian systems. Microautophagy involves the direct engulfment of cytoplasm at the surface of the lysosome by protrusion, invagination, or other deformations of the lysosomal membrane. Macroautophagy is the bulk lysosomal degradation of larger cytoplasmic proteins and organelles such as mitochondria, fractured endoplas-

---

Address for correspondence: Ameeta Kelekar, Department of Laboratory Medicine and Pathology, University of Minnesota, Minneapolis, MN 55455. Voice: 612-625-3204; fax: 612-625-1121.
ameeta@umn.edu

mic reticulum (ER), and peroxisomes within a double-membrane vesicle of nonlysosomal origin that then fuses with lysosomes. Chaperone-mediated autophagy involves formation of a complex between a cytosolic "substrate" protein and "chaperone" protein that interacts with a lysosomal membrane receptor for translocation into the lysosome. This internalization is carried out within the lysosomal lumen by an additional chaperone protein. Among the three, macroautophagy (henceforth referred to simply as autophagy) is the major catabolic pathway for energy generation and for the breakdown of macromolecules and damaged organelles into their essential constituents during periods of stress or nutrient deprivation; this will be the focus for the rest of the chapter.

AUTOPHAGY—THE PROCESS

Autophagosome formation proceeds in morphologically and biochemically distinct phases referred to as the initiation, execution, and maturation phases, all of which are ATP dependent.[2,3] The process begins with the generation of a crescent-shaped isolation membrane or phagophore that sequesters cytoplasm and organelles.[4] The expansion and eventual closing of the phagophore gives rise to the autophagosome (illustrated in FIG. 1). The origins of the double-membrane bilayers are difficult to determine because autophagosomes contain a mixture of markers from the ER, endosomes and lysosomes, as well as bulk cytosolic contents.[5-7] Autophagosomes can be large, ranging between 0.5 and 1.5 mm in diameter in mammalian cells.[8] Maturation of these vesicles involves fusion with endosomes and/or lysosomes and culminates in the release of sequestered contents for degradation and recycling by catabolic enzymes in the lumenal cavity of the lysosome.[9] The endosomal fusion step may provide the "intermediate" autophagosomes with factors that enable them to fuse with lysosomes. Early or immature autophagosomes and mature autolysosomes are distinguishable by electron microscopy: the immature structures contain two or more bilayers while autolysosomes have lost the distinctive inner membrane (FIG. 1). Intermediate filament proteins, cytokeratin and vimentin, are required for sequestration, while fusion with lysosomes requires the micotubular system.[10]

Analysis of mammalian homologues of yeast (Atg or Apg) proteins has revealed a direct association of these proteins with autophagic membranes.[4] Atg12 is covalently conjugated to Atg5 by a ubiquitination reaction during the initiation step. The conjugate behaves as a single molecule and is noncovalently associated with Atg16L, a coiled-coil protein, and the entire 800 kDa complex is found associated with the isolation membrane throughout the elongation process.[11] The membrane-associated light chain 3 protein, MAP LC3 (Atg8), also localizes to the elongation membrane and stays associated with the membrane until autophagosome maturation, unlike the Atg5-Atg12-Atg16 complex that dissociates upon its formation. The cytoplasmic LC3 is posttranslationally modified to LC3-I and then LC3-II in a series of steps that begins with cleavage by Atg4, followed by ubiquitination reactions transiently linking it with Atg7 and Atg3, and finally lipidation.[12] The modified LC3 proteins are found exclusively on autophagosomes. Lysosome-associated membrane proteins, LAMP-1 and LAMP-2, become associated with intermediate autophagosomes, and mature autolysosomes acquire the degradative enzyme, cathepsin, and

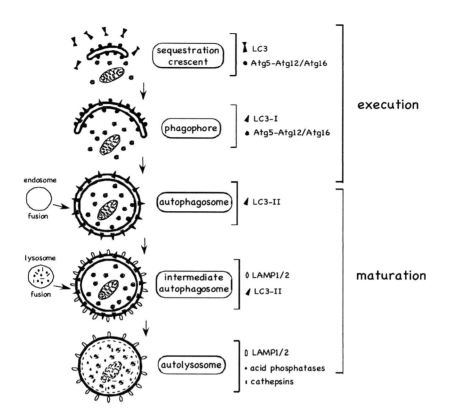

FIGURE 1. Diagrammatic illustration of the execution and maturation phases of autophagy. Markers that are known to be present at each of the steps are also indicated. LC3, microtubule-associated protein light chain 3; LAMP, lysosome-associated membrane protein. The known signaling events that constitute the induction phase are depicted in FIGURE 3.

acid phosphatases.[9] Transmission electron microscopy (FIG. 2), immunoelectron microscopy, and fluorescent labeling of marker proteins, such as Atg5, LC3, Atg7, LAMP1, and LAMP2, in combination with fluorescent and nonfluorescent dyes specific for acidic compartments and lysosomes, such as acridine orange (AO), monodansyl cadaverine (MDC), lysotrackers, and DAMP, have allowed visualization of the steps involved in autophagosome formation.[13] Some of the approaches that are being used to quantify autophagy include flow cytometric analysis of acridine orange uptake, measurement of cytosolic lactate dehydrogenase (LDH) activity in purified membrane fractions,[6,14] determination of the degradation rate of radiolabeled long-lived proteins, and biochemical resolution of LC3 cleavage and lipidation.[12]

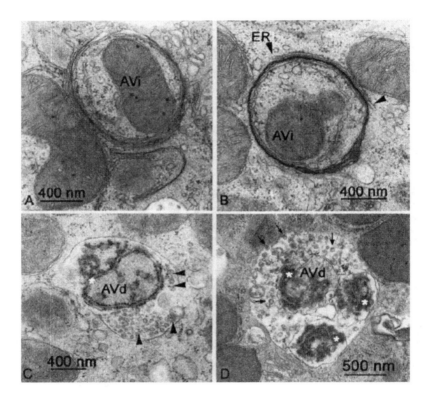

FIGURE 2. Electromicrographs of autophagosomes and autophagic vacuoles in isolated mouse hepatocytes. (**A**) An autophagosome, or initial autophagic vacuole (AVi), containing a mitochondrion, endoplasmic reticulum (ER) membranes, and ribosomes. The two limiting membranes of the autophagosome are visible at the upper rim of the vacuole. Below the AVi is a flat membrane cistern presumably in the process of sequestering a peroxisome. (**B**) Autophagosome membranes are often sandwiched between two cisterns of rough ER. *Arrowheads* point to the ER outside the AVi, and another ER cistern is visible inside the limiting membrane. (**C**) The contents of this late/degradative autophagic vacuole (AVd) look partially degraded, but the remnants of rough ER can still be identified (*asterisk*). The cell was loaded with 6-nm gold particles coated with albumin for 2 h before fixation. As a result of fusion with a multivesicular endosome, the AVd contains both 6-nm gold particles (*arrowheads*) and small vesicles (round structures surrounding the gold particles). (**D**) The degradation of the engulfed rough ER is advanced in this AVd; the remnants of ribosomes form electron-dense partially amorphous masses (*asterisks*). Further, this AV has fused with a multivesicular endosome, as indicated by the content of numerous small vesicles (*arrows*). (From Eskelinen.[68] Reprinted by permission from Eurekah.com and Springer Science + Business Media.)

MOLECULAR SIGNALING IN AUTOPHAGY

A variety of stress conditions can activate autophagy, and a number of signaling pathways are involved in its regulation. Although we are a long way from identifying the precise molecular signals that promote activation and inhibition of autophagy in mammalian cells, recent discoveries of mammalian orthologues of yeast-signaling intermediates have offered significant insights into the underlying pathways. It is abundantly clear, for instance, that phosphatidyl inositol-3 (PI-3) kinases, enzymes synthesizing phosphatidylinositol-3 phosphate (PtdIns 3P) from PtdIns, are major players in mammalian autophagic pathways. While class III PI-3 kinase is required in the early stages of autophagosome generation, class I PI-3 kinase activity has an inhibitory effect mediated, at least partially, through the TOR (target of rapamycin) kinase[15] (autophagic signaling pathways and known intermediates are depicted in FIG. 3).

Inhibition of class III PI3K by wortmannin, 3-methyladenine (3-MA) or LY294002 prevents the generation of autophagic precursors. The class III PI-3 kinase complex on the nascent autophagosome or isolation membrane comprises PDK-1, p150, the membrane-anchored adaptor that tethers the enzyme to cytoplas-

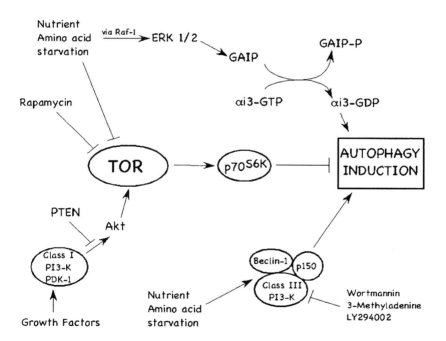

FIGURE 3. Signaling pathways that activate autophagy. The kinase mTOR plays a central role by integrating class I PI3 kinase and amino acid–dependent signaling. The other two major nutrient-sensitive pathways involve class III PI3 kinase and heterotrimeric G proteins (see text for details).

mic membranes, and the Beclin 1 protein (orthologue of yeast Atg6).[16] This complex, through the generation of PtdIns 3P, could play a role in recruiting proteins from the cytosol for autophagic degradation or in supplying the autophagic pathway with membrane components. The expression of Beclin 1 correlates directly with autophagosome formation and is reduced in a large majority of breast and ovarian cancers.[17,18] In addition to interacting with class III PI3 kinase, Beclin 1 is able to bind Bcl-2 family members, Bcl-2 and Bcl-x_L.[17,19] The implications of these interactions in regulating cell death and tumor progression will be discussed later in this chapter.

Amino acids and ATP are negative regulators of autophagic activity—the former, as final breakdown products of degraded proteins and the latter, as an immediate source of energy for metabolism. The inhibition is mediated in large part through the amino acid and ATP "sensor" protein kinase, mTOR,[20] which plays a central role in the class I PI3K/Akt-dependent signaling, and is associated with phosphorylation of proteins in the mTOR pathway, such as p70S6-kinase (p70^{S6K}) and its target ribosomal S6 protein.[21,22] Hypophosphorylated p70^{S6K} promotes one of the early molecular events in sequestration, namely, detachment of ribosomes from the ER.[2] Rapamycin inhibition of TOR leads to the activation of protein phosphatase 2A (PP2A) and induction of autophagy.[23,24] It also leads to translational upregulation of LC3, the protein that plays an essential role in vesicle formation and elongation. The tumor suppressor PTEN, an inhibitor of the PI3K/Akt-dependent signaling pathway, is a positive regulator of autophagy.[24,25]

The formation of autophagic vacuoles also requires GTP hydrolysis at the sequestration step and is controlled by heterotrimeric G proteins and their partners, known to function in membrane transport along the exo-/endocytotic pathway.[26] This pathway is turned off in the presence of amino acids. The autophagic sequestration pathway (FIG. 3) is activated when the trimeric G_{ai3} hydrolyzes GTP and stably binds to the product, GDP, following ERK1/2 phosphorylation of a single serine residue in the Ga-interacting protein (GAIP).[27,28] Since protein synthesis is essential for the expansion of the pre-autophagosomal compartment, as well as for autophagosome maturation after the sequestration step, both starvation-induced and viral autophagy are also regulated by the eIF2a kinase (PKR) signaling pathway in mammalian cells.[29]

PHYSIOLOGICAL ROLE OF AUTOPHAGY

Autophagy occurs at a basal level in most tissues in higher eukaryotes, contributing to turnover and recycling of cytoplasmic components. It also plays essential roles in development, differentiation, and tissue remodeling.

Maintenance of Homeostasis

Autophagy contributes to the maintenance of cellular homeostasis by continually turning over superfluous large proteins and damaged organelles, such as mitochondria, Golgi, ER, and even nuclei, in a manner similar to that of the ubiquitin-proteasomal system of proteolytic recycling.[30] This is a "housekeeping" function that helps maintain an amino acid pool for gluconeogenesis and for the synthesis of essential proteins during periods of starvation. Additionally, autophagy may be the pathway of choice when large-scale cell elimination becomes necessary for maintaining homeostasis, and available professional phagocytes cannot meet the demand. In these situations it functions not only as a cell death pathway, but also as a "scavenging" mechanism for the disposal of dead cell corpses.[31] Autophagy may also serve as an anti-aging mechanism by limiting the harmful effects of oxidative damage caused by reactive oxygen species produced by damaged mitochondria.[32]

Autophagy in Development

Autophagy is increased in cellular tissue undergoing remodeling during differentiation and development, such as in newborn kidney, lung, intestine, fetal duodenum, regressing Mullerian ducts, involuting mammary glands, and keratinizing skin.[33] Since neonates depend on amino acids for survival, autophagy may be critical for sustaining life during this period. For instance, the sudden increase in energy requirements immediately following birth may be addressed, in large part, by the autophagic degradation of glycogen in the newborn liver. Loss of the yeast Atg6 orthologue, Beclin 1, is embryonically lethal in mice, suggesting that autophagy is also essential for the normal development of the mammalian embryo.[19]

Autophagy is also involved in tissue-specific developmental functions such as the elimination of ribosomes and mitochondria from erythrocytes after removal of the nucleus.[34] Curiously, the biconcave shape of the erythrocyte results from the eventual expulsion of this autophagic vacuole.[35] Autophagy also aids in the biosynthesis of neuromelanin in dopaminergic neurons of the substantia nigra[36] and in the biogenesis of surfactants in pneumocytes.[37]

The Adaptive Autophagic Response

As a rule, autophagy is activated on a subcellular scale when either extracellular nutrients or growth factors are limited,[7] and results in the scavenging and recycling of nonessential proteins and organelles for reuse in the cytosol. Such nonspecific autophagy is inhibited under nutrient-rich environmental conditions. The presence of nutrients, however, does not automatically guarantee an energy supply, and it has recently been demonstrated that growth factors that allow cells to take up the extracellular nutrients by increasing the cell surface expression of nutrient transporters are also essential.[38] Autophagy is also activated as a protective measure in response to bacterial and virus infection and serves as an effective way to eliminate infectious agents that enter the cytosolic compartment directly through the plasma membrane or after being internalized in phagosomes.[39]

AUTOPHAGY IN DISEASE

Several human diseases are associated with increased autophagy, particularly in nondividing cells of the nervous or muscular system where turnover of intracellular proteins may be critical. Muscular disorders, known as vacuolar myopathies, are associated with massive accumulation of autophagic or lysosomal vacuoles (reviewed in Ref. 40). Danon disease is characterized by cardiomyopathy and mild mental retardation caused by a deficiency in LAMP-2, the transmembrane protein involved in endosome and lysosome fusion during the late maturation phase of the autophagosome.[41] Other myopathies associated with vacuole accumulation are X-linked myopathy with excessive autophagy (X-MEA),[42] inclusion body myositis,[43] and Marinesco-Sjögren syndrome.[44]

Elevated levels of autophagy also are associated with neurodegenerative diseases such as Parkinson's (PD), Huntington's (HD), and Alzheimer's (AD) diseases,[45–48] and transmissible spongiform encephalopathies (prion diseases).[49] The occurrence of intracellular protein aggregates and altered activity of proteolytic systems is characteristic of neurodegenerative diseases. Autophagy may be activated as a protective mechanism against newly formed aggregates, but increased aggregation decreases susceptibility to degradation and leads to disease progression.[50] The protein that accumulates in PD is mutated α-synuclein,[51] in HD an abnormally expanded form of the huntingtin protein,[52] and in AD, deposits of overproduced β-amyloid precursors that are too large to be removed by the proteasomal system. Excessive autophagy is also associated with chronic liver disease and hepatocellular carcinoma and is due to retention of mutant alpha (1)-anti-trypsin Z (ATZ) protein in the ER and mitochondria of hepatocytes.[53]

Impaired autophagy is associated with some forms of cancer. Beclin 1 is monoallelically deleted in a large number of sporadic breast, ovarian, and prostate cancers, and Beclin 1 haplo-insufficient mice show both reduced autophagy and a measurable increase in epithelial and hematopoietic tumors.[18,19] Furthermore, stable expression of Beclin 1 in breast cancer cells promotes autophagy and reduces the tumorigenic capacity of the transfectants.[19] The reduction in autophagic ability resulting from the deletion of an allele of Beclin-1 in a large number of breast cancers may be essential for maintaining the malignant phenotype and may also help promote tumor progression. The ability of the tumor suppressor PTEN, an inhibitor of the PI3 kinase pathway, to activate autophagy is further indication that a reduction in autophagic capacity contributes to cancer.[25]

Although autophagy functions as a protective mechanism against infection, some microorganisms have developed strategies to manipulate the degradative machinery to their own advantage. After escaping the endophagocytic pathway and preventing fusion with lysosomes, the bacterium *Brucella abortus* shelters and replicates itself in autophagosome-like vesicles,[39] while *Legionella pneumophila* secretes a product that activates autophagy and allows it to replicate itself directly inside autolysosomes.[54] Likewise, a virulence protein, ICP34.5, in *Herpes simplex* virus antagonizes the autophagic pathway allowing the virus to escape degradation.[29]

Active recruitment of autophagy can have preventive or therapeutic potential. Since autophagic elimination of aggregated proteins and damaged organelles preserves cells from further damage, autophagy could serve a protective role. The therapeutic value of activating autophagy in the early stages of neurodegeneration is currently being investigated.[55–57] The mTOR inhibitor and autophagy promoter, rapamycin, has shown promising anticancer activity in clinical trials.[24] Bacterial and viral proteins that modify autophagosome formation could also be harnessed to modulate this degradative pathway for therapeutic purposes. The rate of autophagy decreases with age, and diet restriction-induced autophagy, which takes advantage of the innate adaptive response to effect efficient removal of oxidatively damaged proteins, may contribute to stress resistance and an extended life span.[58,59]

AUTOPHAGY AND APOPTOSIS

Until recently, "programmed cell death" was used synonymously with the word "apoptosis." However, advances in the understanding of the role of autophagy in normal growth and development, and the physiological consequences of knocking down genes that code for intermediates in the autophagic pathway firmly established its involvement in programmed cell death and prompted a reclassification of apoptotic death as Type I PCD and autophagy as Type II PCD. In contrast to apoptosis, subcellular organelles, barring nuclei, are degraded during autophagy while the cytoskeleton is largely preserved. Although the two pathways adopt vastly different means to achieve the same end, they also have much in common. Stimuli that activate autophagy also serve as triggers for apoptosis. Cellular features of apoptosis and autophagy frequently occur together, or in close temporal proximity, in response to stresses such as nutrient deprivation and therapeutic insults. Apoptosis is designed for the rapid deletion of cells using the powerful tool of caspase activation. However, when large-scale tissue histolysis is required during development, autophagy often precedes apoptosis, presumably to deconstruct cells internally, to reduce cell volume, and ease the breakdown of large amounts of apoptotic bodies and dead cells by the phagocytosing host. On the other hand, the intensity of a nonphysiological stress signal may dictate whether apoptosis precedes autophagy during an adaptive response or follows it. The selective sequestration and recycling of damaged mitochondria in autophagosomes following mild stress may have the effect of protecting against apoptosis in normal cells and of promoting tumor progression in cancer cells or in cells with defective apoptotic machinery. Acute starvation or strong doses of damaging drugs activate both pathways although the exposed cell would die an apoptotic death before autophagy was detectable.

Subcellular organelles, mitochondria, ER, and lysosomes may play central roles in integrating the two forms of cell death. The role of damaged mitochondria in activating both caspases and autophagophore generation is well documented. Additionally, some of the endonucleases that participate in the DNA fragmentation associated with apoptosis may originate in lysosomes.[60] Phosphorylation of p70S6K and Akt by PDK-1 (class I PI3-K) inhibits both autophagy and apoptosis.[61,62] Cytochrome c release in TNF-α–treated hepatocytes is attributed to lysosomal destabilization and cathepsin B activation of the Bcl-2 family protein, Bid.[63,64] The apoptosis-inducing ligand TRAIL mediates autophagy in mammary acini.[65] Multidomain Bcl-2 family members, Bcl-2, Bcl-xL, Bax, and Bak have recently been implicated in the regulation of autophagy, both independently and through autophagic intermediate Beclin-1, which interacts with anti-apoptotic proteins Bcl-2 and Bcl-x_L.[66,67] Furthermore, the death-associated protein kinases, DAPk and DRP-1, can induce both apoptotic blebbing and autophagic vacuoles.[67] Thus, it is becoming increasingly evident that the two seemingly independent pathways to cell death are not mutually exclusive.

CONCLUDING REMARKS

Progress in our understanding of the molecular mechanisms underlying this degradative pathway has come in leaps and bounds following the identification of mammalian homologues of autophagy-signaling intermediates in yeast and the analysis of transgenic and knockout animal models. More homologues are certain to be identified in the future, and they will further deepen our understanding of the physiological role of autophagy and its contribution to disease. Excessive autophagy is associated with degenerative diseases of the muscular and nervous system and defective or reduced autophagy may underlie tumor progression in breast and ovarian cancers.

Apoptosis and autophagic cell death can occur simultaneously or conjointly in a single cell. Autophagic cell death or PCD II appears to precede apoptosis when massive cell elimination is required during development. An analysis of the role of autophagy in health and disease suggests it can act both as a protector and a killer of cells. Autophagy is mounted as an adaptive response to pathological stresses, such as nutrient deprivation, therapeutic insults, and bacterial or viral infection. An adaptive response to a moderate stress stimulus would allow for cell survival and recovery, but may culminate in death in the presence of extreme stress and an impaired apoptotic response on the part of the damaged cell. Although the issue of whether autophagy is aimed primarily at cell survival, or at cell death, is likely to remain a subject of debate for some time, for the immediate future the therapeutic uses of autophagy inducers and inhibitors in modulating both its survival and death-promoting properties show promise and continue to be actively explored.

[AUTHOR'S NOTE: We regret the inability to cite many excellent original research articles due to space limitations.]

REFERENCES

1. CUERVO, A.M. 2004. Autophagy: many paths to the same end. Mol. Cell. Biochem. **263:** 55–72.
2. KIM, J. & D.J. KLIONSKY. 2000. Autophagy, cytoplasm-to-vacuole targeting pathway, and pexophagy in yeast and mammalian cells. Annu. Rev. Biochem. **69:** 303–42.
3. KLIONSKY, D.J. & S.D. EMR. 2000. Autophagy as a regulated pathway of cellular degradation. Science **290:** 1717–1721.
4. MIZUSHIMA, N., Y. OHSUMI & T. YOSHIMORI. 2002. Autophagosome formation in mammalian cells. Cell Struct. Funct. **27:** 421–429.
5. KIRKEGAARD, K., M.P. TAYLOR & W.T. JACKSON. 2004. Cellular autophagy: surrender, avoidance and subversion by microorganisms. Nat. Rev. Microbiol. **2:** 301–314.
6. KOPITZ, J. *et al.* 1990. Nonselective autophagy of cytosolic enzymes by isolated rat hepatocytes. J. Cell Biol. **111:** 941–953.
7. MITCHENER, J.S. *et al.* 1976. Cellular autophagocytosis induced by deprivation of serum and amino acids in HeLa cells. Am. J. Pathol. **83:** 485–491.
8. DUNN, W.A., JR. 1990. Studies on the mechanisms of autophagy: formation of the autophagic vacuole. J. Cell Biol. **110:** 1923–1933.
9. PUNNONEN, E.L. *et al.* 1992. Autophagy, cathepsin L transport, and acidification in cultured rat fibroblasts. J. Histochem. Cytochem. **40:** 1579–1587.
10. BURSCH, W. *et al.* 2000. Autophagic and apoptotic types of programmed cell death exhibit different fates of cytoskeletal filaments. J. Cell Sci. **113:** 1189–1198.
11. KUMA, A. *et al.* 2002. Formation of the approximately 350-kDa Apg12-Apg5.Apg16 multimeric complex, mediated by Apg16 oligomerization, is essential for autophagy in yeast. J. Biol. Chem. **277:** 18619–18625.
12. KABEYA, Y. *et al.* 2000. LC3, a mammalian homologue of yeast Apg8p, is localized in autophagosome membranes after processing. EMBO J. **19:** 5720–5728.
13. MIZUSHIMA, N. 2004. Methods for monitoring autophagy. Int. J. Biochem. Cell Biol. **36:** 2491–2502.

14. SEGLEN, P.O. & P.B. GORDON. 1984. Amino acid control of autophagic sequestration and protein degradation in isolated rat hepatocytes. J. Cell Biol. **99:** 435–444.
15. PETIOT, A. et al. 2000. Distinct classes of phosphatidylinositol 3'-kinases are involved in signaling pathways that control macroautophagy in HT-29 cells. J. Biol. Chem. **275:** 992–998.
16. KIHARA, A. et al. 2001. Beclin-phosphatidylinositol 3-kinase complex functions at the trans-Golgi network. EMBO Rep. **2:** 330–335.
17. LIANG, X.H. et al. 1999. Induction of autophagy and inhibition of tumorigenesis by beclin 1. Nature **402:** 672–676.
18. YUE, Z. et al. 2003. Beclin 1, an autophagy gene essential for early embryonic development, is a haploinsufficient tumor suppressor. Proc. Natl. Acad. Sci. USA **100:** 15077–15082.
19. LIANG, X.H. et al. 2001. Beclin 1 contains a leucine-rich nuclear export signal that is required for its autophagy and tumor suppressor function. Cancer Res. **61:** 3443–3449.
20. DENNIS, P.B., S. FUMAGALLI & G. THOMAS. 1999. Target of rapamycin (TOR): balancing the opposing forces of protein synthesis and degradation. Curr. Opin. Genet. Dev. **9:** 49–54.
21. MEJILLANO, M. et al. 2001. Regulation of apoptosis by phosphatidylinositol 4,5-bisphosphate inhibition of caspases, and caspase inactivation of phosphatidylinositol phosphate 5-kinases. J. Biol. Chem. **276:** 1865–1872.
22. SHIGEMITSU, K. et al. 1999. Regulation of translational effectors by amino acid and mammalian target of rapamycin signaling pathways. Possible involvement of autophagy in cultured hepatoma cells. J. Biol. Chem. **274:** 1058–1065.
23. CUTLER, N.S., J. HEITMAN & M.E. CARDENAS. 1999. TOR kinase homologs function in a signal transduction pathway that is conserved from yeast to mammals. Mol. Cell. Endocrinol. **155:** 135–142.
24. GUERTIN, D.A. & D.M. SABATINI. 2005. An expanding role for mTOR in cancer. Trends Mol. Med. **11:** 353–361.
25. ARICO, S. et al. 2001. The tumor suppressor PTEN positively regulates macroautophagy by inhibiting the phosphatidylinositol 3-kinase/protein kinase B pathway. J. Biol. Chem. **276:** 35243–35246.
26. PETIOT, A. et al. 2002. Diversity of signaling controls of macroautophagy in mammalian cells. Cell Struct. Funct. **27:** 431–441.
27. OGIER-DENIS, E. et al. 1996. Guanine nucleotide exchange on heterotrimeric Gi3 protein controls autophagic sequestration in HT-29 cells. J. Biol. Chem. **271:** 28593–28600.
28. OGIER-DENIS, E. et al. 2000. Erk1/2-dependent phosphorylation of G alpha-interacting protein stimulates its GTPase accelerating activity and autophagy in human colon cancer cells. J. Biol. Chem. **275:** 39090–39095.
29. TALLOCZY, Z. et al. 2002. Regulation of starvation- and virus-induced autophagy by the eIF2alpha kinase signaling pathway. Proc. Natl. Acad. Sci. USA **99:** 190–195.
30. LEVINE, B. & D.J. KLIONSKY. 2004. Development by self-digestion: molecular mechanisms and biological functions of autophagy. Dev. Cell **6:** 463–477.
31. YUAN, J., M. LIPINSKI & A. DEGTEREV. 2003. Diversity in the mechanisms of neuronal cell death. Neuron **40:** 401–413.
32. TERMAN, A. & U.T. BRUNK. 2004. Myocyte aging and mitochondrial turnover. Exp. Gerontol. **39:** 701–705.
33. DE DUVE, C. & R. WATTIAUX. 1966. Functions of lysosomes. Annu. Rev. Physiol. **28:** 435–492.
34. TAKANO-OHMURO, H. et al. 2000. Autophagy in embryonic erythroid cells: its role in maturation. Eur. J. Cell Biol. **79:** 759–764.
35. HOLM, T.M. et al. 2002. Failure of red blood cell maturation in mice with defects in the high-density lipoprotein receptor SR-BI. Blood **99:** 1817–1824.
36. SULZER, D. et al. 2000. Neuromelanin biosynthesis is driven by excess cytosolic catecholamines not accumulated by synaptic vesicles. Proc. Natl. Acad. Sci. USA **97:** 11869–11874.

37. HARIRI, M. et al. 2000. Biogenesis of multilamellar bodies via autophagy. Mol. Biol. Cell **11:** 255–268.
38. LUM, J.J. et al. 2005. Growth factor regulation of autophagy and cell survival in the absence of apoptosis. Cell **120:** 237–248.
39. DORN, B.R., W.A. DUNN, JR. & A. PROGULSKE-FOX. 2002. Bacterial interactions with the autophagic pathway. Cell Microbiol. **4:** 1–10.
40. NISHINO, I. 2003. Autophagic vacuolar myopathies. Curr. Neurol. Neurosci. Rep. **3:** 64–69.
41. NISHINO, I. et al. 2000. Primary LAMP-2 deficiency causes X-linked vacuolar cardiomyopathy and myopathy (Danon disease). Nature **406:** 906–910.
42. KALIMO, H. et al. 1988. X-linked myopathy with excessive autophagy: a new hereditary muscle disease. Ann. Neurol. **23:** 258–265.
43. NONAKA, I. 1999. Distal myopathies. Curr. Opin. Neurol. **12:** 493–499.
44. GOTO, Y. et al. 1990. Myopathy in Marinesco-Sjogren syndrome: an ultrastructural study. Acta Neuropathol. (Berl.) **80:** 123–128.
45. ANGLADE, P. et al. 1997. Apoptosis and autophagy in nigral neurons of patients with Parkinson's disease. Histol. Histopathol. **12:** 25–31.
46. CATALDO, A.M. et al. 1996. Properties of the endosomal-lysosomal system in the human central nervous system: disturbances mark most neurons in populations at risk to degenerate in Alzheimer's disease. J. Neurosci. **16:** 186–199.
47. KEGEL, K.B. et al. 2000. Huntingtin expression stimulates endosomal-lysosomal activity, endosome tubulation, and autophagy. J. Neurosci. **20:** 7268–7278.
48. OKAMOTO, K. et al. 1991. Reexamination of granulovacuolar degeneration. Acta Neuropathol. (Berl.). **82:** 340–345.
49. LIBERSKI, P.P. et al. 2004. Neuronal cell death in transmissible spongiform encephalopathies (prion diseases) revisited: from apoptosis to autophagy. Int. J. Biochem. Cell Biol. **36:** 2473–2490.
50. LIPINSKI, M.M. & J. YUAN. 2004. Mechanisms of cell death in polyglutamine expansion diseases. Curr. Opin. Pharmacol. **4:** 85–90.
51. STEFANIS, L. et al. 2001. Expression of A53T mutant but not wild-type alpha-synuclein in PC12 cells induces alterations in the ubiquitin-dependent degradation system, loss of dopamine release, and autophagic cell death. J. Neurosci. **21:** 9549–9560.
52. VENKATRAMAN, P. et al. 2004. Eukaryotic proteasomes cannot digest polyglutamine sequences and release them during degradation of polyglutamine-containing proteins. Mol Cell. **14:** 95–104.
53. TECKMAN, J.H. et al. 2004. Mitochondrial autophagy and injury in the liver in alpha 1-antitrypsin deficiency. Am. J. Physiol. Gastrointest. Liver Physiol. **286:** G851–862.
54. COERS, J. et al. 2000. Identification of Icm protein complexes that play distinct roles in the biogenesis of an organelle permissive for *Legionella pneumophila* intracellular growth. Mol. Microbiol. **38:** 719–736.
55. RAVIKUMAR, B. & D.C. RUBINSZTEIN. 2004. Can autophagy protect against neurodegeneration caused by aggregate-prone proteins? Neuroreport **15:** 2443–2445.
56. QIN, Z.H. et al. 2003. Autophagy regulates the processing of amino terminal huntingtin fragments. Hum . Mol. Genet. **12:** 3231–3244.
57. MERIIN, A.B. & M.Y. SHERMAN. 2005. Role of molecular chaperones in neurodegenerative disorders. Int. J. Hyperthermia **21:** 403–419.
58. MOORE, M.N. 2004. Diet restriction induced autophagy: a lysosomal protective system against oxidative- and pollutant-stress and cell injury. Mar. Environ. Res. **58:** 603–607.
59. BERGAMINI, E. et al. 2004. The role of macroautophagy in the ageing process, anti-ageing intervention and age-associated diseases. Int. J. Biochem. Cell Biol. **36:** 2392–2404.
60. KESSEL, D. et al. 2000. Determinants of the apoptotic response to lysosomal photodamage. Photochem. Photobiol. **71:** 196–200.
61. VANHAESEBROECK, B. & D. R. ALESSI. 2000. The PI3K-PDK1 connection: more than just a road to PKB. Biochem. J. **346:** 561–576.
62. KATSO, R. et al. 2001. Cellular function of phosphoinositide 3-kinases: implications for development, homeostasis, and cancer. Annu. Rev. Cell Dev. Biol. **17:** 615–675.

63. STOKA, V. et al. 2001. Lysosomal protease pathways to apoptosis. Cleavage of bid, not pro-caspases, is the most likely route. J. Biol. Chem. **276:** 3149–3157.
64. GUICCIARDI, M.E. et al. 2000. Cathepsin B contributes to TNF-alpha-mediated hepatocyte apoptosis by promoting mitochondrial release of cytochrome c. J. Clin. Invest. **106:** 1127–1137.
65. MILLS, K.R. et al. 2004. Tumor necrosis factor-related apoptosis-inducing ligand (TRAIL) is required for induction of autophagy during lumen formation in vitro. Proc. Natl. Acad. Sci. USA **101:** 3438–3443.
66. SHIMIZU, S. et al. 2004. Role of Bcl-2 family proteins in a non-apoptotic programmed cell death dependent on autophagy genes. Nat. Cell Biol. **6:** 1221–1228.
67. PATTINGRE, S.A. et al. 2005. Bcl-2 antiapoptotic proteins inhibit Beclin 1–dependent autophagy. Cell **122:** 927–939.
68. INBAL, B. et al. 2002. DAP kinase and DRP-1 mediate membrane blebbing and the formation of autophagic vesicles during programmed cell death. J. Cell Biol. **157:** 455–468.
69. ESKELINEN, E.L. 2005. Macroautophagy in Mammalian Cells. *In* Lysosomes. P. Saftig, Ed. Springer Science+Business Media.

Magnetic Resonance Imaging of Changes in Muscle Tissues after Membrane Trauma

HANNE GISSEL,[a] FLORIN DESPA, JOHN COLLINS, DEVKUMAR MUSTAFI, KATHERINE ROJAHN, GREG KARCZMAR, AND RAPHAEL LEE

Electrical Trauma Research Program, Biological Sciences Division, MC 6035, University of Chicago, Chicago, Illinois 60637, USA

[a]*Department of Physiology and Biophysics, University of Aarhus, DK-8000 Århus C, Denmark*

ABSTRACT: A pure electroporation injury leads to cell membrane disruption and subsequent osmotic swelling of the tissue. The state of water in the injured area of a tissue is changed and differs from a healthy tissue. Magnetic resonance imaging (MRI), which is very sensitive to the quality of the interaction between mobile (water) protons and a restricted (protein) proton pool, is therefore a useful tool to characterize this injury. Here, we present a protocol designed to measure the difference between the values of the transverse magnetic relaxation time (T_2) in MRIs of healthy and electrically injured tissue. In addition, we present a method to evaluate the two main contributions to the MRI contrast, the degree of structural alteration of the cellular components (including a major contribution from membrane pores), and edema. The approach is useful in assessing the level of damage that electric shocks produce in muscle tissues, in that edema will resolve in time whereas structural changes require active repair mechanisms.

KEYWORDS: electroporation; T_2-MRI; electrical injury; edema

INTRODUCTION

It is well recognized that a high-current electrical shock often produces an irreversible damage of biological tissues, followed by necrosis. Damage may occur as a result of local heating to high temperature or as a direct result of breakdown of the cell membrane caused by electroporation.[1–4] Heat may induce significant denaturation of the cellular components, whereas electroporation perturbs the delicate balance of ions and molecules between the intracellular and extracellular compartments. This will, among other things, lead to loss of muscle function, muscle Ca^{2+} overload,[5–7] and most likely an increased production of reactive oxygen species (ROS). Increased production of ROS and muscle Ca^{2+} overload may in turn activate degradative processes intrinsic to the muscle cell, causing further muscle

Address for correspondence: Raphael C. Lee, Department of Surgery, MC 6035, The University of Chicago, 5841 South Maryland Avenue, Chicago, IL 60637. Voice: 773-702-6302; fax: 703-702-1634.
r-lee@uchicago.edu

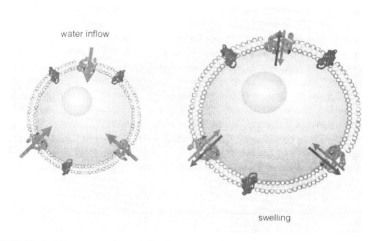

FIGURE 1. Electroporation of the plasma membrane leads to an osmotic water inflow, which in turn leads to cell swelling and increases the hydration of membrane compartments (fatty acids and proteins), which are usually separated from excess amount of water. At quasi-equilibrium, water inflow may be equal to the outer outflow (the reverse arrow).

damage.[8-10] Finally there may be development of edema due to osmotic water inflow into the muscle tissue.

In model studies of electrical trauma, the contribution to the tissue damage from heating and electroporation can be controlled by an appropriate choice of the electrical shock parameters. Sequences of high-voltage pulses in durations of milliseconds, separated by waiting times sufficiently long to allow for heat dissipation and thermal equilibration, may cause predominantly cell electroporation rather than heat (burn) damage.[11] In contrast, high-frequency electric fields might lead primarily to thermal denaturation of the cellular components.

Membrane poration is a frequent injury mechanism in physical trauma of skeletal muscles. This has as a direct effect an alteration of the state of water in the tissue. First, disruption of cellular membranes increases the hydration of membrane compartments (fatty acids and proteins) which are usually separated from excess amounts of water (FIG. 1). Furthermore, structural modifications of cellular components change the dynamics of water molecules in the hydration layers.[12] Secondly, swelling of the cell due to osmotic water inflow (edema) adds extra hydration layers to the cellular components (FIG. 1). These physiological changes affect the magnetic relaxation of the tissue water protons and are likely to be detected and measured by MRI. Water proton MR measurements of protein solutions has recently brought compelling evidence for the change of the MR signal due to protein unfolding.[13-15] In addition, the magnetic relaxation dispersion curves revealed a change between long-lived and short-lived water proton populations during conformational changes of the protein.[13] The physical basis of these observations is that protein unfolding increases the water-accessible surface area (ASA), resulting in change of the number and dynamics of interface water molecules. To properly understand the origin of the change in water dynamics, we recall that a protein structure displays both binding (hydrophilic) and unbinding (hydrophobic) sites for water molecules and that, in the native state, the latter are buried inside the protein structure. Protein unfolding leads to the exposure of the hydrophobic residues to water, which induces a shift of the correlation decay time, that is, the time characterizing the motion of the water molecules in a specific vicinity with respect to the protein structure environment.[13,14] This change in the dynamics of water is expected to occur at the interface with any cellular component subject to structural alterations, including the cell membrane, which seems to be primarily affected by the electric field.

Previous work demonstrated that electrical injury patterns can be visualized in MR images.[16,17] It was shown that muscle subjected to electrical shock showed higher signal intensities in T_2-weighted MRI images. These preliminary results had an immediate clinical impact and lay the ground for applying MRI as a diagnostic method in patients with electric injuries.[18,19] Further investigations focused on edema localization and contrast agent distribution in a rat hind limb electrical injury model.[20]

In this chapter we present a protocol designed to measure the difference between the values of the T_2 relaxation time of healthy and electrically injured tissue. In addition, we present a method to evaluate the two main contributions to the MRI contrast, the degree of structural alteration of the cellular components (including a major contribution from membrane pores), and edema. The approach is useful in assessing the level of damage in muscle tissues produced by the electric shocks, as edema will resolve in time, whereas structural changes require active repair mechanisms.

MEMBRANE DAMAGE BY ELECTROPRATION

The Animal Model

Female Sprague-Dawley rats (weight 300 ± 20 g) were anesthetized using an intraperitoneal injection of a ketamine/xylazine combination (i.e., 90 mg/kg and 10 mg/kg, respectively) dissolved in isotonic saline solution for anaesthesia induction. Additional anaesthetic was administered as needed to maintain a constant deep anaesthetic plane. Ketamine HCl Inj., USP (100 mg/mL, NDC 0856-2013-01) was from Fort Dodge Animal Health (Fort Dodge, IA 50501), and Xylazine sterile solution (20 mg/mL, NDC 11695-4400-1) from the Butler Company (Columbus, OH 43228). The experiments were done in accordance with the standards described in *The Guide for the Care and Use of Laboratory Animals* (NIH Publication No. 86-23, 1985, Department of Health and Human Services), under a protocol approved by the University of Chicago Institutional Animal Care and Use Committee.

All animals were fully anesthetized and medically stabilized before electrical shock. After the anesthetization took effect, catheters were surgically established and secured into the jugular vein and into the peritoneal cavity. A steady maintenance dose of ketamine/xylazine (i.e., 28 mg/mL and 0.20 mg/mL, respectively at a rate of 0.35 mL/h) was administered i.p. throughout the procedure to maintain the anaesthetic plane below pain perception but above respiratory compromise. At the end of the experiment the animals were sacrificed by a Xylazine overdose (0.4 mL, 20 mg/mL) administered intravenously.

The Electroporation Protocol

The electrical conduction path was chosen between the ankle of the left leg and the base of the tail, and thus the electrical current travels along the whole leg of the rat. The skin at the ankle and at the base of the tail was abraded and wrapped in gauze soaked in Parker Electrode Gel, a salt-bridge conducting gel. Stainless-steel cuff electrodes were attached to the salt bridges and connected via high-voltage cables to the pulse generator. A discharge-type rectangular pulse generator (Dialog, Hannover, Germany) was used to generate the electrical shocks. The impedance of this circuit measured at 100 Hz was in the range of 1000 ± 300 Ohms. All experiments were carried out in an electrically insulated chamber.

All rats received 12 shocks. Each shock consisted of a rectangular pulse of 2 kV, ~2A amplitude with duration of 4 ms. There was a 10-second separation period between consecutive shocks to allow thermal relaxation following each pulse. It has

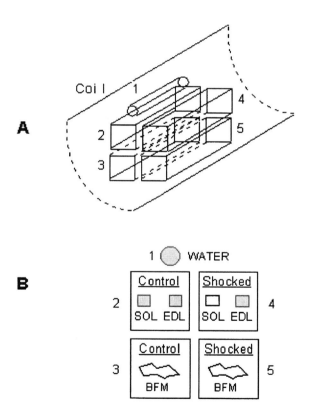

FIGURE 2. A: Illustration of the placing of the four 1-cm cuvettes (2–5) and the water phantom (1) in a 35-mm bird-caged coil for MR imaging. **B**: Cross-sectional view of the placing of the muscles in the cuvettes. Owing to their small size, SOL and EDL could be placed together in the same cuvette. Muscles from the control unshocked leg were placed in 2 and 3, whereas muscles from the shocked leg were placed in 4 and 5 (see text for details).

been shown for this protocol that the temperature rise in the muscle tissue due to the electrical current pulses is less than 0.5°C.[11] The control animals were connected to the pulse generator for the same duration but did not receive electrical shocks.

All animals tolerated the anaesthetic induction and the electroporation injury. A 1-mL bolus of lactated Ringer's solution was administered intravenously immediately after shock and 1 hour post shock. The electroporated thigh muscles could be visually observed to swell over several hours.

Two series of experiments were conducted. In the first series (*In vivo* shock and *ex vivo* imaging) the rats were monitored and then sacrificed at 3 hours post shock. Intact extensor digitorum longus (EDL) and soleus (SOL) muscles were excised, together with a part of the biceps femoris muscle (BFM). Identical muscles isolated from the non-injured leg of the same rat served as controls. The muscles were immediately placed in sealed 1-cm plastic cuvettes to avoid loss of moisture. A small water phantom was placed in the coil along with the muscles as a standard (FIG. 2A).

In the second series (*in vivo* shock and *in vivo* imaging) the rat was prepared for *in vivo* imaging immediately post shock. The shocked leg was imaged from approximately 1 hour post shock through 3 hours post shock. Images obtained separately from unshocked rats were used as controls.

Magnetic Resonance Imaging

Proton magnetic resonance images for T_2 measurements were recorded using a Bruker scanner (Bruker Biospin, Billerica, MA) (200 MHz) equipped with a 4.7 T, 33-cm internal diameter superconducting magnet and actively shielded magnetic field gradients. MRI measurements were acquired using a Bruker bird-caged coil with internal diameter of 35 mm. Images were acquired with a standard multi-slice, multi-echo (MSME) spin echo Carr-Purcell-Meiboom-Gill (CPMG) sequence. The parameters for *in vitro* MRI measurements were as follows: TR, 2000 ms; TE, ten linearly-spaced echoes from 10 to 100 ms; array size, 256 × 256; field of view, 50 mm; flip angle, 0°; readout bandwidth, 50 kHz; slice thickness, 2 mm; slices, 5; average, 2; and in-plane resolution, 195 µm. For each set, ten images with a TR/TE ratio ranging from 1000/10 ms to 1000/100 ms were also acquired. The power and receiver gain of the scanner was kept constant for all experiments to ensure that signal scaling conditions were identical.

For T_2 measurements five sets of samples from five different rats were used. In each set, one leg was shocked and the other leg from the same rat was used as a control. Following MRI measurements the tendons were removed from the explants and muscle wet weight was determined. Then the muscles were placed in a vacuum oven at 60 °C until completely dry (constant weight) in order to obtain the dry weight of the muscle. From this the water content and thus the hydration of the muscle tissue was calculated.

In the second series of experiments the whole animal was placed in the scanner. MRI measurements were carried out using a custom made Helmholtz coil with internal diameter of 25 mm. MR imaging was performed as previously described with the following changes in parameters. The array size is 128 × 128 and ten slices were prescribed, 2 mm thick, and separated by 2 mm over the entire mid-thigh region of the leg. Typically, one imaging sequence took 4.5 minutes to complete. Imaging began at approximately 1 hour after shock and continued until 3 hours post shock.

ASESSING THE MEMBRANE DAMAGE

Ex vivo *MRI Results*

In the first series of experiments the animals were shocked, left for 3 hours before intact SOL and EDL muscles, as well as a part of the BFM, were harvested. The muscles were placed in sealed cuvettes for MR imaging. FIGURE 2A illustrates how the samples were placed in the magnet for MRI measurements and FIGURE 2B illustrates a cross-sectional view of samples inside the coil.

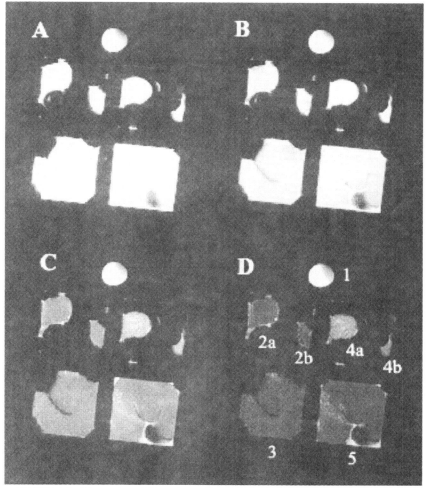

FIGURE 3. Illustration of T_2-weighted MR images. The animal was anaesthetized and shocked. Three hours post shock the animal was sacrificed and SOL, EDL, and BFM were harvested and placed in cuvettes for imaging. Images were acquired at echo times (TE) ranging from 10 to 100 ms. Four sets of images are shown: (**A**) TE = 15 ms; (**B**) TE = 30 ms; (**C**) TE = 60 ms; and (**D**) TE = 90 ms. The numbering scheme (1–5) indicated in panel **D** is the same as in FIGURE 1. The labels **a** and **b** represent SOL and EDL muscles, respectively.

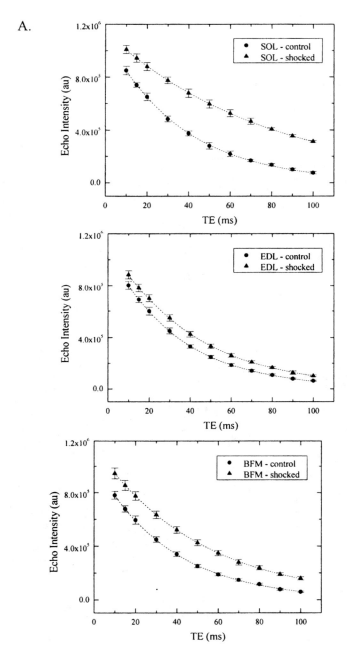

FIGURE 4. The dependence of MR signal intensities on the echo time for normal and electroporated rat muscles. **A**: The data were fitted to an exponential decay curve of the signal amplitude as a function of echo time, as shown by *dashed lines*. Values of T_2 obtained from these plots are shown in TABLE 1.

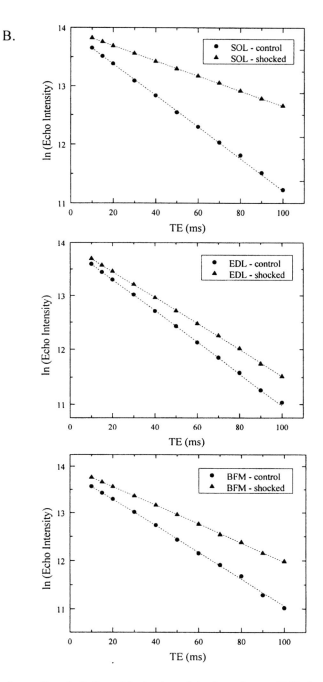

FIGURE 4—*continued*. B: Logarithmic plots of the dependence of MR signal intensities on the echo time. Mean values ± SD are shown; $n = 4–5$ animals.

FIGURE 3 shows a set of T_2 weighted images at echo times ranging from 15 to 90 ms. As seen from the figure, signal intensity decreases as TE increases; however, it is also clear that the contrast between the shocked and the untreated muscles increases with increasing TE.

FIGURE 4A shows the dependence of MR intensities on the echo time for normal and electroporated rat muscles. The data were fitted to an exponential decay curve of the signal amplitude as a function of echo time. It is clear that echo intensity is higher in the shocked muscles at all echo times with the largest difference in SOL and the smallest in EDL. FIGURE 4B shows the log transformation of the data. The slope of each curve represents the expected value of the T_2 relaxation time. T_2 values of control and shocked muscle samples as measured by MRI are listed in TABLE 1. Shocking significantly increased T_2 in all muscles ($P < 0.006$). The largest increase (108%) was observed in SOL. In EDL shocking increased T_2 by 20% and finally in BFM T_2 was increased by 40%.

Muscle Water Content

The hydration of the muscle tissue was measured by determining the water content. As shown in TABLE 1, muscle water content was significantly increased in all shocked muscles compared to untreated controls. In SOL water content was increased by 15%, whereas in EDL and BFM the increase was 4%.

In vivo *MRI Results*

In order to investigate the development of the changes in T_2, MR imaging was performed *in vivo* starting approximately 1 hour after shock and continuing until approximately 3 hours after. FIGURE 5A shows two images of the mid-thigh region of a shocked hind leg from a live rat. The first image is recorded 63 min post shock and the other at 184 min post shock. From these images it can be observed how the signal

TABLE 1. Spin-spin relaxation time (T_2) and water content of three different rat muscles following electroporation *in vivo*

Sample[a]	T_2 (ms)[b]	P Value	Hydration	P Value
SOL (control)	36.90 ± 1.43(5)		0.735 ± 0.022(5)	
SOL (shocked)	76.80 ± 3.07(5)	0.00002	0.845 ± 0.013(5)	0.0005
EDL (control)	34.41 ± 1.41(5)		0.727 ± 0.013(5)	
EDL (shocked)	41.37 ± 1.55(5)	0.006	0.758 ± 0.020(5)	0.017
BFM (control)	36.11 ± 0.72 (4)		0.741 ± 0.017(4)	
BFM (shocked)	50.42 ± 0.017 (4)	0.001	0.771 ± 0.017(4)	0.031

[a]Samples are as follows: SOL, soleus; EDL, extensor digitorum longus; and BFM, biceps femoris muscle. For description of control and shocked muscles and determination of hydration, see text. Mean values ± SD are shown with the number of observations in parentheses. P indicates the significance of the difference between the control and the shocked muscles. Student's paired t test.

[b]T_2 (in millisecond)s were determined by fitting the signal amplitude from the echoes of the multi-slice multi-echo experiments to the spin-spin signal decay equation: $M_{(x,y)} = M_{0(x,y)} \cdot \exp(-TE/T_2)$, where M_0 is the amount of magnetization at thermal equilibrium, TE is the experimental echo time, and T_2 is the spin-spin relaxation time.

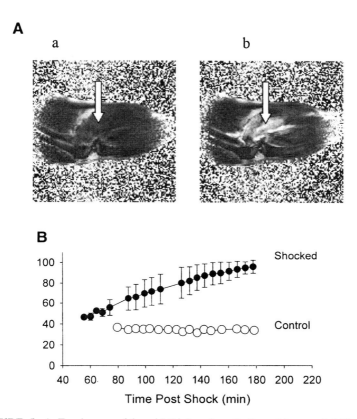

FIGURE 5. A: Two images of the mid-thigh region of a live rat (*a* recorded 63 min post shock, *b* recorded 184 min post shock). The region used for calculation of mean T_2 is depicted by *arrows* **B**: Mean T_2 values for shocked (*solid circles*, SD, $n = 25$ pixels) and unshocked controls (*open circles*, $n = 25$ pixels).

intensity increases with time. The signal intensities at the different TE times were used to fit an exponential decay and measure the T_2 value in each voxel. T_2 values were calculated from the images of each animal for a region of interest (ROI = 25 pixels) and the average value as a function of time is shown in FIGURE 5B. As seen from this figure, T_2 increases progressively to about 180% over a period of 180 min after shock.

Histology

Muscle tissue samples were taken from the electrically injured rat hind limb after applying a different number of pulses to obtain histologic dose-response data. A muscle biopsy was performed at the end of a neurophysiology study (6 hours post shock) and the tissue was fixed in 10% formalin. The samples were paraffin-embedded, cut, slide-mounted, and hematoxylin and eosin–stained (H&E).

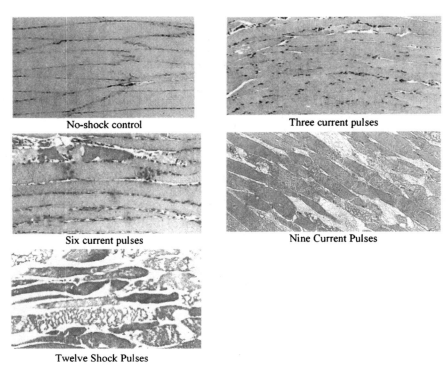

FIGURE 6. H&E stain of histologic sections of hind limb muscle. Muscle has been subjected to multiple DC electric current pulses (2 A, 4 ms, 10-s interval) or no shock control. Images shown were selected from the subsets exposed to 0 (control), 3, 6, 9, and 12 electric shock pulses. Muscle biopsy taken 6 h post shock. (Original magnification × 150.)

A histologic examination of formalin-fixed paraffin-embedded sections of the shocked muscle is displayed in FIGURE 6. This reveals extensive vacuolization, which suggests increased water within damaged myocytes. One can also observe areas of sarcoplasm relatively devoid of eosinophilic contractile proteins and hypercontraction band degeneration, which was not seen in the muscle from the unshocked limb. These results suggest that non-thermal electrical effects can induce structural alteration of the cell.

The Increase of the Hydrophobic Exposure in Injured Cells

The damage of the cell membranes increases the hydrophobic exposure to surrounding water, which marks the magnetic relaxation of this water. The increase of hydrophobic exposure (and, implicitly, the degree of structural alteration of the cells) can be estimated based on a statistical mechanics approach which interprets the T_2-MRI data in terms of dynamical changes and recompartmentalization of water after injury.[12] The net structural change (denaturation) of the cellular components in each muscle tissue is given by the value of α ($0 \leq \alpha \leq 1$), which is a measure of the

TABLE 2. Predicted structural alteration in three different skeletal muscles due to electroporation

Sample[a]	ρ_c (control)	$T_2^{(hyd)}$ (shocked)	$T_2^{(i)} - T_2^{(hyd)}$ (shocked)	α (shocked)
SOL	0.683	71.58 ms	5.22 ms	0.08
EDL	0.673	40.4 ms	0.97 ms	0.03
BFM	0.665	42.49 ms	7.93 ms	0.18

[a]Samples are as follows: SOL, soleus; EDL, extensor digitorum longus; and BFM, biceps femoris muscle. For description of control and shocked muscles, see text.

increase of hydrophobic exposure after injury ($\alpha = 0$, for healthy muscle, and $\alpha = 1$, for the extreme case of denaturated cells). This is extracted by a numerical computation from the difference between values for injured and non-injured muscles. Thus, by replacing η_t for shocked muscles (TABLE 1) in the Zimmerman-Brittin eqs.,[12] we obtain $T_2^{(hyd)}$ values (TABLE 2) which are typically smaller than the measured T_2. This means that there is also an intrinsic contribution to the magnetic relaxation of water in injured muscles ($T_2^{(i)} - T_2^{(hyd)}$), which corresponds to the alteration of the dynamics of water in contact with disrupted cell membranes and denaturated proteins, as we described above. Under such circumstances, we used the measured magnetic relaxation times $T_2^{(i)}$ (TABLE 1) and hydration values η_t (TABLE 1) in the above equations to extract the corresponding values $\alpha^{(SOL)}$, $\alpha^{(EDL)}$, and $\alpha^{(BFM)}$. They predict the degree of alteration of the cellular components following electroporation. Precisely, α represents the fraction of the exposed hydrophobic parts of the cellular components affected by electroporation. Our predictions for $\alpha^{(SOL)}$, $\alpha^{(EDL)}$, and $\alpha^{(BFM)}$ are displayed in TABLE 2.

DISCUSSION

When muscle cells are electroporated *in vivo*, a progressive alteration of the state of water in the tissue is observed (FIG. 5B). Electroporation may lead to the formation of edema due to osmotic water inflow into the muscle tissue. Edema raises the local hydration level over its physiological value. In the absence of significant structural modifications, the amount of immobilized water as well as water in the first hydration layer remains unchanged.[12] The accumulated water adds extra hydration layers to the cellular components and changes the fractions of the spin population. This, in turn, increases the MR signal.

A secondary effect which can influence the long-term pattern of T_2 is the development of secondary structural damage. Upon electroporation of the cell membrane, water can invade membrane compartments that are usually separated from excess amounts of water. This irregular hydration of the cellular membrane may increase the permeability of the membrane leading to influx of Ca^{2+}, thus accelerating tissue damage due to activation of degradative processes intrinsic to the muscle cell.[6,8–10] An increase of the structural damage affects not only the compartmentalization of the spin population, but also modifies spin dynamics during magnetic relaxation, that is, the exposure of the hydrophobic layer of the damaged membrane changes the dynamics of the vicinal water.[12] In this context, the increase in the MRI contrast observed in the injured tissue can be interpreted as a superposition of edema and structural damage at the molecular level.

In this study three different muscles were tested. The muscles differ from each other with regards to the composition of fiber type. SOL is mainly composed of slow-twitch fibers whereas EDL is mainly composed of fast-twitch fibers.[21] No data on fiber type composition was found for BFM. Not surprisingly we find that the muscles respond differently to shocking. However, an interesting aspect of the results is that there is no fixed relation between the degree of edema and the amount of structural damage. SOL clearly experiences the largest degree of edema (15% increase), whereas BFM experiences the largest degree of structural changes (18%). The reason for this difference is not clear; however, anatomic differences may be important. First, the amount of current passing through the tissue depends on the resistance of that tissue, and we would expect this to vary at different locations in the leg. Vasculature and the tightness of the facia surrounding the muscles will be important in determining the degree of edema developing in the muscle. Also, since the electric field is along the length of the fibers, fiber length may be important for the susceptibility to damage. Finally, one might expect that one fiber type is more sensible to electroporation damage than the other. However, in contrast to the findings in this study it has been shown that when isolated muscles are electroporated *ex vivo* (with the electric field perpendicular to the length of the fiber), EDL is more easily damaged than soleus.[7]

The present approach sheds light on the mechanism of development of the electrical trauma and suggests a way to discriminate between the two main contributions to the MRI contrast, edema and the state of integrity of the cellular components.

The major conclusion that can be drawn from this study is that MRI can be used as an efficient non-invasive method for evaluation of muscle electrical trauma. However, the increase in T_2 has two components and it is not necessarily the area showing the largest increase in the water content (edema) which suffers the most structural damage. Future studies will allow us to test the predictions on the structural damage and to locate the damaged components.

ACKNOWLEDGMENTS

The research presented here has been supported by the National Institutes of Health Grant R01 GM61101 and the Danish Medical Research Council Grant 22-02-0523.

REFERENCES

1. HUNT, J.L., R.M. SATO & C. BAXTER. 1980. Acute electric burns: current diagnostic and therapeutic approaches to management. Arch. Surg. **115:** 434–438.
2. LEE, R.C., D.C. GAYLOR, D. BHATT & D.A. ISRAEL. 1988. Role of cell membrane rupture in the pathogenesis of electrical trauma. J. Surg. Res. **44:** 709–719.
3. CHILBERT, M.A. 1992. High-voltage and high-current injuries. *In* Electrical Stimulation and Electropathology. J. Patrick Reilly *et al.* Cambridge University Press. Cambridge, UK.
4. LEE, R.C. 1997. Injury by electrical forces: pathophysiology, manifestations, and therapy. Curr. Prob. Surg. **34:** 677–765.

5. GEHL, J. 2003. Electroporation: theory and methods, perspectives for drug delivery, gene therapy and research. Acta Physiol. Scand. **177:** 437–447.
6. GISSEL. H. & T. CLAUSEN. 2003. Ca^{2+} uptake and cellular integrity in rat EDL muscle exposed to electrostimulation, electroporation, or A23187. Am. J. Physiol. Regul. Integr. Physiol. **285:** R132–R142.
7. CLAUSEN, T. & H. GISSEL. 2005. Role of Na^+,K^+ pumps in restoring contractility following loss of cell membrane integrity in rat skeletal muscle. Acta Physiol. Scand. **183:** 263–271.
8. PUBLICOVER, S.J., C.J. DUNCAN & J.L. SMITH. 1978. The use of A23187 to demonstrate the role of intracellular calcium in causing ultrastructural damage in mammalian muscle. J. Neuropathol. Exp. Neurol. **37:** 554–557.
9. JACKSON, M.J., D.A. JONES & R.H. EDWARDS. 1984. Experimental skeletal muscle damage: the nature of the calcium-activated degenerative processes. Eur. J. Clin. Invest. **14:** 369–374.
10. ARMSTRONG, R.B., G.L. WARREN & J.A. WARREN. 1991. Mechanisms of exercise-induced muscle fibre injury. Sports Med. **12:** 184–207.
11. BLOCK, T.A., J.N. AARSVOLD, K.L. MATHEWS, II, *et al.* 1995. Nonthermally mediated muscle injury and necrosis in electrical trauma. J. Burn Care & Rehabil. **16:** 581–588.
12. DESPA, F., R.S. BERRY & R.C. LEE. 2005. Linking T_2-MRI to protein unfolding. Submitted for publication.
13. DENISOV, V.B. & B. HALLE. 1998. Thermal denaturation of ribonuclease A characterized by water ^{17}O and ^2H magnetic relaxation dispersion. Biochemistry **37:** 9595–9604.
14. DENISOV, V.P., B.H. JONSSON & B. HALLE. 1999. Hydration of denatured and molten globule proteins. Nat. Struct. Biol. **6:** 253–260.
15. BERTRAM, H.C., A.H. KARLSSON, M. RASMUSSEN, *et al.* 2001. Origin of multiexponential T(2) relaxation in muscle myowater. J. Agric. Food Chem. **49:** 3092–3100.
16. KARCZMAR, G.S., L.P. RIVER, J. RIVER, *et al.* 1994. Prospects for assessment of the effects of electrical injury by magnetic resonance. Ann. N.Y. Acad. Sci. **720:** 176–180.
17. FLECKENSTEIN, J.L., D.P. CHASON, F.J. BONTE, *et al.* 1995. High-voltage electric injury: assessment of muscle viability with MR imaging and Tc-99m pyrophosphate scintigraphy. Radiology **195:** 205–210.
18. NETTELBLAD, H., K.A. THUOMAS & F. SJOBERG. 1996. Magnetic resonance imaging: a new diagnostic aid in the care of high-voltage electrical burns. Burns **22:** 117–119.
19. OHASHI, M., J. KOIZUMI, Y. HOSODA, *et al.* 1998. Correlation between magnetic resonance imaging and histopathology of an amputated forearm after an electrical injury. Burns **24:** 362–368.
20. HANNIG, J., D.A. KOVAR, G.S. ABRAMOV, *et al.* 1999. Contrast enhanced MRI of electroporation injury [abstract]. Proceedings of the 1st Joint BMES/EMBS Conference, CD #1078.
21. DESCHENES, M.R., A.A. BRITT & W.C. CHANDLER. 2000. A comparison of the effects of unloading in young adult and aged skeletal muscle. Med. Sci. Sports Exerc. **33:** 1477–1483.

Na$^+$-K$^+$ Pump Stimulation Improves Contractility in Damaged Muscle Fibers

TORBEN CLAUSEN

Institute of Physiology and Biophysics, University of Aarhus, DK-8000 Århus C, Denmark

ABSTRACT: Skeletal muscles have a high content of Na$^+$-K$^+$-ATPase, an enzyme that is identical to the Na$^+$-K$^+$ pump, a transport system mediating active extrusion of Na$^+$ from the cells and accumulation of K$^+$ in the cells. The major function of the Na$^+$-K$^+$ pumps is to maintain the concentration gradients for Na$^+$ and K$^+$ across the plasma membrane. This generates the resting membrane potential, allowing the propagation of action potentials, excitation–contraction coupling and force development. Muscles exposed to (1) high extracellular K$^+$ or (2) low extracellular Na$^+$ show a considerable loss of force. A similar force decline is elicited by (3) increasing Na$^+$ permeability or (4) decreasing K$^+$ permeability. Under all of these four conditions, stimulation of the Na$^+$-K$^+$ pumps can restore contractility. Following exposure to electroporation or fatiguing stimulation, muscle cell membranes develop leaks to Na$^+$ and K$^+$ and a partially reversible loss of force. The restoration of force is abolished by blocking the Na$^+$-K$^+$ pumps and markedly improved by stimulating the Na$^+$-K$^+$ pumps with β$_2$-agonists, calcitonin gene–related peptide, or dbcAMP. These observations indicate that the Na$^+$-K$^+$ pumps are important for the functional compensation of the commonly occurring loss of muscle cell integrity. Stimulation of the Na$^+$-K$^+$ pumps with β$_2$-agonists or other agents may be of therapeutic value in the treatment of muscle cell damage induced by electrical shocks, prolonged exercise, burns, or bruises.

KEYWORDS: Na$^+$-K$^+$pump; skeletal muscle; cell membranes

INTRODUCTION

Like most other cells, skeletal muscle cells contain a high concentration of K$^+$ and a low concentration of Na$^+$. This uneven distribution is maintained by the Na$^+$-K$^+$-ATPase, identical to the Na$^+$-K$^+$ pump, which is situated in the plasma membrane and extrudes 3 Na$^+$ ions against 2 K$^+$ ions in each transport cycle.[1] The action potentials eliciting contractions in skeletal muscle are generated by a sudden influx of Na$^+$, immediately followed by an almost equivalent efflux of K$^+$. These passive Na$^+$-K$^+$ fluxes are large and rapid, and, during intense contractile activity, sufficient to cause progressive rundown of the transmembrane Na$^+$-K$^+$ gradients, depolariza-

---

Address for correspondence: Torben Clausen, Institute of Physiology and Biophysics, University of Aarhus, Ole Worms Alle 160, Universitetsparken, DK-8000 Århus C., Denmark. Voice: +45 8942 2822; fax:+45 8612 6590.

tc@fi.au.dk

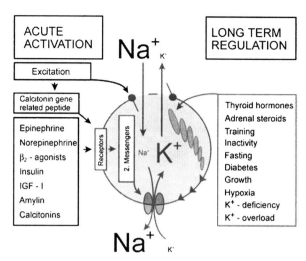

FIGURE 1. Diagram showing factors regulating the Na^+-K^+-pumps in skeletal muscle. *Left:* factors causing acute stimulation of the Na^+,K^+-pumps already present in the muscle. *Right:* list of factors inducing upregulation or downregulation of the total contents of Na^+-K^+-pumps.

tion, and loss of excitability and contractility. Thus it is essential for our ability to move and survive, that the Na^+-K^+ pumps can pump out the Na^+ gained and reaccumulate the K^+ lost, allowing restoration and maintenance of excitability and the ability to carry out contractions.[2,3] Not surprisingly, therefore, skeletal muscle cells contain many Na^+-K^+ pumps situated in the sarcolemma and the T-tubules. As illustrated in FIGURE 1, these Na^+-K^+ pumps are subject to two types of regulation: (1) acute stimulation of the activity of the Na^+-K^+ pumps in the plasma membrane and (2) long-term regulation of the synthesis of Na^+-K^+ pumps.

The acute stimulation of the Na^+-K^+ pumps already present in the muscle is primarily induced by excitation, which in isolated rat soleus muscle increases the pumping rate up to 22-fold in 10 s.[4] Catecholamines, β_2-agonists, amylin, and the calcitonins increase the Na^+-K^+ pumping rate up to about 2.5-fold within minutes, whereas insulin and insulin-like growth factor I produce somewhat smaller stimulation. Thyroid hormones, training, glucocorticoids, and K^+ overload all induce upregulation of the contents of Na^+,K^+-pumps over days or weeks. In contrast, inactivity, diabetes, fasting, hypoxia and K^+ deficiency induce a slow downregulation of the content of Na^+,K^+-pumps.

In the widest sense, the major function of the Na^+-K^+ pumps in skeletal muscles is to counterbalance the passive Na^+-K^+ leaks caused by contractions, increases in the activities of the Na^+-K^+ channels mediating passive Na^+-K^+ fluxes or perhaps also the unspecific Na^+-K^+ leakage seen after membrane damage caused by continued exercise, bruises or electrical shocks.

The possibility that the Na^+-K^+ pump could serve the last-mentioned, hitherto unknown, function was explored in two recent studies.[5,6] The major results, their interpretation, and therapeutic implications are presented and discussed below.

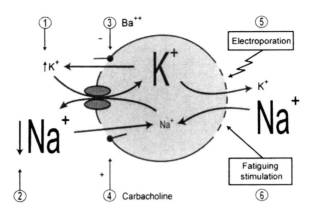

FIGURE 2. Diagram of a typical muscle cell with interventions that cause loss of contractility. In the *left side* of the cell are shown the Na^+-K^+ pump and channels mediating the rapid excitation-induced influx of Na^+ and efflux of K^+. Inhibition of contractile force may be induced by: (1) elevation of extracellular K^+; (2) reduction in extracellular Na^+; (3) addition of Ba^{2+}; (4) carbacholine opening Na^+ channels in the motor end-plate; (5) membrane leakage induced by EP; and (6) membrane leakage induced by fatiguing stimulation.

THE NA^+-K^+ PUMPS RESTORE CONTRACTILITY IN SKELETAL MUSCLE

In skeletal muscle, marked reduction in contractility may be elicited by the following interventions (FIG. 2):

1. Elevation of extracellular K^+;
2. Reduction of extracellular Na^+;
3. Closing K^+ channels with Ba^{2+}, leading to depolarization;
4. Opening Na^+ channels with carbacholine, leading to depolarization;
5. Cell membrane leakage induced by electroporation; and
6. Cell membrane leakage induced by repeated excitation.

Each of these interventions mimic part of the Na^+-K^+ redistribution or depolarization normally taking place in connection with excitation and contraction. In isolated rat soleus and EDL muscles, exposure to high extracellular K^+ (10–12.5 mM) induces a marked loss of contractile force. This can be restored within 10–20 min by stimulating the Na^+,K^+-pumps with catecholamines, insulin, or other agents with a similar action.[3,7]

In other studies, it was shown that exposing muscles to a reduced extracellular Na^+ caused a rapid loss of contractility.[8] This loss of force could be restored by acute stimulation of the Na^+,K^+ pumps with catecholamines or insulin.[9]

The resting membrane potential depends on the permeability to K^+ via K^+ channels. If the K^+ channels are blocked by Ba^{2+}, it is well documented that muscle cells undergo depolarization,[10] which is associated with a loss of contractility. The Ba^{2+}-

induced muscle paralysis can be alleviated by stimulation of the Na^+-K^+ pumps with catecholamines or insulin.[11]

Depolarization may also be induced by increasing the permeability of the plasma membrane to Na^+. Carbacholine, an agent augmenting the permeability of the Na^+ channels in the motor end plates of the muscle cells, increases Na^+ influx, leading to pronounced depolarization and loss of tetanic force. This depolarization and loss of force are both alleviated by stimulating the Na^+-K^+ pumps with catecholamines.[12]

In sum, skeletal muscles clearly lose force when exposed to a marked reduction in the transmembrane concentration gradients for Na^+ or K^+ or to depolarization. This force decline is efficiently counteracted by stimulation of the Na^+-K^+ pumps.

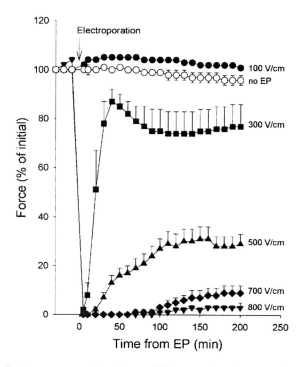

FIGURE 3. Time-course of the effects of EP at varying voltages on force development in rat soleus. Muscles were mounted in force transducers for the measurement of isometric tetanic contractions in Krebs-Ringer bicarbonate buffer at 30°C. Muscles were stimulated using 2-sec pulse trains of 60 Hz (1-ms pulses of 10 V). Following 30-min equilibration, tetanic force was recorded three times and all data presented as percent of this initial level. The muscles were then removed from the transducers and transferred to the EP cuvette and either left unstimulated or exposed to eight pulses of 0.1-msec duration at the indicated voltages (100–800 V). Henceforth, the muscles were remounted on the transducer holders and force recorded at the indicated time intervals. Each point represents the mean of observations on 3–12 muscles with bars denoting SEM. (From Clausen and Gissel.[5] Reproduced by permission.)

EFFECTS OF UNSPECIFIC LEAKAGE OF THE MUSCLE CELL MEMBRANE

The specific channel-mediated leaks to Na^+ and K^+ arising during normal contractions may be mimicked by inducing pores in the membrane. The simplest way of doing this is to expose the muscles to short-lasting high-voltage electrical fields, electroporation[13] (see the paper by R.C. Lee in this volume). By incubating isolated muscles in a small cuvette where two of the walls are aluminium plates, an electrical field can be applied to the muscles. In the cuvette, the muscles are placed in standard Krebs-Ringer bicarbonate buffer (pH 7.35). The force of the muscles is measured before and after the application of a series of eight 0.1-msec pulses varying from 100 to 800 V/cm. Immediately after the pulses, the muscles are remounted in the force transducers and tetanic force is measured at regular intervals.

As shown in FIGURE 3, the lowest voltage tested (100 V/cm) induced no loss of force, whereas higher voltages induced increasing loss of force in isolated rat soleus muscles. It should be noted that the loss of force was followed by a spontaneous re-

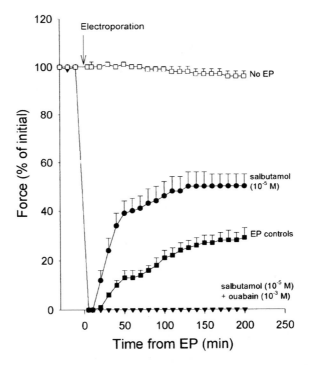

FIGURE 4. Effects of salbutamol and ouabain on force recovery after EP in rat soleus muscle. Experimental conditions as described in the legend to FIGURE 3. After EP, force recovery was followed in buffer without additions or in buffer containing salbutamol (10^{-5} M) or salbutamol (10^{-5} M) with ouabain (10^{-3} M). Each point represents the mean of observations on 4–13 muscles with bars denoting SEM. (From Clausen and Gissel.[5] Reproduced by permission.)

covery. At 300 V/cm, force recovered to around 80% of the level measured before electroporation. After exposure to pulses of 800 V/cm, there was only a few percent recovery, even after 200 min. Using 500 V/cm, around 30% force recovery was reached in 200 min, and this stimulation paradigm was used as a standard for the examination of the effects of Na^+-K^+ pump stimulation. As shown in FIGURE 4, the addition of the β_2-adrenoceptor agonist salbutamol (10^{-5} M) induced a considerable improvement of the force recovery after exposure to an electroporation field of 500 V/cm. The initial rate of force recovery as measured over the first 30 min increased by 433%, and at steady state, force recovery was doubled.

The spontaneous force recovery as well as that seen in the presence of salbutamol were abolished by the addition of ouabain (10^{-3} M), indicating that the Na^+-K^+ pumps are important for the restoration of force. This may be related to the electrogenic action of the Na^+-K^+ pumps. After electroporation, rat soleus muscle showed a depolarization of around 50 mV, followed by a spontaneous repolarization. Salbutamol increased this repolarization by 10–15 mV.[5]

Other agents known to stimulate the activity of the Na^+,K^+-pumps (epinephrine, calcitonin gene–related peptide [CGRP], and cAMP, the second messenger for the action of epinephrine and CGRP) all induced comparable increases in the initial rate of force recovery as well as the steady-state force level.

Also in the isolated rat extensor digitorum longus (EDL) muscle, salbutamol (10^{-5} M) induced a doubling of the steady-state force level, an effect that was also suppressed by ouabain.[5] These experiments performed with two muscles containing predominantly slow-twitch (soleus) or fast-twitch fibers (EDL) indicate that the functional defect induced by acutely induced cell leakage is clearly compensated by Na^+-K^+ pump stimulation in both of the major types of muscle fibers.

EFFECTS OF CELL LEAKAGE INDUCED BY FATIGUING STIMULATION

Prolonged fatiguing stimulation may lead to depolarization,[14,15] loss of excitability, contractile performance,[16] and cellular integrity.[5,17–20]

Work-induced cell damage probably represents the most common cause of cell membrane leakage in skeletal muscle.[20] It was of considerable interest, therefore, to define the conditions causing this effect and to identify mechanisms of possible restoration of the force.

Studies on isolated rat EDL muscles[21] showed that intermittent stimulation (1-msec 10-V pulses delivered at 40 Hz — 0 sec on, 30 sec off) caused approximately 90% loss of force within 10 min. When this stimulation paradigm was applied for 30 or 60 min, the muscles showed respectively a three- and nine-fold increase in the release of the intracellular enzyme lactic acid dehydrogenase (LDH). This was associated with a 40 and 70% increase in the sucrose space, respectively, further evidence that the fatiguing stimulation induced cell leakage.

After the fatiguing stimulation, the muscles showed a spontaneous slow force recovery from 10% to 30% of the initial force. As shown in FIGURE 5, this force recovery was clearly improved by stimulating the Na^+-K^+ pump with salbutamol (10^{-5} M) or epinephrine (10^{-5} M).[21] Further studies[5] showed that also at lower concentrations, salbutamol (10^{-7} M), and epinephrine (10^{-8}–10^{-6} M) induced a significant im-

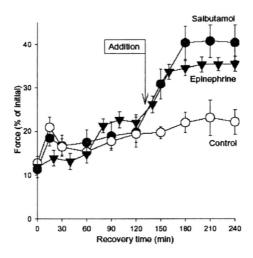

FIGURE 5. Effects of salbutamol and epinephrine on force recovery after fatiguing electrical stimulation in rat EDL. Force is given as percent of mean initial force measured using 90 Hz stimulation for 0.5 s. Salbutamol (10^{-5} M) or epinephrine (10^{-5} M) were added at the *arrow*. Each point represents the mean of observations on 6–8 muscles with bars denoting 2 × SEM.

provement of the force recovery. Similar effects were obtained by stimulating the Na^+-K^+ pump with CGRP (10^{-7} M) or dibutyryl cyclic AMP (1 mM), providing the second messenger (cyclic AMP) for the action of salbutamol, epinephrine, or CGRP. The effect of epinephrine, but not that of CGRP, was suppressed by the β-blocker propranolol (10^{-5} M). Inhibition of the Na^+-K^+ pump with ouabain (10^{-5}–10^{-3} M) suppressed the spontaneous force recovery as well as that induced by stimulation of the Na^+-K^+ pump with salbutamol.

It is well-known that $β_2$-agonists exert a trophic action on skeletal muscle. This effect develops over weeks and is expressed in increased muscle size and protein content. In healthy athletes, chronic intake of salbutamol has been shown to increase the time to exhaustion by 29%.[22] The $β_2$-agonist fenoterol was found to enhance the functional repair of regenerating rat muscle in the first weeks after injury.[23] However, no studies of acute effects of $β_2$-agonists on contractility in damaged muscles have been reported.

In dog diaphragm muscle fatigued with repeated 20 Hz contractions to around 50% of the control level, intravenous administration of the $β_2$-agonist broxaterol increased contractile performance by up to 36%.[24]

In hypoxic rat diaphragm muscles, both twitch and tetanic force are reduced. Under these conditions, $β_2$-agonists induced a significantly improved twitch force, but caused no change in tetanic force.[25] It should be noted, however, that the $β_2$-agonist clenbuterol, when injected at doses around 5 mg/kg, was found to induce cell death in around 1% of soleus muscle fibers after 2–12 hours.[26] This suggests that the use of $β_2$-agonists for the restoration of muscle force carries a risk of cell damage.

CLINICAL RELEVANCE, CONCLUSIONS AND PERSPECTIVES

In isolated skeletal muscles that have been exposed to loss of cellular integrity by electroporation or fatiguing stimulation, the Na^+-K^+ pumps seem to be important for the spontaneous recovery of force. Moreover, acute stimulation of the Na^+-K^+ pump activity leads to further force recovery. These observations would suggest that β_2-agonists or other agents known to stimulate the Na^+-K^+ pump are potential tools for the treatment of muscle cell damage arising from exercise, bruises, rhabdomyolysis, or electrical schocks. Restoration of the contractility in a damaged muscle, due to the marked stimulating effect of contractions on perfusion, might favor the supply of oxygen and nutrients. This, in turn, could improve the recovery of energy supplies, protein synthesis, and cell structure. It should be noted, however, that β_2-agonists may also cause cellular damage, both in skeletal muscles and in the heart. Up to now, no clinical trials have assessed the possible benefits or risks of Na^+-K^+ pump stimulation in the restoration of muscle damage. β_2-agonists are used by millions of patients every day, including many with muscle cell damage arising from hypoxia or fatiguing work of the respiratory muscles. A focused clinical study is warranted to explore the possible benefits and risks of this treatment in relation to muscle performance. Another unexplored aspect is the possible effects of the content of Na^+-K^+ pumps on restoration of function after muscle cell damage. The present data and the information available on the regulation of Na-K pump contents (FIG. 1) indicate that the potential for restoring muscle force after cell damage depends on their content of Na^+-K^+ pumps.

ACKNOWLEDGMENTS

This study was supported by grants from the Danish Center for Biomembrane Research, the Lundbeck Foundation and the Danish Medical Research Council (J. No 22-04-0241).

REFERENCES

1. SKOU, J.C. 1965. Enzymatic basis for active transport of Na^+ and K^+ across cell membrane. Physiol. Rev. **45:** 596–617.
2. CLAUSEN, T. 2003. The sodium pump keeps us going. Ann. N.Y. Acad. Sci. **986:** 595–602.
3. CLAUSEN, T. 2003. Na^+-K^+ pump regulation and skeletal muscle contractility. Physiol. Rev. **83:** 1270–1324.
4. NIELSEN, O.B. & T. CLAUSEN. 1997. Regulation of Na^+-K^+ pump activity in contracting rat muscle. J. Physiol. **503:** 571–581.
5. CLAUSEN, T. & H. GISSEL. 2005. Role of Na^+,K^+ pumps in restoring contractility following loss of cell membrane integrity in rat skeletal muscle. Acta Physiol. Scand. **183:** 263–271.
6. MIKKELSEN, U.R., A. FREDSTED, H. GISSEL & T. CLAUSEN. 2005. Excitation indunded cell damage and β_2-adrenoceptor agonist stimulated force recovery in rat skeletal muscle. Am. J. Physiol. In press.
7. CLAUSEN, T., S.L.V. ANDERSEN & J.A. FLATMAN. 1993. Na^+-K^+ pump stimulation elicits recovery of contractility in K^+-paralysed rat muscle. J. Physiol. **472:** 521–536.

8. BOUCLIN, R., E. CHARBONNEAU & J. M. RENAUD. 1995. Na^+ and K^+ effect on contractility of frog sartorius muscle: implication for the mechanism of fatigue. Am. J. Physiol. **268:** C1528–C1536.
9. OVERGAARD, K., O.B. NIELSEN & T. CLAUSEN. 1997. Effects of reduced electrochemical Na^+ gradient on the contractility in skeletal muscle: role of the Na^+-K^+ pump. Pflügers Arch. **434:** 457–465.
10. GALLANT, E.M. 1983. Barium-treated mammalian skeletal muscle: similarities to hypokalaemic periodic paralysis. J. Physiol. **335:** 577–590.
11. CLAUSEN, T. & K. OVERGAARD. 2000. The role of K^+ channels in the force recovery elicited by Na^+-K^+ pump stimulation in Ba^{2+}-paralysed rat skeletal muscle. J. Physiol. **527:** 325–332.
12. MACDONALD, W., O.B. NIELSEN & T. CLAUSEN. 2005. Na^+-K^+ pump stimulation restores carbacholine-induced loss of excitabilituy and contractility in rat skeletal muscle. J. Physiol. **563:** 459–469.
13. BHATT, D.L., D.C. GAYLOR & R.C. LEE. 1990. Rhabdomyolysis due to pulsed electric fields. Plast. Reconstr. Surg. **86:** 1–11.
14. LOCKE, S. & H.C. SOLOMON. 1967. Relation of resting potential of rat gastrocnemius and soleus muscles to innervation, activity and the Na-K pump. J. Exp. Zool. **166:** 377–386.
15. BALOG, E.M., L.V. THOMPSON & R.H. FITTS. 1994. Role of sarcolemma action potentials and excitability in muscle fatigue. J. Appl. Physiol. **76:** 2157–2162.
16. LINDSTRÖM, L., R. MAGNUSSON, R. & I. PETERSEN. 1970. Muscular fatigue and action potential conduction velocity changes studied with frequency analysis of EMG signals. Electromyography **10:** 341–356.
17. CLARKSON, P.M. & M.J. HUBAL. 2002. Exercise-induced muscle damage in humans. Am. J. Physiol. Med. Rehabil. **81:** S52–S69.
18. BELCASTRO, A.N. 1993. Skeletal muscle calcium-activated protease (calpain) with exercise. J. Appl. Physiol. **74:** 1381–1386.
19. MIKKELSEN, U.R., A. FREDSTED, H. GISSEL & T. CLAUSEN. 2004. Excitation-induced Ca^{2+} influx and muscle damage in the rat: loss of membrane integrity and impaired force recovery. J. Physiol. **559:** 271–285.
20. ALLEN, D.G., N.P. WHITEHEAD & E.W. YEUNG. 2005. Mechanisms of stretch-induced muscle damage in normal and dystrophic muscle: role of ionic changes. J. Physiol., published online, July 2005.
21. MIKKELSEN, U.R., A. FREDSTED, H. GISSEL & T. CLAUSEN. 2004. Muscle cell damage and β_2-agonist stimulated force recovery in rat [abstract]. The Physiologist **47:** 307.
22. COLLOMP, K., R. CANDAU, F. LASNE, et al. 2000. Effects of short-term oral salbutamol on exercise endurance and metabolism. J. Appl. Physiol. **89:** 430–436.
23. BEITZEL, F., P. GREGOREVIC, J. G. RYALL, et al. 2003. β_2-adrenoceptor agonist fenoterol enhances functional repair of regenerating rat skeletal muscle after injury. J. Appl. Physiol. **96:** 1385–1392.
24. BURNISTON, J. G., L. B. TAN & D. F. GOLDSPINK. 2005. β_2-adrenergic receptor stimulation in vivo induces apoptosis in the rat heart and soleus muscle. J. Appl. Physiol. **98:** 1379–1386.
25. DEROM, E., S. JANSSENS, G. GURRIERI, et al. 1992. Am. Rev. Respir. Dis. **146:** 22–25.
26. VAN DER HEIJDEN, H. F., L. M. HEUNKS, H. FOLGERING, et al. 1999. $beta_2$-adrenoceptor agonists reduce the decline of rat diaphragm twitch force during severe hypoxia. Am. J. Physiol. **276:** L474–L480.

Multimodal Strategies for Resuscitating Injured Cells

JAYANT AGARWAL,[a] ALEXANDRA WALSH,[a] AND RAPHAEL C. LEE[a,b]

[a]*Section of Plastic and Reconstructive Surgery, the University of Chicago Hospitals, Chicago, Illinois, USA*

[b]*Department of Surgery, Medicine and Organismal Biology (Biomechanics), University of Chicago, Chicago, Illinois, USA*

ABSTRACT: Our cells and tissues are challenged constantly by exposure to extreme conditions that cause acute and chronic stress. Wounding at the cellular level is a common event, and results from cell exposure to supra-physiologic forces, or is the consequence of action by reactive chemical agents. An individual cellular wound results from either the alteration of protein or DNA structure, or the disruption of molecular assemblies, the most important of which is the cell's membranes. Tissue healing at the macroscopic level is a complex and coordinated process involving many different cell types while, in contrast, the wounds of individual cells heal primarily via biomolecular interactions. Like tissue wound healing, cellular wound healing involves the upregulation or acceleration of processes that are constitutively expressed in routine physiologic repair of cellular structures In addition, recent advances have been made in the identification of pharmaceutical strategies to aid the cellular repair response. Many of these strategies offer promise for augmenting the already present cellular repair mechanisms.

KEYWORDS: cell injury; cell repair; resuscitating injured cells

INTRODUCTION

Biologists commonly consider a wound to be an acquired defect in the structural integrity of tissues. The healing of tissue wounds represents a complex, well orchestrated, multi-cellular process that results in repair or replacement. The signals that control and coordinate this repair process are not all known, but this is an active area of investigation. When a tissue experiences traumatic injury (a "wound"), this trauma also necessarily affects its component cells. Our cells and tissues are challenged constantly by exposure to extreme conditions that cause acute and chronic stress. However, while much is known about various aspects, healing responses by injured cells are a less well conceptualized and a coordinated process.

Wounding at the cellular level is a common event, and results from cell exposure to supra-physiologic forces, or as the consequence of action by reactive chemical agents.[1,2] An individual cellular wound results from either the alteration of protein

Address for correspondence: Raphael C. Lee, M.D., Sc.D., University of Chicago Hospitals, 5841 S. Maryland Ave., MC 6035, Chicago, IL 60637. Voice: 773-702-3320; fx: 773-702-1634 r-lee@uchicago.edu or DNA structure, or the disruption of molecular assemblies, the most important of which is the cell's membranes. Tissue healing is a complex and coordinated process involving many different cell types while, in contrast, the wounds at the level of individual cells occur primarily via biomolecular interactions. However, like tissue wound healing, cellular wound healing involves accelerated processes that are constitutively expressed in routine physiologic repair of cellular structures.[3]

This review first will aim to explain the causes and consequences of cellular wounds, and then will discuss pharmaceutical strategies that may be useful for augmenting the cellular healing response in clinical situations such as microvascular surgery. Wounding of an individual cell can occur either as a disruption of the phospholipid bilayer of the cell's membrane, or can be due to a change in the structure of individual proteins or DNA. An understanding of these mechanisms leads to a greater appreciation for the complexity of our cellular repair processes.

LOSS OF MEMBRANE STRUCTURE

Mechanisms of Membrane Breakdown

The bilayer phospholipid membrane surrounding the cell is necessary for protection of the cell from the outside environment. The cell's membrane is about 5 nm thick and primarily consists of phospholipids. The lipid bilayer is dynamically stabilized by its ability to self-assemble in an aqueous environment if the local concentration of the component phospholipids is above a crucial concentration level. In a pure surfactant membrane (like a soap bubble), a sufficiently large defect will expand until the entire assembly ruptures.[4] Surface tension is responsible for causing the membrane defect to expand. Unlike a soap bubble, cellular membranes contain large proteins, some of which are anchored together in the intracellular and extracellular spaces by other proteins, and thus a breach in the membrane does not lead always to complete cellular rupture. Studies using free-fracture electron microscopy suggest that stable membrane defect dimensions might be in the range of 0.1 micron, large enough for most biological macromolecules to pass.[5]

Basically, cellular membranes function as solute transport barriers. Lipid bilayers are very suitable for this function. The energy required to move solvated ions across pure phospholipid bilayers in an aqueous environment approaches 80–100 k_BT, indicative of the strong impediment to passive ion-diffusion across the lipid bilayer.[6] However, cell membranes are typically 20–30% protein, which renders the cell membrane approximately 10^6 times more conductive to ions than the pure lipid bilayer.[7] Structural integrity of the cell membrane, including its constituent proteins, is essential for making possible the transmembrane physiological ionic concentration gradients at a metabolic energy cost that is affordable. Despite the effectiveness of the membrane barrier, approximately 40–85% of the metabolic energy expended by cells is used to maintain normal transmembrane ion gradients. Cellular wounds involving disruption of the membrane structure quickly lead to cell necrosis and a disruption of the ionic gradient (FIG. 1). If a cell is to survive a plasma membrane disruption, the resealing process must take place within minutes of injury.[8]

Membranes of many cell types are significantly injured in the course of normal physiologic conditions.[2] For example, it has been estimated that up to 30% of skel-

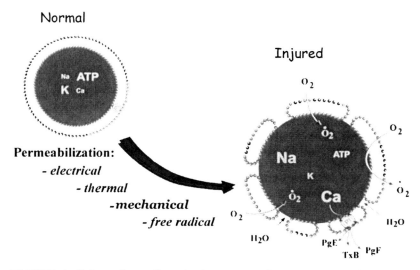

FIGURE 1. Cell membrane disruption is a common feature linking cell injury and cell death. Disruption of the plasma membrane can initiate metabolic energy depletion that leads to cell death, or plasma membrane disruption can manifest later in the pathogenesis of cell death initiated by other insults.

etal muscle cells are wounded during strenuous exercise.[8,9] Loss of cell membrane integrity has been documented under many conditions besides thermal trauma as well: in frostbite injuries,[10] in free-radical mediated radiation injury,[11] in barometric trauma,[2,12] and in electric shock,[13,14] and mechanical shear or crush forces.[3,15,16] Ischemia–reperfusion injury, which is mediated by the toxic effects of reactive oxygen species (ROS), is a common cause of membrane disruption and is an important factor in many common medical problems.[11,17]

Although the effect remains the same—a defect in the cell membrane—injuries to cellular membranes can occur in many different ways. ROS mediated wounding of the cell membrane through peroxidation of phospholipids and the oxidative deamination of membrane proteins. These effects alter lipid conformation and results in blebbing which eventually ruptures. Membrane electroporation results from the pull of water into the membrane by supraphysiologic electric fields. Heating increases the kinetic energy of membrane amphiphilic lipids until their momentum overcomes the forces of hydration that act to hold the lipids within the membrane lamella. Under cold conditions, ice nucleation in the cytoplasm can lead to factors that are very destructive to the cell membrane, including the mechanical disruption of the membrane by the ice crystal growth and the damaging effects of increasing salt concentration as the ice spreads and excludes ions.[18] Abrupt barometric pressures can lead to acoustic wave disruption of the cell membrane.[12]

The sensitivity of the membrane to all types of injury may be due to the fact that the membrane is exceptionally fragile. Using differential scanning calorimetry (DSC) we have shown that membrane components are more sensitive to fluctuations in temperature than other cellular components,[19] and it seems likely that membrane breakdown is one of the first effects of supraphysiologic temperatures.[10] Clearly, for cells to survive the stresses that are imposed upon them constantly, there must be cellular strategies for membrane repair.

Physiological Membrane Resealing

Sealing of disrupted membranes is an important naturally occurring process in cell types, as disruption of the cell membrane is a common event. Recent experiments have shed light on cellular membrane sealing. The review by Dr. Steinhardt in this symposium cover the essential aspects (this volume, pages 152–165). There seem to be two different but related mechanisms for membrane sealing after disruption: tension reduction and patching by intracellular vesicle fusion.[21] The transmembrane Ca^{2+} influx that occurs when a membrane is damaged causes vesicles to be transported by contractile proteins to the membrane defect site and to fuse with the damaged membrane, thus causing a decrease in membrane tension, which leads to sealing.[20] Studies have shown that even in the absence of Ca^{2+}, disrupted membranes are able to reseal if the membrane tension decreases profoundly enough.[20] In addition, it has been shown that there are pools of vesicles available in the cell, awaiting activation by the influx of Ca^{2+}, which fuse together to form a large "patch." The patch then fuses with the defect, enabling the sealing of holes of up to 50 μm in diameter.[21]

DENATURATION OF CELLULAR PROTEINS

At physiological temperatures, proteins in solution experience thermally driven fluctuations in their three-dimensional folded conformations over time; most of them are irrelevant to the function of the molecule. The molecular free energy difference between an active conformation and an inactive one is usually quite small, typically ranging between 5–10 kcal/mol.[14] In order for biomacromolecules to function correctly, however, the spatial relationship of the constituent reactive side chains must be precisely positioned. Particularly for enzymes, molecular activity is governed by the conformational state and the activity of enzymes is tightly controlled within the cellular milieu. This control is often achieved by slight changes in the position of the enzyme's active site caused by phosphorylation.

Alteration of molecular conformation beyond physiological range is termed denaturation and results from exposure to traumatic stresses.[22] Physical forces—whether heat, cold or other trauma—cause the proteins to unfold, altering their function. Examples include protein melting (thermal denaturation) in response to higher than physiologic temperature exposure,[14,23] electroconformational denaturation of ion channels and lipid bilayer electroporation following exposure to large transmembrane electrical potentials,[24] and freeze-induced protein damage.[10] Loss of protein architecture can also result from exposure to reactive chemical agents such as what commonly occurs as a result of excessive intracellular generation of reactive oxygen species (ROS).[11,17] Reactive chemical injuries have in common the ability to alter the primary structure of macromolecules. In comparison to thermal burn and electrical injury, the major damaging effect on cells from ionizing radiation injury are initiated at the primary protein structure level.[22] Altered macromolecules are not able to assemble into functional cellular structures, thus leading to cellular damage.

TABLE 1. Synopsis of chaperone categories that work to properly fold newly synthesized proteins and refold those proteins that lose their native conformation

Chaperone Category	Chaperone Name	Function	Notes
Prevent aggregation	Small HSPs	Attach to unfolded protein's hydrophobic domains and prevent aggregation	Non-ATP-dependent
Protein folding	GroEL/GroES	Protein folding under normal and stress conditions	Essential for cell function; double helix configuration
Chaperones of the endoplasmic reticulum	Calnexin/calreticulin	Initial correct folding of newly synthesized proteins	
Signal transduction	Hsp90	Involved in signal transduction, cell cycle control, and transcriptional regulation	Most abundant; 1–2% of cell's dry weight; highly conserved among species
Disaggregation	ClpB/HSP104 (also known as AAA+)		Double "windmill" configuration
Refolding of aggregated proteins	Hsp70 and Hsp40	Necessary for the refolding of previously aggregated proteins	

Natural Chaperones

In order to minimize the damage to cells from protein damage, molecular chaperones are continually active in cells, folding and refolding damaged proteins. Following cell injury, the production of a certain subset of these proteins, the heat-shock proteins (HSPs), is unregulated, causing an increase in the number available to help the proteins recover. These HSPs chaperone the proper refolding of denatured intracellular proteins or enable their quick removal. Induced overexpression of heat-shock proteins (chaperones) is often sufficient to protect cells from otherwise lethal exposures to environmental stresses ranging from hydrogen peroxide to extreme temperatures.[25] This process of increasing the cell stress tolerance is termed "preconditioning."

In the native state, proteins are folded such that their hydrophobic domains are buried within the center of the molecule, while the hydrophilic domains are at the surface of the molecule. When a protein unfolds due to an environmental stress, the interior hydrophobic domains are suddenly exposed and the hydrophobic domains from neighboring denatured molecules are attracted to each other, forming aggregates. If a protein manages to escape aggregation, it is likely that the newly refolded state will be functionally inactive and have a different molecular conformation than the native protein. Thus, there are two components to the repair of denatured proteins: disaggregation and refolding. Within a cell, proteases are present to break down proteins which are permanently damaged and not refoldable. Chaperones also suppress protein production and capture and maintain intermediate folded states, facilitating their refolding or degradation so that the misfolded proteins do not aggregate.[25]

In general, cells that have recently survived stressful conditions are more likely to survive another stressful incident due to the upregulation of molecular chaperones. For example, skeletal muscle cells which had been pretreated with hyperthermia were better equipped to handle the effects of calcium ionophores and mitochondrial uncouplers, corresponding to the increased production of HSPs.[26] Similarly, yeast cells treated with hydrogen peroxide or cold shock were better prepared to survive barotraumas.[27]

DNA DAMAGE AND REPAIR

Cells are constantly subjected to conditions which damage not only the cellular membranes and proteins, but also the building blocks of life: DNA. Nucleic acids are particularly vulnerable to attack by reactive oxygen species (ROS). Exposure to ROS comes about because of ROS generation by normal cellular metabolism or exposure to extrinsic sources such as chemical peroxides and ionizing irradiation. In addition, any trauma or chemical that leads to metabolic exhaustion will trigger formation of high levels of ROS in the cell that overwhelm the natural radical buffering mechanisms. This this leads to DNA and RNA damage. Typical examples would be cell membrane disruption by mechanical stress or transient oxygen depletion.

If the DNA is not repaired in a timely fashion, the consequences may be mutation, malfunction, or cellular death. These effects can usually be prevented by cellular DNA repair mechanisms which remove the damaged nucleotides and thus restore the original DNA sequence.[28] DNA repair mechanism failure is the hallmark of many types of cancer as well as diseases such as ataxia telangectasia.[29] Different injuries cause differing types of damage to cells; for example, UV light causes dimerization of adjacent pyrimidine in DNA, while other types of injury can cause single- or double-stranded breaks in the DNA.[30] The cell, if it is to survive these insults, must rapidly detect the damage and efficiently repair it. Therefore, there are more than 130 DNA repair gene to handle this matter for mammalian cells. Many of the DNA repair processes are ATP consuming, and thus cells that are metabolic energy– depleted are not likely to succeed in DNA repair.[30]

Augmenting repair of DNA and other nucleic acid repair is an important consideration in cell injury resuscitation. A review of the basic mechanisms of DNA repair and their operation conditions can found in the article by Kao *et al.* in this volume.[30]

POTENTIAL MEDICAL THERAPIES

Membrane Sealing Polymers

Therapeutic strategies to accelerate or facilitate membrane sealing are primarily based on the use of either highly polar and strongly hydrophilic polymers or amphiphilic surfactant copolymers with highly polar regions. Polyethylene glycol

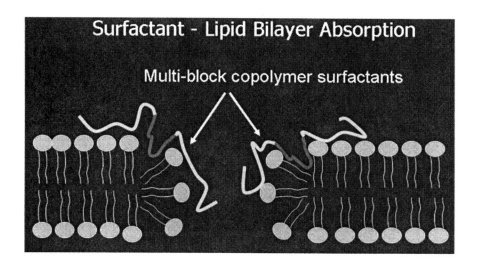
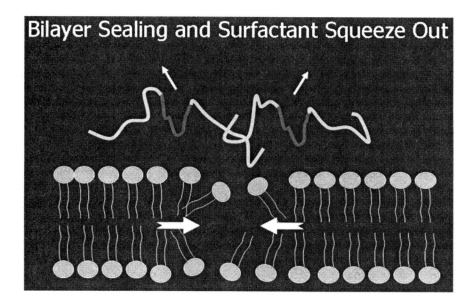

FIGURE 2. Sealing damaged cell membranes with biocompatible surfactants can augment natural membrane sealing strategies. (**a, top**) The hydrophobic domain adheres to exposed hydrophobic regions in disrupted membranes, which bring along large, highly polar blocks that alter local water structure and reduce membrane tension. (**b, bottom**) Once the surfactant is concentrated in the bilayer defect, the surface tension is altered, allowing the water to escape and the lipids to self-assemble and squeeze out the surfactant, which results in defect sealing.

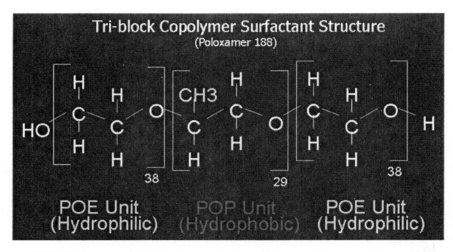

FIGURE 2—*continued.* (c) The chemical structure of a biocompatible surfactant, poloxamer 188, which is effective in membrane sealing, is detailed. The chemistry of membrane sealing is reviewed in detail by Lee *et al.* in this volume.

(PEG) is a purely hydrophilic molecule with a long history in cell-to-cell fusion applications. PEG is also very well investigated in the fusion of model membranes.[31] It is hypothesized that PEG can force very close contact between vesicle membranes by lowering the activity of water adjacent to the membrane.[32] Studies done at Purdue University have shown that PEG injected into dogs suffering from spinal cord injury caused rapid recovery of function.[33,34] In addition, when cardiac myocytes were exposed to damagingly low Ca^{2+} environments, PEG at 9% weight/volume was able to abolish the cellular damage, indicating sealing of leaks from gap junctions.[35]

As promising as PEG-mediated treatments appear to be, the high concentrations necessary for the desired benefit may limit the application of this therapy to site-specific application as opposed to intravascular administration in an injured trauma victim. For intravascular delivery, an alternative to PEG is the amphiphilic surfactant copolymers with low detergency. These agents are large-molecular-weight tri- or high block copolymers having both hydrophilic and hydrophobic domains (see structural diagram in FIGURE 2). One member of this family, poloxamer 188 (P188), has been shown to have membrane-sealing effects at low concentrations, similar to those observed upon administration of a much higher concentration of PEG.

Poloxamer 188 (P188) was initially shown in 1992 in our laboratory to seal cells against loss of a fluorescent dye after electroporation, as seen in FIGURE 3.[1] In the years following this initial finding, it has been demonstrated that P188 can also seal membrane pores in skeletal muscle cells after heat shock,[36] enhance the functional recovery of lethally heat-shocked fibroblasts,[37] and decrease damage due to high-dose ionizing radiation.[38] P188 has also been shown to protect embryonic hippocampal neurons against death due to a neurotoxin-induced loss of membrane integrity.[39]

The capability of these amphiphilic copolymers to repair cell membranes at millimolar concentrations distinguishes the sealing capability of this molecule from

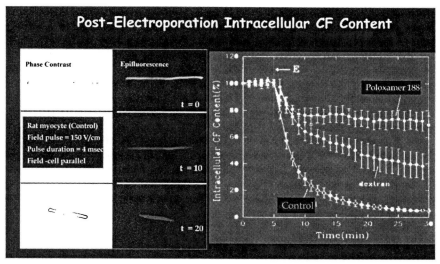

FIGURE 3. The curves on the *right* reflect the intracellular content of carboxyfluorcein (CF) before and after electroporation of adult skeletal muscle cells using a single 150 V/cm, 4-millisecond duration field pulse. The relative efficacy of poloxamer 188 in inducing membrane sealing and preventing dye loss is demonstrated by a comparison with treatment of an equimolar concentration of dextran (10 kD). Controls were treated by sampling changing the buffered saline media.

purely hydrophilic polymers such as PEG, which require molar concentrations for the same effect.[16] It is our hypothesis that the tri-block copolymers—with their hydrophobic central core—act to concentrate the hydrophilic tails to the regions of damage. This is due to the observation that damaged membranes have more hydrophobic regions exposed, and the hydrophobic region of the triblock molecule would be attracted to these exposed hydrophobic regions. The triblock molecule would thus act as a defect-targeted PEG molecule, thus requiring much lower concentrations to achieve fusion (sealing) of a permeabilized cell membrane.[40,41]

Precisely how triblock copolymers are able to seal membranes is as of yet unknown. Experiments done in our laboratory suggest that P188 is able to insert itself into damaged membranes, thus increasing the fluidity of the membrane and decreasing membrane tension. When the membrane reseals, the membrane tension increases, which seems to push P188 out of the resealed membrane.[42,43]

Surfactants to Disaggregate Proteins

Effective strategies to augment cell wound healing also involve developing therapeutic molecular templates to speed refolding of denatured proteins. Studies involving small-angle X-ray scattering performed at the Advance Photon Source at Argonne National Laboratories have allowed the visualization of individual molecules as they undergo heating and cooling cycles with and without various surfactants added.[44] In these studies, poloxamers had a remarkable ability to break apart small aggregates of denatured proteins. Light-scattering experiments, which detect larger groups of aggregates, have confirmed this finding, and current work is being done to optimize the concentration and characteristics of poloxamers in decreasing aggregation after heat denaturation. A cell is an intensely crowded environment; approximately 20% of a typical cell's weight is due to protein. Therefore, when proteins denature (from whatever cause) within a cell, the predominating event will be aggregation as the individual protein's hydrophobic domains are exposed and stick together. Any strategy to combat wounding at the cellular level will have to address the issue of aggregation.

As previously described, the membrane-sealing capability of P188 is well established. When P188 was intravenously injected in animal models of burn injury, the area of coagulation surrounding the burn decreased by 40%.[37,45] We are unsure as of yet whether these experimental results are the result of the membrane-sealing properties of P188, or whether these reflect the protein-folding and disaggregative effects of the molecule as well.

Metabolic Support

Antioxidants

Reactive species of oxygen (ROS) are ubiquitous within and across aerobic organisms. An increasingly complex picture is emerging about their physiological role in signal transduction, the cytoprotective mechanisms that prevent toxic side-effects and the deleterious effects that result when the rate of generation of these molecules exceed the buffering capability of the cells.[46] Unfortunately, excessive production of ROS can both cause cellular damage and result from other forms of cell damage that, in turn, accelerate the damage. Following cell membrane damage, the production of reactive oxygen species increases by many orders of magnitude,[47] leading to membrane injury as well as protein denaturation. It would seem to naturally follow that antioxidant administration would be a logical therapeutic modality in traumatic injury, but experimental results have been mixed. On the positive side, superoxide dismutase plus catalase was shown in 1984 to reduce the area of myocardial infarct size in dogs[48] and dimethylsulfoxide (DMSO) added to a model of ischemia–reperfusion injury was shown to decrease the release of CPK when administered prior to injury.[49] However, clinical trials in myocardial infarctions have shown that antioxidants merely delay the eventual extent of ischemic injury.

Data from our own laboratory indicates that ascorbate (vitamin C, a powerful antioxidant) as well as the combination of P188 with ascorbate may provide synergistic therapeutic benefits after electrical injury.[50] Another naturally occurring antioxidant, vitamin E, has also been shown to be protective against free radical–mediated lipid peroxidation of biological membranes.[51] Evidence is accumulating that surfactants themselves may have some antioxidant properties. Polyethylene glycol (at a concentration of 50% weight/volume) was shown to inhibit free radical production in a model of neuronal membrane damage.[52] However, although there is tantalizing evidence that free radical scavengers and antioxidants might decrease cellular injury after trauma, clinical trials of antioxidant administration have been mostly negative to this point.[53] This reality hints at the complexity of the ROS system, and future therapeutic possibilities will have to be more specific and targeted, both temporally and spatially, to the areas of the injury. In addition, as the extent of membrane damage due to free radicals becomes more evident, targeted membrane-sealing therapeutics will probably need to be included as an adjunct to the antioxidants. If a cell's membrane is injured beyond repair, antioxidants would not be able to do any good.

Energy Substrates

Loss of cell membrane integrity quickly drains the cell's energy resources as the cell attempts to maintain the normal transmembrane ion gradients using ATP-dependent membrane transport enzymes. In an animal model of extremity ischemia (causing primarily free radical–mediated membrane injury) the administration of high-energy phosphates, in particular adenine, was effective in reducing ischemic damage.[54] In a trauma causing membrane-damaging injury, a temporarily large extracellular concentration of these co-enzymes could perhaps provide transient energy charging of the cell, thus leading to decreased cell damage. In addition, it has been shown that extracellular magnesium protects cells against energy-dependent cell damage.[55] Administration of exogenous ATP-$MgCl_2$ has been shown to be beneficial for the survival of experimental animals subjected to hemorrhagic shock[56,57] and ischemia.[17] Moreover, ATP-$MgCl_2$ has been found to improve mitochondrial function following shock and ischemia.[58]

Our laboratory has observed that a combination of P188, ATP, and vitamin C has been shown to significantly enhance the survival of post-mitotic muscle cells permeabilized by exposure to intense (40 Gray) ionizing radiation.[50] We are currently examining the effect of post-injury membrane sealing on DNA repair function. With molecular repair of cell membranes and resuscitation of cellular metabolism, it appears that acute necrosis following severe membrane injury can be blocked. It remains to be seen whether additional measures are needed to block later cell entry into apoptotic pathways.

HYPOTHERMIA

Lowering tissue temperature to temperatures between 30–32°C after injury has been found to substantially reduce cell death and tissue loss following a wide range of injuries. Particular attention has been paid to the protective effects following cerebrovascular stroke[64–66] and myocardial infarction.[67,68] Because of the biochemical nature of biological systems, it is not surprising that lowering the temperature would alter the outcomes. The progress of cell injury responses should be no exception.

Recently, there has been quite a bit of interest in determining the most important mechanisms underlying hypothermic protection. It is been shown that the generation of reactive oxygen species during reperfusion injury is attenuated by cooling.[69] Mitochondria have an important role in this generation of damaging reactive oxygen species after injury. In addition, hypothermia has been found to significantly inhibit the opening of the mitochondrial permeability transition pore following an ischemic insult.[70] Given the extraordinary temperature dependence of lipid membrane dynamics, it can be expected that the gating of most membrane proteins will be slowed by cooling.

In addition to slowing down certain processes, it appears that hypothermia stimulates the biosynthesis of certain protective stress proteins as well so that cells have evolved mechanisms that provide some general cytoprotection against hypothermia.

SUMMARY

Trauma at the cellular level, from whatever cause, leads to the common pathways of membrane disruption and protein denaturation. As our knowledge of cellular membrane repair and chaperone systems increases, we begin to discern how cells have evolved strategies to alleviate these problems. With increasing knowledge of the biomechanics of cellular injury and repair, new therapeutic strategies may present themselves. Triblock copolymer compounds (such as P188) have the ability to seal membranes under some traumatic conditions, and evidence is increasing that they may also have a role as artificial chaperones in the refolding of denatured proteins, and in the breaking up of aggregated proteins. Our laboratory has ongoing investigations into the mechanisms of these effects, which will hopefully shed light on potential treatments for burn, as well as strategies to effectively reduce ischemia-/reperfusion-based membrane injury in free-flaps.[19]

REFERENCES

1. LEE R.C., L.P. RIVER, F.S. PAN, *et al.* 1992. Surfactant-induced sealing of electropermeabilized skeletal muscle membranes in vivo. Proc. Natl. Acad. Sci. USA. **89:** 4524.
2. MCNEIL P.L.& R.A. STEINHARDT. 1997. Loss, restoration, and maintenance of plasma membrane integrity. J. Cell Biol. **137:** 1.
3. MCNEIL P.L., S.S. VOGEL, K. MIYAKE, *et al.* 2000. Patching plasma membrane disruptions with cytoplasmic membrane. J. Cell. Sci. **113 (Pt. 11):** 1891.
4. TAYLOR G.I. & D.H. MICHAEL. 1973. On making holes in a sheet of fluid. J. Fluid Mechanics **58:** 625.
5. CHANG D.C. & T.S. REESE. 1990. Changes in membrane structure induced by electroporation as revealed by rapid-freezing electron microscopy. Biophys. J. **58:** 1.
6. PARSEGIAN, A. 1969. Energy of an ion crossing a low dielectric membrane: solutions to four relevant electrostatic problems. Nature **221:** 844.
7. POWELL, K.T. & J.C. WEAVER. 1986. Transient aqueous pore in bilayer membranes: a statistical theory. Bioelectrochem. Bioenerg. **15:** 211.
8. MIYAKE, K. & P.L. MCNEIL. 2003. Mechanical injury and repair of cells. Crit. Care Med. **31:** S496.
9. BISCHOF J.C., J. PADANILAM, W.H. HOLMES, *et al.* 1995. Dynamics of cell membrane permeability changes at supraphysiological temperatures. Biophys. J. **68:** 2608.
10. RUBINSKY, B., A. ARAV & A.L. DEVRIES. 1992. The cryoprotective effect of antifreeze glycopeptides from antarctic fishes. Cryobiology **29:** 69.
11. PALMER, J.S., W.J. CROMIE & R.C. LEE. 1998. Surfactant administration reduces testicular ischemia-reperfusion injury. J. Urol. **159:** 2136.
12. FISCHER, T.A., P.L. MCNEIL, R. KHAKEE, *et al.* 1997. Cardiac myocyte membrane wounding in the abruptly pressure-overloaded rat heart under high wall stress. Hypertension **30:** 1041.
13. GAYLOR, D.C., K. PRAKAH-ASANTE & R.C. LEE. 1988. Significance of cell size and tissue structure in electrical trauma. J. Theor. Biol. **133:** 223.
14. TSONG, T.Y. & Z.D. SU. 1999. Biological effects of electric shock and heat denaturation and oxidation of molecules, membranes, and cellular functions. Ann. N.Y. Acad. Sci. **888:** 211.

15. ANNO, G.H., R.W. YOUNG, R.M. BLOOM, et al. 2003. Dose response relationships for acute ionizing-radiation lethality. Health Phys. **84:** 565.
16. SHI, R., R.B. BORGENS & A.R. BLIGHT. 1999. Functional reconnection of severed mammalian spinal cord axons with polyethylene glycol. J. Neurotrauma **16:** 727.
17. HICKEY, M.J., K.R. KNIGHT, D.A. LEPORE, et al. 1996. Influence of postischemic administration of oxyradical antagonists on ischemic injury to rabbit skeletal muscle. Microsurgery **17:** 517.
18. KARLSSON, J.O., E.G. CRAVALHO, I.H. BOREL RINKES, et al. 1993. Nucleation and growth of ice crystals inside cultured hepatocytes during freezing in the presence of dimethyl sulfoxide. Biophys. J. **65:** 2524.
19. DESPA, F., D.P. ORGILL, J. NEUWALDER, et al. 2005. The relative thermal stability of tissue macromolecules and cellular structure in burn injury. Burns **31:** 568–577.
20. TOGO, T., T.B. KRASIEVA & R.A. STEINHARDT. 2000. A decrease in membrane tension precedes successful cell-membrane repair. Mol. Biol. Cell. **11:** 4339.
21. MCNEIL P.L. & R.A. STEINHARDT. 2003. Plasma membrane disruption: repair, prevention, adaptation. Annu. Rev. Cell Dev. Biol. **19:** 697.
22. Lee, R.C. & R.D. Astumian. 1996. The physicochemical basis for thermal and nonthermal "burn" injuries. Burns **22:** 509.
23. TSONG, T.Y. & C.J. GROSS. 1994. Reversibility of thermally induced denaturation of cellular proteins. Ann. N. Y. Acad. Sci. **720:** 65.
24. CHEN, W. & R.C. LEE. 1994. Evidence for electrical shock-induced conformational damage of voltage-gated ionic channels. Ann. N.Y. Acad. Sci. **720:** 124.
25. MORIMOTO, R.I. 1998. Regulation of the heat shock transcriptional response: cross talk between a family of heat shock factors, molecular chaperones, and negative regulators. Genes. Dev. **12:** 3788.
26. MAGLARA, A.A., A. VASILAKI, M.J. JACKSON, et al. 2003. Damage to developing mouse skeletal muscle myotubes in culture: protective effect of heat shock proteins. J. Physiol. 548: 837.
27. PALHANO, F.L., M.T. ORLANDO & P.M. FERNANDES. 2004. Induction of baroresistance by hydrogen peroxide, ethanol and cold-shock in *Saccharomyces cerevisiae*. F.E.M.S. Microbiol. Lett. **233:** 139.
28. LIVNEH, Z. 2001. DNA damage control by novel DNA polymerases: translesion replication and mutagenesis. J. Biol. Chem. **276:** 25639.
29. Zou, L. & S.J. Elledge. 2001. Sensing and signaling DNA damage: roles of rad17 and rad9 complexes in the cellular response to DNA damage. Harvey Lect. **97:** 1.
30. KAO, J., B.S. ROSENSTEIN, M.T. MILANO & S.J. KRON. 2005. Cellular response to DNA damage. Ann. N.Y. Acad. Sci. **1066:** 243–268 (this volume).
31. LEE, J. & B.R. LENTZ. 1997. Evolution of lipidic structures during model membrane fusion and the relation of this process to cell membrane fusion. Biochemistry **36:** 6251.
32. Arnold, K., O. Zschoernig, D. Barthel, et al. 1990. Exclusion of poly(ethylene glycol) from liposome surfaces. Biochim. Biophys. Acta **1022:** 303.
33. BORGENS, R.B. & D. BOHNERT. 2001. Rapid recovery from spinal cord injury after subcutaneously administered polyethylene glycol. J. Neurosci. Res. **66:** 1179.
34. BORGENS, R.B., D. BOHNERT, B. DUERSTOCK, et al. 2004. Subcutaneous tri-block copolymer produces recovery from spinal cord injury. J. Neuroscience Res. **76:** 141.
35. DIEDERICHS, F. 1997. A decrease of both $[Ca^{2+}]e$ and $[H^+]e$ produces cell damage in the perfused rat heart. Cell Calcium **22:** 487.
36. PADANILAM, J.T., J.C. BISCHOF, R.C. LEE, et al. 1994. Effectiveness of poloxamer 188 in arresting calcein leakage from thermally damaged isolated skeletal muscle cells. Ann. N. Y. Acad.Sci. **720:** 111.
37. MERCHANT, F.A., W.H. HOLMES, M. CAPELLI-SCHELLPFEFFER, et al. 1998. Poloxamer 188 enhances functional recovery of lethally heat-shocked fibroblasts. J. Surg. Res. **74:** 131.
38. GREENEBAUM, B., K. BLOSSFIELD, J. HANNIG, et al. 2004. Poloxamer 188 prevents acute necrosis of adult skeletal muscle cells following high-dose irradiation. Burns **30:** 539.
39. MARKS, J.D., C.Y. PAN, T. BUSHELL, et al. 2001. Amphiphilic, tri-block copolymers provide potent membrane-targeted neuroprotection. FASEB J. **15:** 1107.

40. BAEKMARK, T.R., S. PEDERSEN, K. JORGENSEN, et al. 1997. The effects of ethylene oxide containing lipopolymers and tri-block copolymers on lipid bilayers of dipalmitoylphosphatidylcholine. Biophys. J. **73:** 1479.
41. SHARMA, V., K. STEBE, J.C. MURPHY, et al. 1996. Poloxamer 188 decreases susceptibility of artificial lipid membranes to electroporation. Biophys. J. **71:** 3229.
42. MASKARINEC, S.A., J. HANNIG, R.C. LEE, et al. 2002. Direct observation of poloxamer 188 insertion into lipid monolayers. Biophys. J. **82:** 1453.
43. MASKARINEC S.A. & K.Y. LEE. 2003. Comparative study of poloxamer insertion into lipid monolayers. Langmuir **19:** 1809.
44. LEE, R.C., F. DESPA, L. GUO, et al. 2005. Surfactant copolymers facilitate refolding of *Heat Denatured Lys*ozyme. Ann. Biomed. Eng. In press.
45. BASKARAN, H., M. TONER, M.L. YARMUSH, et al. 2001. Poloxamer-188 improves capillary blood flow and tissue viability in a cutaneous burn wound. J. Surg. Res. **101:** 56.
46. BECKER, L.B. 2004. New concepts in reactive oxygen species and cardiovascular reperfusion physiology. Cardiovasc Res. **61:** 461.
47. GABRIEL, B. & J. TEISSIE. 1994. Generation of reactive-oxygen species induced by electropermeabilization of Chinese hamster ovary cells and their consequence on cell viability. Eur. J. Biochem. **223:** 25.
48. JOLLY, S.R., W.J. KANE, M.B. BAILIE, et al. 1984. Canine myocardial reperfusion injury: its reduction by the combined administration of superoxide dismutase and catalase. Circ. Res. **54:** 277.
49. GANOTE, C.E., M. SIMS & S. SAFAVI. 1982. Effects of dimethylsulfoxide (DMSO) on the oxygen paradox in perfused rat hearts. Am. J. Pathol. **109:** 270.
50. KINGSLEY, S. & R.C. LEE. 2005. Cofactor enhanced poloxamer 188 treatment increases viability of irradiated skeletal muscle cells.
51. YAMAMOTO, Y., A. FUJISAWA, A. HARA, et al. 2001. An unusual vitamin E constituent (alpha-tocomonoenol) provides enhanced antioxidant protection in marine organisms adapted to cold-water environments. Proc. Natl. Acad. Sci. USA **98:** 13144.
52. LUO, J., R. BORGENS & R. SHI. 2002. Polyethylene glycol immediately repairs neuronal membranes and inhibits free radical production after acute spinal cord injury. J. Neurochem. **83:** 471.
53. BECKER, L.B. 2004. New concepts in reactive oxygen species and cardiovascular reperfusion physiology. Cardiovasc. Res. **61:** 461.
54. ABLOVE, R.H., O.J. MOY, C.A. PEIMER, et al. 1996. Effect of high-energy phosphates and free radical scavengers on replant survival in an ischemic extremity model. Microsurgery **17:** 481.
55. KRISTENSEN, S.R. & M. HORDER. 1989. Effect of extracellular Ca^{2+} and Mg^{2+} on enzyme release from quiescent fibroblasts during various exposures. Enzyme **41:** 209.
56. CHAUDRY, I.H., M.M. SAYEED & A.E. BAUE. 1974. Effect of adenosine triphosphate-magnesium chloride administration in shock. Surgery **75:** 220.
57. HIRASAWA, H., S. ODA, H. HAYASHI, et al. 1983. Improved survival and reticuloendothelial function with intravenous ATP-$MgCl_2$ following hemorrhagic shock. Circ. Shock **11:** 141.
58. MACHIEDO, G.W., S. GHUMAN, B.F. RUSH, JR., et al. The effect of ATP-$MgCl_2$ infusion on hepatic cell permeability and metabolism after hemorrhagic shock. Surgery **90:** 328.
59. CARROLL, C.M., S.M. CARROLL, M.L. OVERGOOR, et al. 1997. Acute ischemic preconditioning of skeletal muscle prior to flap elevation augments muscle-flap survival. Plast. Reconstr. Surg. **100:** 58.
60. MOUNSEY, R.A., C.Y. PANG & C. FORREST. 1992. Preconditioning: a new technique for improved muscle flap survival. Otolaryngol. Head Neck Surg. **107:** 549.
61. CODNER, M.A., J. BOSTWICK, 3RD, F. NAHAI, et al. 1995. Tram flap vascular delay for high-risk breast reconstruction. Plast. Reconstr. Surg. **96:** 1615.
62. JENSEN, J.A., N. HANDEL, M.J. SILVERSTEIN, et al. 1995. Extended skin island delay of the unipedicle tram flap: experience in 35 patients. Plast. Reconstr. Surg. **96:** 1341.
63. KUNTSCHER, M.V., B. HARTMANN & G. GERMANN. 2005. Remote ischemic preconditioning of flaps: a review. Microsurgery **25:** 346–352.

64. ZEINER, A., M. HOLZER, F. STERZ, et al. 2000. Mild resuscitative hypothermia to improve neurological outcome after cardiac arrest: a clinical feasibility trial: Hypothermia After Cardiac Arrest (HACA) Study Group. Stroke **31:** 86–94.
65. DIETRICH, W.D., R. BUSTO, O. ALONSO, et al. 1993. Intraischemic but not postischemic brain hypothermia protects chronically following global forebrain ischemia in rats. J. Cereb. Blood Flow Metab. **13:** 541–549.
66. MARKARIAN, G.Z., J.H. LEE, D.J. STEIN, et al. 1996. Mild hypothermia: therapeutic window after experimental cerebral ischemia. Neurosurgery **38:** 542–550.
67. ABELLA, B.S., D. ZHAO, J. ALVARADO, et al. 2004. Intra-Arrest cooling improves outcomes in a murine cardiac arrest model. Circulation **109:** 2786-2790
68. HASSOUN, H.T., R.A. KOZAR, B.C. KONE, et al. 2002. Intraischemic hypothermia differentially modulates oxidative stress proteins during mesenteric ischemia/reperfusion. Surgery **132:** 369–376.
69. SIMKHOVICH, B.Z., S.L. HALE & R.A. KLONER. 2004. Metabolic mechanism by which mild regional hypothermia preserves ischemic tissue. J. Cardiovasc. Pharmacol. Ther. **9:** 83–90.
70. RYTTER, A., C.M.P. CARDOSO, P. JOHANSSON, et al. 2005. The temperature dependence and involvement of mitochondria permeability transition and caspase activation in damage to organotypic hippocampal slices following in vitro ischemia. J. Neurochem. **95:** 1108–1117.

Membrane Sealing by Polymers

STACEY A. MASKARINEC,[a] GUOHUI WU,[a-c] AND KA YEE C. LEE[a-c]

[a]*Department of Chemistry,* [b]*James Franck Institute, and* [c]*the Institute for Biophysical Dynamics, The University of Chicago, Chicago, Illinois, USA*

ABSTRACT: An intact cell membrane serves as a barrier, controlling the traffic of materials going into and out of the cell. When the integrity of the membrane is compromised, its transport barrier function is also disrupted, leaving the cell vulnerable to necrosis. It has been shown that triblock copolymer surfactants can help seal structurally damaged membranes, arresting the leakage of intracellular materials. Using model lipid monolayers along with concurrent Langmuir isotherm and fluorescence microscopy measurements as well as surface X-ray scattering techniques, the nature of the interaction between lipids and a particular family of triblock copolymers in the form poly(ethylene oxide)-poly(propylene oxide)-poly(ethylene oxide) is examined. The polymer is found to selectively insert into membranes where the lipid packing density is below that of an intact cell membrane, thus localizing its sealing effect on damaged portions of the membrane. The inserted polymer is "squeezed out" of the lipid film when the lipid packing density is increased, suggesting a mechanism for the cell to be rid of the polymer when the membrane integrity is restored.

KEYWORDS: cell membrane; lipid bilayer; poloxamer; poloxamine; surfactant

THE CELL MEMBRANE

The cell membrane separates materials inside the cell from those in the environment. In essence, it plays the crucial role of a gatekeeper, acting as a permeable barrier for transport into and out of the cell, thus regulating the molecular and ionic content of the intracellular medium. The majority of the energy required to sustain cellular function is expended in maintaining large differences in electrolyte ion concentrations across the cell membrane. The lipid bilayer constituting the membrane provides the necessary ionic diffusion barrier that makes it energetically possible to maintain large transmembrane ion concentration gradients. The lipid bilayer serves this role remarkably well by establishing a nonpolar region through which an ion must pass to cross the membrane. However, cell membranes consist typically of 30% proteins, many of which facilitate and regulate membrane ion transport. These membrane protein effects combine to make the mammalian cell membrane roughly 10^6 times more conductive to ions than the pure lipid bilayer.[1]

The mammalian cell membrane is essentially a two-dimensional sheet-like structure with a typical thickness of 60 to 100 Å. Forces that hold the lipid and protein molecules together in this assembly are not strong covalent or ionic bonds, but rather

---

Address for correspondence: Ka Yee C. Lee, Department of Chemistry, The University of Chicago, 5735 S. Ellis Avenue, Chicago, IL 60637. Voice: 773-702-7068 fax: 773-702-0805.
kayeelee@uchicago.edu

the much weaker forces, such as van der Waals, hydrophobic, hydrogen bonding, and screened electrostatic interactions. The Fluid Mosaic model proposed in 1972[2] depicts the membrane lipid bilayer as a passive entity, serving no special purpose other than providing a solvent for the membrane proteins to freely diffuse within the membrane. Over the last decade, however, evidence has emerged to suggest that compositional heterogeneity in the lipid bilayer within the membrane is important for membrane trafficking, signal transduction, selective protein attachments, and biomolecular reactions. Such membrane heterogeneity has been proposed to be the result of self-organization of various lipid species into domains[3] or rafts.[4] Irrespective of where the lipids reside, they all exhibit rotational and lateral diffusion within the membrane. Occasionally, small separations in the lipid packing order occur, producing transient structural defects with lifetimes on the order of nanoseconds. This lifetime is sufficient to permit passage of small solutes including water. The lifetime and size of these transient pores are influenced by temperature, electric field strength in the membrane, and polymers absorbed onto the membrane interface. The integrity of the membrane bilayer is essential for maintaining physiological transmembrane ionic concentration gradients at an affordable metabolic energy cost.

CELL MEMBRANE DAMAGE

Despite its critical role in supporting life, the lipid bilayer is quite fragile compared to other biological macromolecular structures. Many forms of trauma can disrupt the transport barrier function of the cell membrane. Loss of cell membrane integrity occurs in tissues at supraphysiologic temperatures as in the case of thermal burns, with very intense ionizing radiation exposure, in frostbite, in barometric trauma, and with exposure to strong electrical forces in electrical shock. Electrical shock is the paradigm for necrosis primarily mediated by membrane permeabilization. Skeletal muscle and nerve tissue exposed to strong electrical fields (greater than 50 V/cm) can experience membrane damage by at least three distinct physiochemical processes—electroporation, heat-mediated membrane poration, and electroconformational membrane protein denaturation.

When the bilayer structure is damaged, ion pumps cannot keep pace with the increased diffusion of ions across the membrane. Under these circumstances, the metabolic energy of the cell is quickly exhausted, leading to biochemical arrest and necrosis. Defects formed in the membrane can be stabilized by membrane proteins anchored in the intra- or extracelluar space. Chang and Reese[5] have demonstrated that stable structural defects—"pores" in the range of 0.1 µm—occur in electroporated cell membranes. In other cases, the translateral motion of the lipids, normally restricted by anchored proteins may cause the membrane to form bubbles as a result of the expansion of electroporated cell membranes, compromising the local lipid packing and leading to an enhanced permeability.

SURFACTANT SEALING OF CELL MEMBRANES

Sealing of porated or permeabilized cell membranes is an important, naturally occurring process. Fusogenic proteins induce sealing of membranes following exocytosis by creating a low-energy pathway for the flow of phospholipids across the defect or to induce fusion of transport vesicles to plasma membranes. Membrane sealing has also been accomplished using surfactants. The amphiphilic properties of poloxamers, a group of triblock copolymers, are able to interact with the lipid bilayer to restore its integrity. Poloxamer 188 (P188) has been used widely in medical applications since 1957, mainly as an emulsifier and anti-sludge agent in the blood.[6] Thus, most investigations on the sealing capabilities of synthetic surfactants have focused on P188 due to its already established medical safety record.

The first demonstration was that P188 could seal cells against loss of carboxyfluorescein dye after electroporation.[7] Low molecular weight (10 kDa) neutral dextran was unsuccessful in producing the same effect. In the following years, it has been shown that P188 can also seal membrane pores in skeletal muscle cells after heat shock[8] and enhance the functional recovery of lethally heat-shocked fibroblasts.[9] More recently, P188 has been shown to protect against glutamate toxicity in the rat brain[10] and protect embryonic hippocampal neurons against death due to neurotoxic-induced loss of membrane integrity[11,12] and reduce the leakage of normally membrane impermeant calcein dye from high-dose irradiated primary isolated skeletal muscle cells.[13] Other surfactants, such as poloxamine 1107 (P1107), have been shown to reduce testicular ischemia-reperfusion injury,[14] hemoglobin leakage from erythrocytes after ionizing radiation,[15,16] and propidium iodine uptake of lymphocytes after high-dose ionizing irradiation.[17] In all the aforementioned investigations, the observed phenomena were attributed to sealing of permeabilized cell membranes by the surfactants. In addition, the effect of P188 infusions in reducing duration and severity of acute painful episodes of sickle cell disease is presently also explained by beneficial surfactant-erythrocyte membrane interactions.[18]

TRIBLOCK COPOLYMER SURFACTANTS

Poloxamers and poloxamines belong to a class of water-soluble triblock copolymers often abbreviated as PEO-PPO-PEO, with PEO and PPO representing poly(ethylene oxide) and poly(propylene oxide), respectively. The PEO chains are hydrophilic due to their short carbon unit between the oxygen bridges, whereas the PPO center is hydrophobic due to the larger propylene unit (FIG. 1). Commercially available poloxamers and poloxamines have both PEO chains of similar length in a

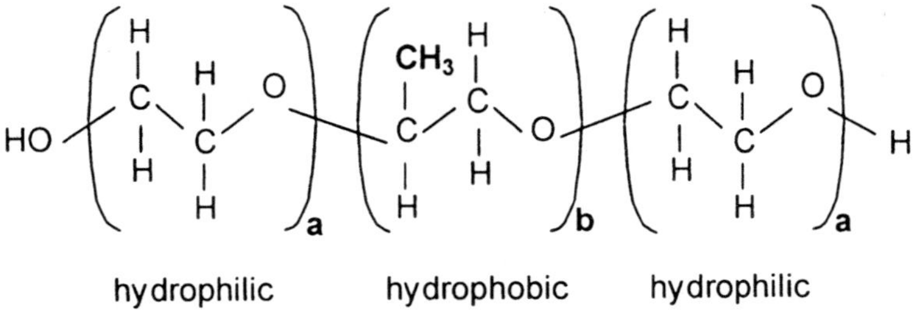

FIGURE 1. Chemical structure of poloxamers. The series of different poloxamers is constituted through varying numbers and ratios for **a** and **b**.

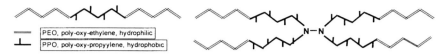

FIGURE 2. Schematic drawing to illustrate structural differences between poloxamers (*left*) and poloxamines (*right*). The PEO and PPO chain lengths vary among the members of the surfactant families.

particular copolymer. The lengths of the hydrophilic and the hydrophobic chains and their lengths ratios (FIG. 1, a vs. b) can vary tremendously, forming a large group of copolymers widely used in industrial applications as emulsifying, wetting, thickening, coating, solubilizing, stabilizing, dispersing, lubricating, and foaming agents.[21] The poloxamine series is slightly different from the poloxamer series in that the hydrophobic center consists of two tertiary amino groups each carrying two hydrophobic PPO chains of equal length each followed by a hydrophilic PEO chain. Thus, it still is a triblock copolymer but it is much bulkier than poloxamers (FIG. 2).

The poloxamer series covers a range of liquids, pastes, and solids, with molecular weights varying from 1100 to about 14,000 Da. The ethylene oxide:propylene oxide weight ratios range from about 1:9 to about 8:2. P188 has an average molecular weight of about 8400 kDa. It is prepared from a 1750-Da average molecular weight hydrophobe (29 propylene oxide units), and its hydrophile (76 ethylene oxide units) comprises about 80% of the total molecular weight. In the nomenclature of the poloxamers, the last digit (here 8) indicates the weight percentage of the hydrophilic part of a surfactant (here 80%). Thus, the poloxamers 108, 188, 238, and 288 are a series with increasing overall chain lengths but constant 80% hydrophile weight percentage. Among the group of poloxamers named P183, P185, and P188, the length of the hydrophobic chains stay constant at about 1800 Da (indicated by the first two digits, here 18) but the hydrophile weight percentage varies from 30% to 80%.

Physicochemical Properties of Triblock Copolymer Surfactants

A characteristic physicochemical parameter of surfactants is their critical micelle concentration (CMC). Above their CMC, surfactants self-aggregate to micelles causing the (active) surfactant monomer concentration to remain constant (= CMC) independent of the total surfactant concentration. Triblock copolymer surfactants, unlike conventional nonionic surfactants, do not form micelles at a critical micelle concentration. Instead, aggregation occurs over a broad concentration range, which is referred to as the aggregation concentration range. The limiting aggregation concentration (LAC) is the point at which the surfactant reaches saturation, which would correspond to the more conventional CMC (BASF Corporation 1999). This aggregation behavior of the triblock copolymers most likely accounts for the widespread values of CMCs reported in the literature, reflecting its dependence on the particular determination method used. For example, the CMC of P188 at 30° has been given as ≥100 mg/mL by Kabanov *et al.*[19] and 12.5–51.7 mg/mL by Alexandritis and Hatton.[20] In *in vitro* membrane sealing applications, P188 is typically used at concentrations well below the CMC of 0.1–1.0 mM corresponding to about 1–10 mg/mL. On the basis of these results, the surfactant monomer is presumed to be the active agent not the surfactant micelle.

LIPID–POLOXAMER INTERACTIONS

Given the membrane-sealing capability of poloxamer, one can envision using the poloxamer as a membrane sealant for therapeutic purposes. The design of an effective therapy for membrane sealing requires a good understanding of the nature of the interaction between lipid membranes and poloxamers. Ideally, the poloxamer should be able to discriminate between damaged and healthy cell membranes, interacting only with the former and not interfering with the latter. Moreover, once its presence is no longer needed (i.e., when the membrane structural integrity is restored), an exit mechanism for the poloxamer from the previously damaged membrane should be in place so that the poloxamer would not inhibit the cell healing process. Furthermore, elucidation of the mechanism by which the poloxamer helps seal the damaged membrane should aid the design of suitable polymer or polymers for therapeutic purposes.

Poloxamer and Lipid Monolayers

Although cell membranes are made up of lipid bilayers, the monolayer system provides a good mimic of the outer leaflet, with the aqueous subphase acting as the

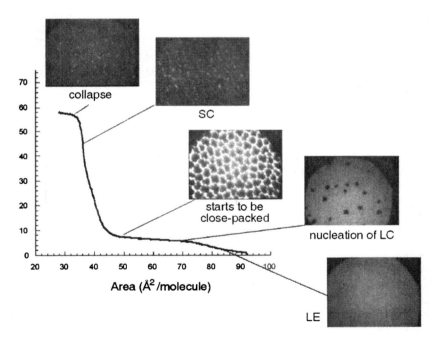

FIGURE 3. Surface pressure–area isotherm for DPPG at 30°C on pure water. The corresponding fluorescence micrographs are shown.

extracellular matrix. A Langmuir monolayer is two-dimensional (2D) film formed by a single layer of insoluble lipid molecules at the air–liquid interface. Using surface pressure–area isotherms,[22] one can observe that decreasing the lipid's surface area at the interface induces a series of 2D phase transitions.[23–29] At very high areas per lipid molecule, the molecules at the air–water interface exist in a 2D gas-like (G) state. Upon reduction of surface area by lateral compression, the monolayer condenses from the G state to an isotropic 2D fluid state known as the liquid-expanded (LE) phase. A further decrease in surface area causes a transition from the LE phase to the anisotropic condensed (C) phase. Compression beyond the minimum surface area needed for each molecule destabilizes the 2D monolayer film, resulting in the eventual collapse of the film. FIGURE 3 shows a typical surface pressure–area isotherm for dipalmitoylphosphatidylglycerol (DPPG) with the corresponding surface morphology obtained by fluorescence microscopy. The morphological images were obtained by incorporating a small amount of dye into the monolayer. Due to steric hindrance, the dye molecules preferentially partition into the disordered phase, rendering it bright and leaving the ordered phase dark.

Langmuir lipid monolayers have been extensively used as model biological membranes,[27] with the monolayer acting a good 2D model for studying interactions between different surfactants residing in the aqueous subphase and various lipids or lipid mixtures constituting the outer leaflet of the membrane surface. Langmuir troughs can be used to alter the surface area for a known amount of spread lipid accumulated at the air–water interface. The packing density of the lipid can thus be

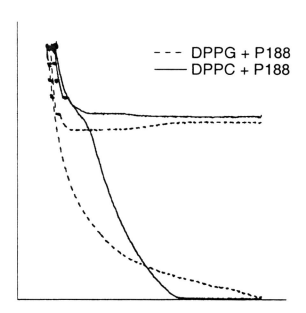

FIGURE 4. DPPC and DPPG monolayers on pure water at 30°C with P188 injected into the subphase. No change in the area was found for both cases until the surface pressure was lowered to several mN/m below the bilayer equivalent pressure.

easily controlled to simulate cell membrane damage. By measuring the extent to which these transitions are affected by the presence of poloxamers, we can gain insight into the incorporation of P188 into the monolayer.

Do Poloxamers Interact Preferentially with Damaged Membranes?

To address the question as to whether poloxamers interact preferentially with damaged membranes, we have examined the interaction of poloxamer P188 with both anionic phospholipids DPPG and zwitterionic dipalmitoylphosphatidylcholine (DPPC). An intact membrane was mimicked by compressing a spread lipid film to the bilayer equivalent surface pressure of 30 mN/m; the pressure was held constant by adjusting the surface area via a feedback mechanism. P188 was then injected into the subphase at this pressure, and the surface area of the film was monitored. Insertion of the poloxamer into the lipid film would result in an area increase while desorption of lipids into the subphase by the poloxamer would lead to a decrease in the area. For a lipid film at the bilayer equivalent pressure, no immediate change in the area per molecule (FIG. 4) or morphology was observed for a period of 10 min. Subsequently, the surface pressure was lowered to 28 mN/m, but still no observable change was detected. A pressure step-down procedure was then adopted until a low level of P188 insertion was observed at 22 mN/m. Because this change in the effective area per lipid molecule was only approximately 3 Å^2 for DPPG after 10 min, the surface pressure was lowered again to 20 mN/m. Rapid insertion of P188 into the DPPG monolayer was detected at this pressure with an overall change in an area per molecule of 74 Å^2, or until the barriers were expanded to their original position (see expansion in FIG. 4).

FIGURE 5A–C shows the morphology of a monolayer of DPPG on a water subphase at 30°C before and after P188 injection. Before injection, the condensed flower-shaped domains of DPPG occupy a much higher area fraction than the LE phase at 30 mN/m (FIG. 5A). Upon the insertion of P188 at 20 mN/m, the condensed domains become elongated, forming a more network-like structure with various-sized domains linked (FIG. 5B). In addition, there is a drastic increase in the percentage of LE or disordered phase, indicating the disordering of lipid molecules by the incorporation of P188. An additional phase of intermediate brightness is also observable (FIG. 5C).

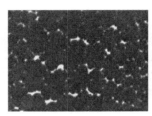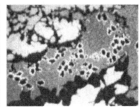

FIGURE 5. Fluorescence images showing the effect of P188 insertion into a DPPG monolayer at 30°C on pure water. P188 cannot pack well with the ordered lipid phase and preferentially associate with the disordered lipids.

Constant surface pressure injection experiment with DPPC gave similar results, with no observable change in area at the bilayer equivalent pressure after P188 administration but with substantial polymer insertion when the surface pressure was lowered to 22 mN/m (see FIG. 4). Similar morphological changes were observed upon P188 insertion.

Together these experiments suggest that P188 would only interact with compromised bilayers where the local lipid packing density is reduced and would not nonspecifically insert into membranes that were not affected. Moreover, as similar injection results were obtained for DPPC and DPPG monolayers, the insertion of the poloxamer is not influenced by the electrostatics of the lipid head group.

What is the Fate of the Poloxamer upon Cell Healing?

To determine whether there exists a mechanism for the poloxamer inserted in the damaged membrane to leave the membrane when the integrity of the once structurally compromised membrane is restored, we have examined the ability of the inserted poloxamer to retain in the model membrane at high lipid packing densities. Just as in the the injection experiments described above, the monolayer material was spread at the interface at a low surface density ($\pi \cong 0$ mN/m), but unlike in the previous case, P188 was introduced to the subphase before the lipid monolayer was compressed. The entire assembly was left undisturbed for five minutes before lateral compression commenced.

The addition of P188 to each lipid monolayer instantly displayed a drastic increase in surface pressure, from 0 mN/m to approximately 20 mN/m, close to the

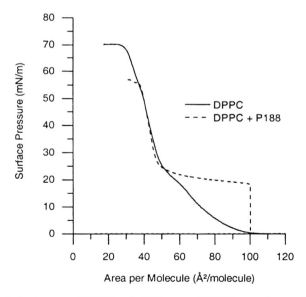

FIGURE 6. Isotherms of DPPC and P188-treated DPPC monolayers. The two isotherms overlap at surface pressures beyond 25 mN/m, indicating that the poloxamer is "squeezed-out" at high pressures.

equilibrium spreading pressure of pure P188. This high surface activity probably aids in its absorption and facilitates its insertion into lipid monolayers. The heterogeneous lipid-poloxamer system was then compressed fully. FIGURE 6 shows that as DPPC was compressed to high surface pressures, the isotherms of the poloxamer-pretreated monolayers reverted to those of the pure lipids, suggesting that P188 had been eliminated from the system. Similar results have been obtained for DPPG. These observations suggest that P188 activity is localized, capable of incorporating itself into the monolayers only when the film pressure is several mN/m below the bilayer equivalent pressure. When the lipids regain the tighter packing density found in intact cells, however, P188 cannot maintain its position within the lipid film and is "squeezed out" or eliminated as its association with the lipid layer is no longer detectable. The incapability of P188 to sustain its involvement in the system at high surface pressures can be beneficial in terms of its application. After insults of traumas such as electroporation that damage the barrier function of the cell membrane, the cell may activate a self-healing process that eventually restores the structural integrity of the bilayer. Consequently, as the cell heals and the lipid packing of the membrane is regained, P188 can be easily removed from the cell membrane.

What is the Underlying Mechanism for This Sealing Action?

A hint about the mechanism for the poloxamer-sealing action comes from experiments in which the area, instead of the surface pressure, was held constant. Here, the pure lipid monolayer was first compressed to 20 mN/m, and P188 was injected at a constant area allowing the surface pressure to increase should the polymer inserts. In the case of both DPPG and DPPC, there were dramatic surface pressure increases as a result of P188 administration (FIG. 5). Such an increase in surface pressure is indicative of tighter packing. These results therefore point to the ability of P188 to effectively insert into the damaged region of the membrane where the local lipid-packing density is reduced. By so doing, the poloxamer helps increase the local packing density.

We have recently reported that P188 changes the phase behavior and morphology of both zwitterionic DPPC and anionic DPPG monolayers.[30] P188 is found to insert into both films at surface pressures equal to and lower than ~ 22 mN/m at 30°C; this pressure corresponds to the maximal surface pressure attained by P188 on a pure water subphase. Similar results for the two phospholipids indicate that P188 insertion is not influenced by head-group electrostatics, which is not surprising as the polymer is nonionic. Because the equivalent pressure of a normal bilayer is on the order of 30 mN/m, the lack of P188 insertion above 22 mN/m further suggests that the poloxamer selectively adsorbs into damaged portions of the membrane, thereby localizing its effect. P188 is also found to be squeezed out of monolayers at high surface pressures, suggesting a mechanism for the cell to be rid of the poloxamer when the membrane is restored.[30] This squeeze-out hypothesis has also been proposed previously by Weingarten *et al.*,[31] based on their P188/PC-monolayer compression experiments.

Recent surface X-ray diffraction experiments further demonstrate that the insertion of poloxamers into a lipid film with low packing density indeed leads to a tighter packing of the lipid molecules.[33] By physically occupying part of the surface area, the adsorbed poloxamers leave the lipid molecules a smaller surface area to span and hence help tighten their packing. X-ray reflectivity results, on the other hand, show that at high surface pressures lipid films with and without poloxamers in the subphase exhibit identical electron density profiles.[34] This signifies the absence of any poloxamer in the lipid matrix and corroborates the "squeeze-out" of poloxamers at high surface pressures (or when normal lipid packing density is restored) revealed by isotherm measurements.

Polymer Design

Do surfactant monomers interact only with disrupted parts of the membrane, sealing the pores? Do they interact with the entire bilayer, altering certain membrane properties that result in its restoration (e.g., decreased fluidity)?[35,36] Does the glycocalix play a role in the sealing process? Because there is a large variety of surfactants with different hydrophilic/hydrophobic proportions, a surfactant other than P188 might have different interactions with bilayer membranes that may be better suited to seal membranes of a specific cell or useful for a particular type of injury. In considering a transmembrane scenario for its sealing mechanism, the chain length of the hydrophobic center part, including its 3D folding, can be expected to accommodate within the thickness of the lipid bilayer. The length of the hydrophilic chain might influence the strength of the interaction between the permeabilized membrane and the surfactant and thereby influence the polymer's effectiveness as a membrane *sealant*. Poloxamines, might be more effective in restoring the membrane integrity in some instances due to their overall bulkier hydrophobic center and four hydrophilic chains, providing a stronger anchor to the membrane through increased interactions with hydrophilic lipid head groups. A thorough understanding of the structure–activity of these polymers is clearly needed to better design and develop them as sealing agents.

REFERENCES

1. SCHANNE, P.F. & E.R.P. CERETTI, 1978. Impedance Measurements in Biological Cells. Wiley. New York.
2. SINGER & NICOLSON, 1972. Fluid mosaic model of structure of cell-membranes. Science **175**: 720.
3. MOURITSEN, O.G. 1998. Biol. Skr. Dan. Vid. Selsk. **49**: 47.
4. SIMONS, K. & E. IKONEN. 1997. Nature **387**: 569.
5. CHANG, D.C. & T.S. REESE. 1990. Changes in membrane structure induced by electroporation as revealed by rapid-freezing electron microscopy. Biophys. J. **58**: 1–12,
6. SCHMOLKA, I.R. 1994. Physical basis for poloxamer interactions. Ann. N.Y. Acad. Sci. **720**: 92–97.
7. LEE, R.C., P. RIVER, F.-S. PAN, L. JI & R.L. WOLLMANN. 1992. Surfactant-induced sealing of electropermeabilized skeletal muscle membranes in vivo. Proc. Natl. Acad. Sci. USA **89**: 4524–4528.
8. PADANILAM, J.T., J.C. BISCHOF, R.C. LEE, *et al.* 1994. Effectiveness of poloxamer 188 in arresting calcein leakage from thermally damaged isolated skeletal muscle cells. Ann. N.Y. Acad. Sci. **720**: 111–123.
9. MERCHANT, F.A., W.H. HOLMES, M. CAPELLI-SCHELLPFEFFER, *et al.* 1998. Poloxamer 188 enhances functional recovery of lethally heat-shocked fibroblasts. J. Surg. Res. **74**: 131–140.
10. FRIM, D.M., D.A. WRIGHT, D.J. CURRY, *et al.* 2004. The surfactant poloxamer-188 protects against glutamate toxicity in the rat brain. NeuroReport **15**: 171–174.
11. MARKS, J.D., W. CROMIE & R.C. LEE. 1998. Nonionic surfactant prevents NMDA-induced death in cultured hippocampal neurons. Soc. Neurosci. Abs. **24**: 462.

12. MARKS, J.D., C-Y PAN, T. BUSHELL, et al. 2001. Amphiphilic tri-block copolymers provide potent, membrane-targeted neuroprotection, FASEB J. doi:10.1096/fj.00-0547fje, 2001.
13. HANNIG, J. & R.C. LEE. 2000. Structural changes in cell membranes after ionizing electromagnetic field exposure. IEEE Trans. Plasma Sci. **28:** 97–101.
14. PALMER, J.S., W.J. CROMIE & R.C. LEE. 1998. Surfactant administration reduces testicular ischemia-reperfusion injury. J. Urol. **159:** 2136–2139.
15. HANNIG, J., D.J. CANADAY, M. BECKETT, et al. 2000. Sealing of membranes permeabilized by ionizing radiation. Radiat. Res. **154:** 171–177.
16. GREENEBAUM. B., K. BLOSSFIELD, J. HANNIG, et al. 2004. Poloxamer 188 prevents acute necrosis of adult skeletal muscle cells following high-dose irradiation, Burns **30:** 539–547.
17. TERRY, M.A., J. HANNIG, C.S. CARRILLO, et al. 1999. Oxidative cell membrane alteration: evidence for surfactant mediated sealing. Ann. N.Y. Acad. Sci, **888:** 274–284.
18. ADAMS-GRAVES, P., A. KEDAR, M. KOSHY, et al. 1997. RheothRx (Poloxamer 188) injection for the acute painful episode of sickle cell disease: a pilot study. Blood **90:** 2041–2046.
19. KABANOV, A.V., I.R. NAZAROVA, I.V. ASTAFIEVA, et al. 1995. Micelle formation and solubilization of fluorescent probes in poly(oxyethylene-β-oxypropylene) solutions. Macromolecules **28:** 2303–2314.
20. ALEXANDRIDIS, P. & T.A. HATTON. 1995. Poly(ethylene oxide)-poly(propylene oxide)-poly(ethylene oxide) block copolymer surfactants in aqueous solutions and at interfaces: thermodynamics, structure, dynamics, and modeling. Colloids Surf. **96:** 1–46. 1995.
21. CHU, B. & Z. ZHOU, 1996. Physical chemistry of polyoxyalkylene block copolymer surfactants. Surf. Sci. Ser. **60:** 67–144.
22. GAINES, G.L. 1966. Insoluble Monolayers at Liquid-Gas Interface. Interscience. New York.
23. ANDELMAN, D., F. BROCHARD, C. KNOBLER & F. RONDELEZ. 1994. Structures and phase transitions in Langmuir monolayers. *In* Micelles, Membranes, Microemulsions and Monolayers. W. Gelbart, A. Ben-Shaul & D. Roux, Eds.: 559–602. Springer-Verlag. New York.
24. KAGANER, V.M., H. MÖHWALD & P. DUTTA. 1999. Structure and phase transitions in Langmuir monolayers. Rev. Modern Phys. **71:** 779–819,
25. KNOBLER, C.M. & R.C. DESAI. 1992. Phase-transitions in monolayers. Annu. Rev. Phys. Chem. **43:**207–236.
26. MCCONNELL, H.M, 1991. Structures and transitions in lipid monolayers at the air-water interface. Annu. Rev. Phys. Chem. **42:** 171–195.
27. MÖHWALD, H, 1990. Annu. Rev. Phys. Chem. **41:** 441–476.
28. MÖHWALD, H, 1993. Surfactant layers at water surfaces. Rep. Prog. Phys. **56(5):** 653–685.
29. WEIS, R.M, 1991. Fluorescence microscopy of phospholipid monolayer phase-transitions. Chem. Phys. Lipids **57:** 227–239.
30. MASKARINEC, S.A., J. HANNIG, R.C. LEE & K.Y.C. LEE. 2002. Direct observation of Poloxamer 188 insertion into lipid monolayers. Biophys. J. **82:** 1453–1459.
31. MASKARINEC, S.A. & K.Y.C. LEE. 2003. Comparative study of poloxamer insertion into lipid monolayers. Langmuir **19:** 1809–1815.
32. WEINGARTEN, C., N.S.S. MAGALHAES, A. BASZKIN, et al. 1991. Interaction of non-ionic APA copolymer surfactant with phospholipid monolayers. Int. J. Pharmacol. **75:** 171–179.
33. WU, G., C. EGE, J. MAJEWSKI, et al. 2004. Lipid corralling and poloxamer squeeze-out in membranes. Phys. Rev. Lett. **93:** 02810.
34. WU, G, J. MAJEWSKI, C. EGE, et al. 2005. Interaction between lipid monolayers and Poloxamer 188: an X-ray reflectivity and diffraction study. Biophys. J. **89:** 3159–3173.
35. SHARMA, S, K. STEBE. K. MURPHY & L. TUNG. 1996. Poloxamer 188 decreases the susceptibility of artificial lipid membranes to electroporation. Biophys. J. **71:** 3229–3241.
36. BAEKMARK, T.R., S. PEDERSEN, K. JØRGENSEN & O.G. MOURITSEN, 1997. The effects of ethylene oxide containing lipopolymers and tri-block copolymers on lipid bilayers of dipalmitoylphosphatidylcholine. Biophys. J. **73:** 1479–1491.

A Surfactant Copolymer Facilitates Functional Recovery of Heat-Denatured Lysozyme

ALEXANDRA M. WALSH,[a] DEVKUMAR MUSTAFI,[b] MARVIN W. MAKINEN,[b] AND RAPHAEL C. LEE[a]

[a]*Department of Surgery, University of Chicago Hospitals, Chicago, Illinois 60637, USA*

[b]*Department of Biochemistry and Molecular Biology, Center for Integrative Science, University of Chicago, Chicago, Illinois 60637, USA*

ABSTRACT: The triblock copolymer poloxamer 188 is a non-cytotoxic, nonionic surfactant with both hydrophobic and hydrophilic domains. We show that P188 is able to facilitate the recovery of catalytic activity of heat-denatured lysozyme in dilute solution at low molar ratios of P188:enzyme. Heat-denatured enzyme retained 55% of native activity. After treatment with P188, the enzyme's activity was 85% of native. Because of the low molar ratios used and the non-cytotoxic nature of the compound, P188 may be of potential use in burn therapy.

KEYWORDS: artificial chaperone; protein folding; triblock copolymer

INTRODUCTION

The addition of inert macromolecules to dilute solutions of enzymes has been shown to have important thermodynamic effects, causing increased aggregation, enhancement of catalytic reactivity, and decreased solubility.[1] Studies using 12–20% weight/volume of polyethylene glycol or dextran have shown increased spectrin self-association[2] and increased refolding of hen egg white lysozyme.[3] These artificial chaperone effects are thought to be due to macromolecular crowding kinetically facilitating recovery of the folded protein.[4] The nonionic triblock copolymer poloxamer 188 (P188, MW of 8400) has recently been shown to function as an artificial chaperone to induce refolding of heat-denatured proteins.[5] In addition, when administered intravenously in an animal model of burn injury, P188 was shown to decrease the area of coagulation surrounding the burn by 40% 24 hours after injury.[6] In this investigation we have sought to identify the molecular basis of the interaction of nonionic, triblock copolymers with heat-denatured proteins in dilute solution. We

ABBREVIATIONS: HEW, hen egg white; HEWL, hen egg white lyzozyme; NCLD, *Micrococcus lysodeikticus*; MW, molecular weight; NT, untreated group; P188, poloxamer 188; PEG, polyethylene glycol.

Address for correspondence: Raphael C. Lee, Department of Surgery, University of Chicago Hospitals, MC 6035, 5841 S. Maryland Ave., Chicago, IL 60637. Fax: 773-702-1634.
r-lee@uchicago.edu

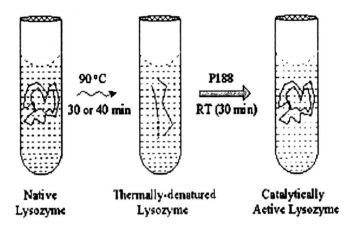

FIGURE 1. Experimental scheme.

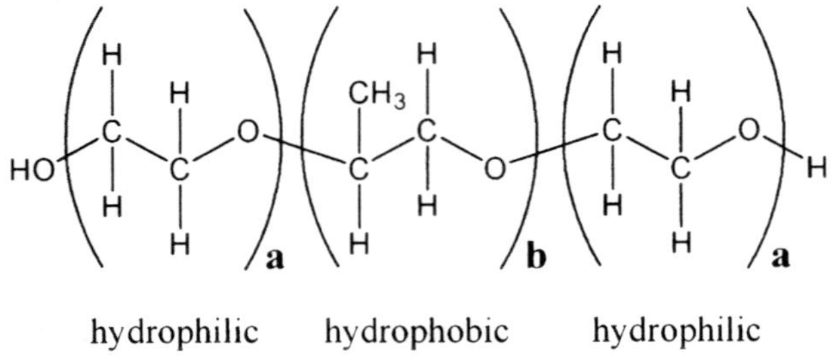

FIGURE 2. Poloxamer 188 structure.

observed that addition of P188 to a solution of heat-denatured hen egg white lysozyme (HEWL) facilitated the recovery of enzymatic activity (FIG. 1).

In contrast, the addition of polyethylene glycol of comparable molecular weight had no effect. These results suggest that the molecular basis of the chaperone-like activity of P188 may differ from that associated with other inert macromolecules and that P188 or similar molecules may have potential clinical applications.

The basic structure of P188 is shown in FIGURE 2.

The family of poloxamers consist of differing ratios of polyoxyethylene (a) and polyoxypropylene (b) chains (for P188, a:b:a=80:27:80). P188 is non-cytotoxic and recent phase I trials have shown that P188 is safe for human administration.[7] HEWL has been used extensively as a model protein for protein-folding research[8,9] and was thus chosen for these initial studies. Lysozyme is composed of 129 residues containing one α-domain, one β-domain, and four disulfide bonds.[10] The folding of HEWL has been characterized in detail, and consists of both fast and slow processes.[8-12]

ENZYMATIC ACTIVITY ASSAY FOR LYSOZYME

We used a standard lysozyme activity assay to study the catalytic effects of P188 on heat-denatured HEWL in dilute solution.[13] HEWL at an initial concentration of 50 µM was heated at 90°C for 40 min and then cooled to 25°C for 30 min. The resulting solution was then treated with either P188 or polyethylene glycol (PEG, MW 8000) and the assay begun as quickly as possible (within 3–4 minutes). Enzymatic activity was measured by following spectrophotometrically the lysis of *Micrococcus lysodeikticus* (MCLD) cells.

The Experimental Protocol

HEWL, MCLD cells, and PEG were obtained from Sigma (Milwaukee, WI); and P188 was obtained from Lutrol, BASF Pharma (Mount Olive, NJ) and used as received without further purification. We used the protocol for the turbidimetric determination of lysozyme with MCLD cells optimized by Morsky.[13] HEWL was dissolved in 55 mM phosphate buffer at pH 6.2 and the concentration was determined spectrophotometrically using the extinction coefficient of 3.65×10^4 $M^{-1}cm^{-1}$ at 280 nm. Final concentrations of MCLD and lysozyme were 410 µg/ml and 1.67×10^{-7} M, respectively, suitable for steady-state kinetics. Kinetic data were acquired at 700 nm using UV-VIS spectrophotometer for 1, 3, and 5 min. For each trial at least three scans were collected. Kinetic data were acquired with use of a Cary 15 spectrophotometer modified by On-Line Instruments Systems, Inc. (Jefferson, GA) for microprocessor controlled data acquisition, as previously described.[14]

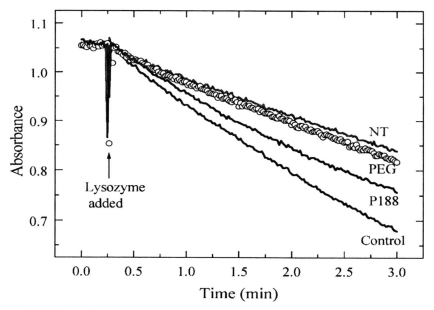

FIGURE 3. Time-dependent change in absorbance of MCLD cells at 700 nm in the presence of HEWL.

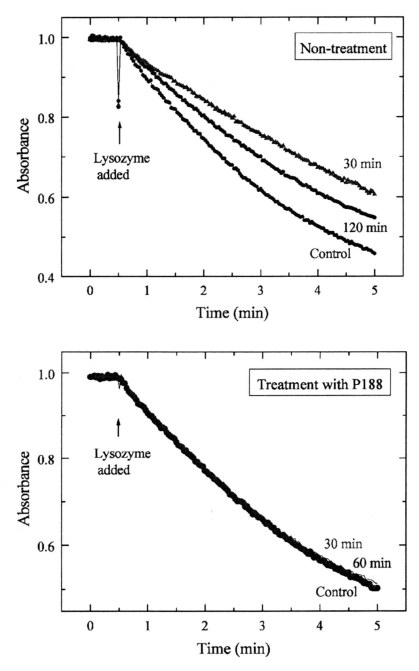

FIGURE 4. Kinetic progress curves for the turbidimetric determination of the activity of hen egg white lysozyme (HEWL) with *Micrococcus lysodeikticus* (MCLD) cells.

RESULTS

FIGURE 3 illustrates initial velocity data for the turbidimetric determination of HEWL against MCLD. P188 was able to significantly increase the fraction of catalytically active HEWL, while PEG had no statistically significant effect.

P188 or PEG was added at a 2:1 molar ratio of polymer:protein and the assay was started as quickly as possible. Untreated samples (NT) underwent the same heating–cooling cycle, but with the addition of buffer, without P188 or PEG. The control sample was kept at 25 °C throughout. The reaction was initiated by addition of HEWL to the suspension of MCLD after 15 sec, as indicated by the arrow.[13–15]

In FIGURE 4 we can see kinetic progress curves for the turbidimetric determination of the activity of HEWL with MCLD cells. The final concentration of MCLD was 410 µg/mL, and the final HEWL concentration was 1.67×10^{-7} M. In the left panel, the change in absorbance with time is compared for the native enzyme and thermally denatured enzyme. The enzyme solution was heated at 90°C for 30 min, followed by cooling to 25°C for a period of 30 min. Then the enzyme activity was measured at various time intervals following re-equilibration to 2 °C (T = 30, 60, 90, and 120 min). The reaction was initiated by adding either the native or thermally denatured HEWL to the solution of MCLD, as indicated by the arrow in the figure. Untreated HEWL regained partial activity up to T = 60 min and thereafter no further recovery of activity was observed. As seen in the left panel, even after 120 min, the catalytic activity of the thermally denatured enzyme was less than 100% when com-

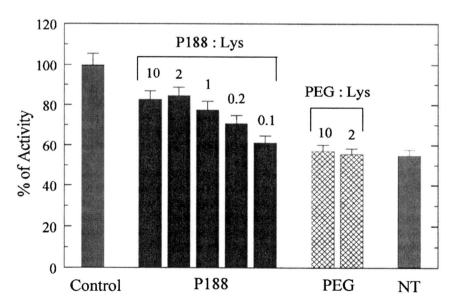

FIGURE 5. Fraction of recovered activity of heat-denatured HEWL upon addition of P188 surfactant or PEG at various molar ratios. Error bars indicate standard deviations of the measurements with the numbers above each bar indicating the molar ratio of P188 or PEG with respect to HEWL.

pared with the native enzyme, as indicated by the trace labeled as control (blue in online version). In the right panel, similar scans are shown but with P188-treated enzyme. P188 was added at 10:1 molar ratio of P188:HEWL to the thermally denatured enzyme after it reached 25°C. As seen in this panel, the enzyme activity was restored to 100% within 30 min after adding P188.

FIGURE 5 shows the percent of enzymatic activity recovered upon addition of PEG and P188, comparing the results to samples of control and heat-exposed but untreated enzyme. Untreated enzyme retained ~55% activity, while samples treated with P188 recovered up to 85% catalytic activity. The maximum effect of P188 was observed at a P188:HEWL molar ration of 2:1. The effect of PEG on the restoration of enzymatic activity was tested at 10:1 and 2:1 molar ratios of PEG:HEWL.

As seen in FIGURE 5, the percent of recovered enzyme activity of PEG-treated enzyme with a 10-fold molar excess of PEG is lower than that of P188-treated enzyme at a P188:HEWL molar ratio of 0.1:1, and is similar to that of the untreated protein. The results demonstrate that P188 acts as a molecular chaperone for the renaturation of thermally damaged lysozyme at low molar ratios, while PEG is essentially ineffective.

When the HEWL solutions were heated to 90 °C for 30 min followed by cooling to 25°C for a period of 30 min, addition of P188 in a 2:1 molar ratio resulted in full recovery of catalytic activity while only 58% of the catalytic activity was recovered in untreated samples. We have also performed enzyme activity assays of TEM-1 β-lactamase and observed similar chaperone-like effects of P188.[16]

DISCUSSION AND CONCLUSIONS

Analysis of the steady-state kinetic parameters of native HEWL showed that P188 at molar ratios of 2:1 poloxamer:HEWL exhibited no inhibition of enzymatic activity. Preliminary investigations in our laboratory utilizing light scattering at 600 nm indicate that the presence of P188 decreases aggregation.[16] Interestingly, the effect of P188 was maximal upon our first measurement of activity. If P188-facilitated recovery of enzyme activity involved only disaggregation, we would expect a time-dependent increase in enzymatic activity instead of the observed immediate maximal effect. Our observations suggest, therefore, that P188's action is primarily as an artificial chaperone.

P188 has been shown to provide protection for numerous cell types subjected to various types of stress including radiation, heat, mechanical impact, and oxidative injury.[17–21] This protective effect has been linked to membrane-sealing properties of the amphiphilic molecule.[22] The amphiphilic nature of the P188 molecule, which is necessary for its observed membrane-sealing activity,[23] may also underlie its capacity to act as an artificial chaperone in dilute solution. We hypothesize that the exposed hydrophobic portions of the P188 molecule is attracted to the hydrophobic portion of the denatured protein and displaces solvent, allowing the native conformation to be regained. The low poloxamer:protein ratios sufficient for the observed chaperone-like effect suggest that P188 or similar molecules may have potential industrial as well as medical applications. Compounds such as cyclodextrin and PEG[24,25] are used at significantly higher concentrations to promote refolding of recombinant proteins obtained from inclusion bodies.

In summary, we have employed a classical enzyme assay to evaluate the recovery of catalytic activity of heat-denatured enzymes facilitated by a nonionic, synthetic, triblock copolymer which has the unique characteristic of being safe for human administration. Since PEG (a completely hydrophilic polymer of ethylene glycol monomers) of similar molecular weight was not associated with recovery of catalytic activity, the hydrophobic domain of P188 must underlie the capacity for chaperone-like action at low molar concentrations. The ability to refold proteins at low molar concentrations may lead to potential breakthroughs in the treatment of burn injuries or protein folding diseases.

ACKNOWLEDGMENTS

This work was supported by Grant GM 64757 from the National Institutes of Health.

REFERENCES

1. A.P. MINTON. 1083. Mol. Cell. Biochem. **55:** 119–140.
2. LINDNER, R. & G. G. RALSTON. 1994. Biophys. Chem. **57:** 15–25;
3. VAN DEN BERG, B. R.J. ELLIS & C.M. DOBSON.1999. **18:** 6927–6933.
4. MINTON, A.P. 2001. J. Biol. Chem. **276:** 10577–10580.
5. KUO, F., K. BLOSSFIELD, F. DESPA, et al. 2004. Proc. Soc. Phys. Reg Biol. Med **2:** 1.
6. BASKARAN, H.M. TONER, M.L. YARMUSH & F.J. BERTHIAUME. 2001. Surg. Res. **101:** 56–61.
7. JEWELL, R.C., S.P. KHOR, D.F. KISOR, et al. 1997. J Pharm. Sci. **86:** 808–812.
8. KIEFHABER, T. 1995. Proc. Natl. Acad. Sci. USA **92:** 9029–9033.
9. MIRANKER, A., S.E. RADFORD, M. KARPLUS & C.M. DOBSON. 1991. Nature **349:** 633–636.
10. MATAGNE, A., E.W. CHUNG, L.J. BALL, et al. 1998. J. Mol. Biol. **277:** 997–1005.
11. GU, Z., X. ZHU, S. NI, et al. 2004. Int. J. Biochem. Cell. Biol. **36:** 795–805.
12. ROUX, P., M. DELEPIERRE, M.E. GOLDBERG & A.F. CHAFFOTTE. 1997. J. Biol. Chem. **272:** 24843–24849.
13. MORSKY, P. 1983. Anal. Biochem. **128:** 77–85.
14. MUSTAFI, D., M.M. KNOCK, R.W. SHAW & M.W. MAKINEN. 1997. J. Am. Chem. Soc. **119:** 12619–12628.
15. MUSTAFI, D., A. SOSA-PEINADO & M.W. MAKINEN. 2001. Biochemistry **40:** 2397–2409.
16. MUSTAFI, D., A.M. WALSH, M.W. MAKINEN & R.C. LEE. Unpublished observations.
17. GREENEBAUM, B., K. BLOSSFIELD, J. HANNIG, et al. 2004. Burns **30:** 539–547.
18. LEE, R.C.. L.P. RIVER, F.S. PAN, et al. Proc. Natl. Acad. Sci. USA **89:** 4524–4528.
19. MARKS, J.D. C.Y. PAN, T. BUSHELL, et al. 2001. FASEB J. **15:** 1107–1109.
20. MERCHANT, F.A., W.H. HOLMES, M. CAPELLI-SCHELLPFEFFER, et al. 1998. Surg. Res. **74:** 131–140.
21. PHILLIPS, D.M. & R.C. HAUT. 2004. J. Orthop. Res. **22:** 1135–1142.
22. LEE, R.C. 2002. Ann. N.Y. Acad. Sci. **961:** 271–275.
23. MASKARINEC, S.A., J. HANNIG, R.C. LEE & K.Y. LEE. 2002. Biophys. J. **82:** 1453–1459.
24. COUTHON, F., E. CLOTTES & C. VIAL. 1996. Biochem. Biophys. Commun. **227:** 854–860.
25. CLELAND, J.L., C. HEDGEPETH & D.I.C. WANG. 1992. J. Biol. Chem. **267:** 13327–13334.

Index of Contributors

Agarwal, J., 295–309

Barbee, K.A., 67–84
Bass, J., 222–242
Berry, R.S., 34–53
Bischof, J.C., 12–33

Chen, W., 92–105
Clausen, T., 286–294
Collins, J., 272–285

Despa, F., ix–x, 1–11, 54–66, 272–285
Diller, K.R., 222–242

Feng, Y., 222–242
Fowler, A., 119–135
Fridlyand, L.E., 136–151

Gissel, H., 166–180, 272–285
Guo, W., 34–53

Hamann, K.J., ix–x
He, X., 12–33
Hunt, T.K., vii–viii

Kao, J., 243–258
Karczmar, G., 272–285
Kelekar, A., 259–271

Kron, S.J., 243–258

Lee, K.Y.C., 310–320
Lee, R.C., ix–x, 54–66, 85–91, 272–285, 295–309, 321–327

Makinen, M.W., 321–327
Maskarinec, S.A., 310–320
Meredith, S.C., 181–221
Milano, M.T., 243–258
Mustafi, D., 272–285, 321–327

Orgill, D.P., 54–66, 106–118

Peters, S., 243–258
Philipson, L.H., 136–151
Porter, S.A., 106–118

Rojahn, K., 272–285
Rosenstein, B.S., 243–258
Rylander, M.N., 222–242

Shea, J.-E., 34–53
Steinhardt, R.A., 152–165

Taylor, H.O., 106–118
Toner, M., 119–135

Walsh, A.M., 295–309, 321–327
Wu, G., 310–320

Printed in the United Kingdom
by Lightning Source UK Ltd.
129304UK00001B/310-327/P